T0188839

Better Land Husbandry

From Soil Conservation to Holistic Land Management

Land Reconstruction and Management
Vol 4, 2006

Land Reconstruction and Management Series
Series Editor: Martin J. Haigh

Vol. 1: Reclaimed Land, Erosion Control, Soils and Ecology
Vol. 2: Ecological Effects of Roads
Vol. 3: Stone Deterioration in Polluted Urban Environments
Vol. 4: Better Land Husbandry

Better Land Husbandry

From Soil Conservation to Holistic Land Management

Jon Hellin

International Maize and Wheat Improvement Center (CIMMYT)
El Batán, Mexico

Science Publishers

Enfield (NH)   Jersey   Plymouth

CIP data will be provided on request.

SCIENCE PUBLISHERS
An imprint of Edenbridge Ltd., British Isles.
Post Office Box 699
Enfield, New Hampshire 03748
United States of America

Website: *http://www.scipub.net*

sales@scipub.net (marketing department)
editor@scipub.net (editorial department)
info@scipub.net (for all other enquiries)

ISBN 1-57808-244-7

Published by Science Publishers, Enfield, NH, USA
An imprint of Edenbridge Ltd.
Printed in India

For Sophie, Maya and Rosa

In memory of my father
Heinz Hellin
1930-2004

Land Reconstruction and Management
Series Preface

This is the fourth volume in an international series of monographs and thematic collections of review and research papers based around the broad theme of *'Land Reconstruction and Management'* and which deal with aspects of environmental management in constructed landscapes. Previous volumes have examined the ecological effects of roads, the impacts of air pollution on the urban fabric and the reclamation of surface coal-mined lands. This volume tackles a topic even more fundamental: the management of soil and water on agricultural steeplands in the humid tropics. It is also special because it is one of the first monographic expositions of a new paradigm for international soil and water conservation known as "better land husbandry".

The Better Land Husbandry (BLH) movement has grown in response to the failure of conventional soil and water conservation to deal effectively either with the fundamental causes of soil erosion and accelerated runoff or, indeed, many of its serious symptoms. In its search for a better approach, BLH turns many of the ideas of conventional soil and water conservation on their head. For a start, BLH supporters argue that soil and water loss is less the issue than the quality of the soil and soil moisture that remains in the field. They emphasise the importance of organic matter management in and on the soil rather than the construction of soil conservation structures. BLH theorists argue that accelerated soil erosion is caused by land degradation and that this is a land management problem. The point is that farmers, more than external technical specialists, are the better advisors where it comes to the sustainable management of the lands they farm, and that most farmers aim to make the best decisions for their families and their land within their envelope of operational opportunities—the realities that they perceive. Consequently, the greater part of most BLH projects involves discovering what farmers' realities are, determining where lie the problems in their operational systems that allow land degradation, and then seeking strategies that help those farmers rise above the obstacles by improving their welfare and by helping them conserve and enhance their land. The trick is to find land management solutions that combine economic uplift to the farmers and that provide both incentives for and the practical capability of managing the land self-sustainably. This monograph lights a path and also illustrates some of the shortcomings of conventional thinking.

This new volume joins a *'Land Reconstruction and Management'* series that is devoted to works that offer insights, deeper than those of the advanced textbooks, which are of immediate value to the researcher and of lasting practical value to the innovative practitioner. The series aspires to do more than provide an overview of the state of play in a key area of applicable research. It also aims to address the conceptual and contextual underpinning of the applied sciences and technologies involved in practical landscape management and reconstruction. It hopes to provide a platform for new philosophies and the ideals of particular international communities, 'Schools' of applied researchers and practitioners. These include the Better Land Husbandry Movement, which seeks to transform the profession of agricultural soil and water conservation and the Headwater Control Movement, which promotes new thinking in integrated watershed management through emphasising land management ahead of engineering solutions. These two new schools will both be featured by future volumes in the series.

The '*Land Reconstruction and Management*' series focuses on 'greening' the impacts of development. It deals with the management of lands that are, or have been, intensively used by human societies. Its philosophy is a delicate shade of green; its core concerns are sustainability, or better self-sustainability, in landscapes that are consequent upon, or very strongly affected by, human activities.

To this end, the series invites suggestions for new contributions. If you are the keeper of a body of knowledge that fits within this remit, if you are committed to the production of knowledge that is useful for both the researcher and environmental management practitioner, then this series may be the right home for your work?

Martin Haigh
(Series Editor)
Oxford, UK
2006
mhaigh@brookes.ac.uk

Preface

Land shortages are forcing increasing numbers of smallholder farmers to cultivate tropical steeplands. Resulting accelerated soil erosion is often countered by the promotion of soil conservation technologies, such as cross-slope barriers, which aim to reduce soil loss and preserve land productivity. This conventional approach, focusing on controlling soil erosion rather than maintaining or improving the quality of soil that remains in farmers' fields, has several drawbacks. Firstly, it addresses the symptoms of soil degradation rather than its causes and secondly, it seldom tackles the complex social and economic constraints that affect farmers' decision-making *vis-à-vis* land management. Most seriously of all, conventional soil conservation approaches tend to downplay the significance of the central purpose of farmers' activities, which is to generate maximum livelihood benefit from the land through agricultural production.

Land degradation is a social, economic, political, and technical problem requiring multi- and inter-disciplinary solutions. In this context, there is clearly a need for a new way ahead, one that shakes off the negativity of the soil loss prevention ethos and focuses, instead, on the positive aspects of land management. The new approach must place the farmer first and also address the issues of soil quality and land productivity. It is precisely these ideas that provide the foundations and starting point for the better land husbandry approach.

The better land husbandry approach aspires to discover an integrated framework within which land management issues can be analysed and practical, productivity-enhancing, strategies be formulated and implemented. The focus of better land husbandry is to improve soil quality, particularly soil architecture via soil protection, incorporation of organic matter, and the use of soil organisms. In this respect, it is in sharp contrast to the conventional soil conservation approach. The conventional approach attempts to 'combat' soil erosion head-on by capturing soil once it has been eroded. The better land husbandry approach, on the other hand, aims to maintain optimum soil conditions—in physical, chemical, biological and hydrologic terms—for the acceptance, transmission and retention of water, and for root growth and crop production.

Whilst it is not a panacea to land degradation, an approach that encourages better land husbandry has the potential to draw social and natural scientists into productive dialogue with the land users, leading to practical and realistically sustainable land management development initiatives that can play a critical role in sustainable agriculture intensification and the sound management of natural resources. Despite a growing number of successful better land husbandry examples worldwide, the aficionados of the better land husbandry approach continue to face a challenge in getting their message across. There is, therefore, a need to win over the sceptics and those who are either disinterested or uninterested. Informed debate backed up by rigorous research is an effective way to do this. This book is designed to contribute to this debate.

Jon Hellin
El Batán, Mexico

Acknowledgements

I am very grateful to a considerable number of people for all their help and assistance while I carried out research in Honduras in the mid- and late-1990s and while I wrote this book over the subsequent years. Four people deserve special mention. First and foremost, I would like to thank Professor Martin Haigh (Land Reconstruction and Management Series Editor) for his enormous patience and guidance over the last decade. I am also very grateful to Sergio Larrea and Samuel Carranza, both of whom contributed enormously to the research process in Honduras. Over the years, Francis Shaxson's insight has assisted me immeasurably during my intellectual journey from soil conservation to better land husbandry.

In Honduras, I am very grateful to everyone who was working with the *Proyecto de Conservación y Silvicultura de Especies Forestales de Honduras* (CONSEFORH) for granting me access to Santa Rosa and for making the research run much more smoothly than might otherwise have been the case. I would particularly like to thank Ernesto Ponce, Edgardo Padilla, Andrew Pinney, Kevin Crockford and Jos Wheatley. In addition, I am very grateful to Adrian Barrance, Roland Bunch, Gaye Burpee, Harri Carranza, Roberto Carranza, Ian Cherrett, Anne Dickinson, Guadalupe Durón, Karen Dvorak, Steen Johansen, Don Kass, Gabino López, Horacio Martínez, Guillermo Mendoza, Carlos Padilla, Myriam Paredes, John Pender, Sara Scherr, Steve Sherwood, Gloria de Rojas, Ian Walker, Rees Warne, Mario Zavala, José Zepeda, Humbero Zepeda, and all the farmers who participated in the focus group meetings, semi-structured interviews and the questionnaire.

In the United Kingdom, I would like to thank my former colleagues at Oxford Brookes University, especially Lesley Ambrose, Tim Blackman, Wakefield Carter, Sin Yi Cheung, Abbey Halcli, Chris McDonaugh and Elaine Welsh. I am also very grateful to Colin Hughes and John Palmer.

The United Kingdom's Department for International Development (DFID) through the Forestry Research Programme (research project R6292CB - *The use of vegetation for better land husbandry in Honduras*) part-funded the writing of this book and also funded field research in Honduras from 1995 to 1997. In 1998, research in Honduras was supported in part by the United States Agency for International Development (USAID) Soil Management Collaborative Research Support Program (Grant No. LAG-G-00-97-00002-00). I am extremely grateful for this support although the views expressed here are not necessarily those of DFID or USAID.

I am very grateful to the Food and Agriculture Organization of the United Nations for permission to reproduce figures from: Shaxson, T.F. 1999. *New Concepts and Approaches to Land Management in the Tropics with Emphasis on Steeplands*. Soils Bulletin 75: Figures 1, 3, 4 and 5. FAO, 1995. *Agricultural*

Investment to Promote Improved Capture and Use of Rainfall in Dryland Farming.
FAO Investment Centre Technical Paper 10: Figure 1.
    Lastly, I would like to thank Sophie for all her love and support and for
never doubting that I would eventually finish this book.

Contents

Series Preface   *vii*
Preface   *ix*
Acknowledgements   *xi*

1. **Introduction to Better Land Husbandry**   **1**
   1.1 Land and Soil Degradation   1
   1.2 Soil Conservation Technologies   6
   1.3 Better Land Husbandry Approach   10
   1.4 Winning the Debate?   18

2. **Social and Economic Components of Better Land Husbandry**   **20**
   2.1 Conventional Soil Conservation Approach   20
   2.2 Farmers' Realities   26
   2.3 Putting Farmers First   35
   2.4 Unlocking the Complexity of Farmers' Realities   42
   2.5 Farmers' Priority Needs   54
   2.6 Obstacles to Better Land Management   62
   2.7 Farmers' Relationship with Outsiders   72
   2.8 Tapping into Farmers' Resourcefulness   76

3. **Agro-ecological Components of Better Land Husbandry**   **81**
   3.1 Questioning the Soil Conservation Approach   81
   3.2 Soil Erosion Processes   82
   3.3 Are the Data Reliable?   87
   3.4 Testing the Soil Conservation Approach   95
   3.5 Measuring Rainfall   116
   3.6 Impacts of Conventional Agriculture   120
   3.7 Effectiveness of Soil Conservation Technologies   127
   3.8 Soil Conservation and Site Variability   135
   3.9 Time for a Rethink   138

4. **Better Land Husbandry: A New Paradigm**   **140**
   4.1 Soil Conservation Versus Better Land Husbandry   140
   4.2 New Perspectives on Soil Erosion and Land Degradation   141
   4.3 Importance of Soil Quality   149
   4.4 Maintaining and Improving Soil Quality   162
   4.5 Lessons for Outsiders   177
   4.6 Better Land Husbandry: Why has It Taken so Long?   181

5. **Better Land Husbandry in Practice**   **185**
   5.1 Introduction to Case Studies   185
   5.2 Zero Tillage in South America   185

5.3   Soil Cover in Honduras   192
5.4   Farmer Participation in Honduras   195
5.5   Soil Moisture and Yields in Africa   205
5.6   Hurricane Mitch: From Soil Erosion to Landslides   210

6. **Better Land Husbandry and Policy**   **218**
6.1   A New Approach to Land Management   218
6.2   Future of Smallholder Farming   219
6.3   A New Professionalism   222
6.4   From Production to Markets   228
6.5   Broadening the Policy Agenda   233
6.6   Better Land Husbandry and Future Research   238

Appendix 1—Impact Monitoring and Assessment   **244**

References   **268**

Index   **310**

Colour Plates 1–8   **317**

1

Introduction to Better Land Husbandry

"Human-induced agricultural land degradation is widespread in irrigated and rainfed land and in both tropical and temperate zones. Land degradation represents a challenge to the sustainability in all regions, even those of low population densities" (Shaxson and Barber, 2003:2).

1.1 LAND AND SOIL DEGRADATION

Since 1700, and as populations have increased, the cultivated area of the earth has increased by nearly 6-fold (Pimentel et al. 1995). By the year 2020, the world will have around 7.7 billion people, of whom 84 percent will be in developing countries (Scherr, 1999:31). These people will need to be fed and as populations grow and inequalities in land distribution continue, the amount of cultivable land per capita will decline. If the world's presently cultivated arable land area is maintained at about 1.45 billion hectares (ha), the per capita arable land will fall from approximately 0.25 ha to 0.15 by 2050 (Lal, 1991).

One of the challenges in the tropics and subtropics is, therefore, to intensify the output from the land without destroying the land resource (soil, water and land) upon which it all depends (Shaxson, 1993; Dumanski et al. 1998:108; Pieri et al. 1995:1; El-Ashry, 1988). In the front-line of this struggle are steeplands. These can be defined as land with an average slope of more than 20 percent. In these areas, economically disadvantaged farmers are bringing fragile and erodible lands into cultivation (Stocking, 1995). It is an immense challenge and our achievement to date has not been particularly encouraging. There are, however, parts of the world where peoples have lived sustainably in mountainous regions for centuries and where they have developed agricultural practices that are well suited to the region. Examples include communities living in the Andes in South America.

In general, though, the result of attempts to increase production in steeplands has often been soil and land degradation. Land degradation is a composite term for the aggregate diminution of the productive potential of the land, including its major uses (rainfed arable, irrigated, rangeland, forestry), its farming systems (e.g. smallholder subsistence) and its value as an economic resource. Soil degradation, which includes soil erosion, is any process that reduces the current and/or future productivity of soils (see Box 1.1).

Box 1.1 Soil degradation

Two categories of soil degradation processes are recognised. The first category is soil erosion, defined as the displacement of soil material either by water or wind. Soil erosion is one of the main processes of soil degradation and the bulk of research and practical attempts to alleviate the problem of land degradation has been directed at controlling soil loss. Erosion of exposed soil leads to the loss of fine particles, rich in organic matter and basic plant nutrients such as nitrogen, phosphorous, potassium, and calcium, which are attached to the eroded sediments and which are essential for crop production (Norman and Douglas, 1994:5; Lavelle et al. 1992). Erosion decreases soil's water-holding capacity, reduces effective rooting depth and plant available water reserves, exposes relatively infertile subsoil, and adversely affects plant growth and vigour. Erosion is a serious problem and Stocking (1995) estimates that since 1945, 1.2 billion ha—an area roughly the size of China and India combined—have been eroded to the point where their original biotic functions are impaired.

The second group deals with soil deterioration *in situ*. This can involve biological, chemical or physical soil degradation processes that lead to a decline in soil quality (Hurni et al. 1996:11). These processes include:

- Compaction and reduced water-holding capacity
- Impairment or elimination of one or more important populations of microorganisms
- Reduction in soil organic matter
- Reduced structural stability of soil crumb structure, which becomes more vulnerable to collapse on wetting or when subjected to mechanical stress.
- Excessive leaching of cations in soils with low activity which leads to a decline in soil pH and a reduction in base saturation.

Soil degradation is a major component of land degradation and human-induced soil degradation worldwide has affected 1,966 million ha. This represents 15 percent of the total land area or 38 percent of agricultural land worldwide. Furthermore, it has been suggested that, worldwide, approximately 12×10^6 ha of arable land are destroyed and abandoned annually because of unsustainable farming practices (Pimentel et al. 1995).

The major causes of both land and soil degradation are poor land husbandry and unsustainable agriculture practice. Natural rates of subsoil formation in the tropics, by the slow process of rock weathering, are estimated to be in the region of 0.5-1 t ha^{-1} yr^{-1} (Stocking, 1995). Since this is approximately equivalent to natural rates of soil erosion, all agricultural and land-use systems are, theoretically, unsustainable because almost any disturbance will cause an increase in soil loss above the natural rate of subsoil formation.

However, soil formation is a two-way process and is equally due to the addition of organic matter through the decay of roots in the soil, through the churning actions of soil organisms including earthworms, and the accumulation of litter at the soil surface. This is why, as we will see in Chapter 4, the addition of organic matter, humification and tillage can lead to the formation of soils from the top downwards (Shaxson, 1999:9).

Human activities or factors responsible for soil and land degradation include deforestation, cultivation of marginal lands, excessive and indiscriminate use of chemicals, and excessive grazing with high stocking rates (Lal and Stewart, 1990). Soil and land degradation are often greater problems in many developing countries because, in the tropics, soils commonly have inherent fertility problems. Typically, they are low in nutrients and organic matter (because the higher temperatures favour the more rapid decomposition of organic residues) and have low water-holding capacity due to a coarse texture and low clay content (Douglas, 1993).

In recent years, many smallholder farmers have been displaced from their traditional lands by way of various kinds of pressures, such as inequalities in access to land, population growth and the expansion in area of large-scale commercial agriculture (Hawkesworth and Garcia Pérez, 2003; Tadesse and Belay, 2004). Many of these smallholder farmers have taken to the cultivation of ever-steeper hillsides (see Box 1.2 and Plate 1).

Box 1.2   Political, social and economic dimensions to soil and land degradation

Throughout the world, cash crops have claimed the better lands and, having been displaced from these lands, smallholder farmers have few alternatives but to cultivate steep and marginal areas (El-Swaify, 1994; Stonich, 1991). Countries such as Honduras in Central America are no exception to this rule. In Honduras, 90 percent of the land designated as agricultural is in the hands of 10 percent of the producers (*Secretaría de Coordinación, Planificación y Presupuesto*, 1989). Furthermore, 2.8 m ha is classified as agricultural land (25 percent of the land area) but almost 4.0 m ha are farmed. The difference of 1.2 m ha is largely made up by appropriation of hillsides.

The inequitable distribution of land in Honduras that forces farmers onto steeplands increased after the Second World War, commensurate with the growth of agricultural export crops and a high population growth rate (3.2 percent per annum according to Leonard, 1987:38). In the 1950s and 1960s, parts of Honduras were for the first time drawn intimately into national and international markets. There was a growth in non-traditional agricultural exports such as cotton, sugar and beef (Stonich, 1995; Stonich and DeWalt, 1989; Durham, 1979:117).

In the early 1980s, declines in the prices of most agricultural exports encouraged large landowners to switch from labour-intensive crops such as sugar to less labour-intensive and more land-extensive alternatives such as cattle production (Moran, 1987; Stonich and DeWalt, 1989; Stonich 1991). In more recent years, the export of shrimp and fresh melons from southern Honduras has increased in importance in comparison with beef (Stonich, 1993:64).

Plate 1   Due to inequalities in access to land, population growth and the expansion in area of large-scale commercial agriculture, many smallholder farmers have been forced to cultivate ever-steeper hillsides. The result is often deforestation and soil and land degradation. Honduras. (Hellin, J.)

The cultivation of steeplands is particularly problematic because the encroachment onto hillsides represents a move to an area of lower resilience (the resistance to degradation) and higher sensitivity (the degree to which soils degrade when subjected to degradation processes) (Stocking, 1995). Sloping lands are particularly susceptible to rapid soil degradation caused by physical, chemical and biological processes (Lal, 1988). Many steeplands are in semi-arid areas and during the long dry season, plant cover is very sparse. The soil is exposed to the first intense rains of the wet season and on steep slopes with high rainfall, cultivation will inevitably cause more erosion. Steeplands also have soil fertility problems in terms of low nutrient status and aluminium toxicity (Cook, 1988).

Soil degradation can be mitigated if the land is left for sufficiently long periods in 'fallow'. During this recuperation phase, biological processes are responsible for restoring soil characteristics that have become degraded, notably porosity, water holding capacity, and nutrient content (Shaxson, 1993). Throughout the world, however, inequalities in land distribution mean that land shortages are reducing the time that land can be allowed to remain in fallow. Consequently, the rate of degradation, and the loss of underlying productive potential, is increasing. These land shortages exist despite the fact that there is much rural-urban migration throughout the developing world. This is because rapid rural population growth means that in many countries, there is no *net* out-migration from marginal rural areas such as steeplands.

Steeplands have been identified as the area where soil degradation will have the greatest and most adverse impact on food security by 2020. Degradation of irrigated lands through salinization and waterlogging poses the second greatest threat because these lands play a central role in commercial food supply in Asia (Scherr, 1999:43). It is all too easy to see land and soil degradation as a new phenomena. The reality is that these problems have plagued mankind for thousands of years (see Box 1.3).

Box 1.3   Historic land degradation in South America

The abandoned cities and agricultural terraces found throughout Mexico and Central America are testament to the fact that land degradation in the region is not, uniquely, a recent phenomenon. The ancient Maya learnt to reduce soil loss with a variety of conservation techniques including terraces (Hughes, 1999; Wilken, 1987). There is, however, some debate as to the effectiveness of these techniques (Beech and Dunning, 1995; O'Hara et al. 1993). Evidence suggests that a collapse of the shifting cultivation system and increased soil erosion was partly to blame for the demise of the Mayan empire (Abrams and Rue, 1988; Deevey et al. 1979; Kellman and Tackaberry, 1997:236). Land degradation has also contributed to the demise of other ancient civilisations in South America, despite their investment in complex systems of irrigation, aqueducts and raised fields (Hallsworth, 1987:69; Zimmerer, 1995).

Most of the lands in the Mediterranean basin owe their rugged appearance to centuries of erosion and land degradation. Even in northern Europe, many soils are called 'beheaded' because their topsoil layer was damaged long ago and reduced by misuse. According to Hillel (1991:62), the dating of fluvial sediments in river valleys in England, for example, suggests that they were the products of erosion caused by anthropogenic clearings in the originally closed deciduous forest during the Late Paleolithic period. Meanwhile, Nyssen et al. (2004) cite human impact as a possible cause of ancient soil erosion in the Ethiopian and Eritrean highlands.

That having been said, soil and land degradation have been exacerbated in the last 100 years and researchers and policy makers have long recognised the dangers that accelerated soil degradation is undermining efforts to increase agricultural productivity on a sustainable basis. Their response has often been the promotion of soil conservation technologies.

1.2   SOIL CONSERVATION TECHNOLOGIES

1.2.1   Controlling Runoff and Preventing Erosion

Generalisations about the extent of land degradation are abundant and are readily used by planners and policy-makers. There is little disagreement over *how* agricultural activities on steeplands lead to soil and land degradation, or that such activities accelerate soil loss. What is more controversial is the debate surrounding *why* farmers are cultivating steep slopes, often on an non-sustainable basis, and *what* ought to be done to alleviate the problem. In the last 40-50 years, and in response to the problem, soil conservation programmes have been initiated in many parts of the developing and developed world (Hudson, 1995:354).

For many years, soil conservation programmes have been based on the assumption that runoff is the main cause of erosion, and that runoff and erosion are inevitable consequences of farming and the principle causes of land degradation. Declines in soil productivity are often equated with losses of soil particles and plant nutrients to erosion (Food and Agriculture Organization of the United Nations (FAO), 1995:5). The main objective of soil conservation programmes has, therefore, been to control runoff on agricultural lands in order to prevent loss of soil through accelerated erosion (Douglas, 1993).

Based on this premise, and following patterns of work in the United States, much research in the developing world has been directed at finding out what the current rate of erosion actually is and, more importantly, how the factors responsible can be modified so as to reduce that rate. The goal is often to bring the calculated rate of soil loss to within 'acceptable levels'. The so-called *soil loss tolerance* is the level of soil erosion at which soil fertility can be maintained, with realistic inputs of technology, during a time frame of about 30 years (Stocking, 1995). These days, however, it is to be hoped that thinking on sustainable development encompasses a rather longer timescale.

In order to combat the perceived threat to soil productivity, and backed up by a huge amount of field and laboratory research data, soil conservation specialists have provided farmers with technical advice, assistance and technologies designed to control runoff and restrict soil losses (Suresh, 2000). Soil conservation initiatives have generally adopted a 'top-down' physical planning approach. Government and non-governmental organisations often implement national and regional soil conservation programmes. In general, their work aims to educate and involve the uninformed farming communities

(Norman and Douglas, 1994:55; Young, 1989) and the focus has often been on the concept of transfer of technology where a small array of soil conservation techniques are seen as having universal application.

Soil erosion control methods usually take the form of some combination of practices that do one or more of the following:

- Reduce the susceptibility of the soil surface to detachment
- Reduce the application of detaching forces to erodible surfaces by providing soil cover
- Reduce the ability of erosion processes to transport detached materials
- Induce deposition of transported materials

The conventional approach to soil conservation has involved cross-slope technologies such as live barriers[1], rock walls, terraces, and/or earth bunds, along with other physical structures such as drainage channels and vegetated waterways (Suresh, 2000) (see Plates CP 1 and 2). As we shall see below, however, the vertical component of rainfall precedes and is quite different in effect from the lateral component of runoff. They need attention in different ways. Despite this, in soil conservation programmes, there is much conflation of the two components. This has often led to sub-optimal approaches to dealing with soil degradation.

Plate 2   In response to soil and land degradation, soil conservation specialists have recommended technologies such as live barriers. Honduras. (Hellin, J.)

According to Hudson (1995:214-275), soil conservation technologies can be grouped into five categories according to their main objective, although there is overlap across the divisions (see Box 1.4). A more general division of erosion control measures is into two categories. Mechanical protection describes all those practices that involve moving the earth and includes digging drains and building terraces. All other practices, such as live barriers,

---

[1]Live barriers of woody species are sometimes referred to as contour hedgerows (e.g. Young, 1997). The principle of the technology—capturing eroded material and forming terraces over time—is the same irrespective of the species being used.

are known as biological methods. As we will see in Chapter 3, however, many biological actions are aimed at maintaining and improving soil quality and productivity rather than at controlling erosion *per se*.

Box 1.4   Five categories of soil conservation technologies (based on Hudson, 1995 pp. 214-275)

- Hold the rain where it falls or to transfer the runoff onto arable land (e.g. absorption ridges such as the *Murundum* terraces in Brazil; irrigation terraces such as the *Banaue* rice terraces in the Philippines).
- Retain as much rainfall as possible but to cater for infrequent overflows (e.g. contour cultivation; strip cropping and tied-ridging).
- Manage unavoidable runoff (e.g. a combination of cut-off drains, graded channel terraces and grassed waterways).
- Modify slopes by 'one-shot' terracing (e.g. bench terraces that convert a steep slope into a series of steps with horizontal ledges and vertical rises between the ledges).
- Modify slopes progressively over time by trapping soil on-site (e.g. earth banks such as *fanya juu* in Kenya; trash lines from crop residues; live barriers of grass and woody species).

Recently, more attention has been directed at the use of cover crops such as *Mucuna* spp. to protect the surface of soil from the impact of high-intensity raindrops (Anderson et al. 2001). Since the early 1980s, these biological methods have been seen as particularly suitable for the increasing number of farmers cultivating steep slopes, because of relatively low establishment and maintenance costs. Civil engineers have also demonstrated a change in emphasis in erosion control from mechanical to biological measures (or a combination of the two) (Clark and Hellin, 1996; Morgan and Rickson, 1995; Barker, 1995).

1.2.2   Farmer Adoption and Adaptation of Technologies

One of the problems in judging the success or failure of soil conservation initiatives is deciding which criteria to use. Since it is usually assumed, not always correctly, that the technical recommendations offered are effective, the most commonly used measure of success is the level of farmer adoption. Adoption may reflect genuine farmer interest in a particular technology, though there may be many reasons for adoption other than that of halting erosion (cf. Bunch, 1982). For example, reported farmer adoption rates may be enhanced by temporary incentives, such as subsidies. These tend to encourage temporary farmer participation but are often discarded once the external support is withdrawn, proving that the incentive, rather than erosion control, provided the motivation for participation (Dvorak, 1991; Lutz et al. 1994; Pretty, 1998b: 293).

Adoption rates may also be affected by negative incentives such as the threat of resettlement, social ostracism or peer pressures, which 'encourage' farmers to participate in soil conservation programmes against their better judgement (Enters, 1996). It has, therefore, been argued that 'success' should partly be based on the degree of adaptation rather than adoption of technologies (Hinchcliffe et al. 1995:15). The salutary fact is that where the criterion of farmer adoption/adaptation of technologies and practices is used, the results of many soil conservation projects to date have been disappointing (Hudson, 1992; Tenge et al. 2004; Bewket and Sterk, 2002).

The record demonstrates, however, that there is a widespread 'reluctance' on farmers' parts to adopt recommended soil conservation technologies. So, where does this leave researchers and development practitioners, especially in the context of an abundance of literature that fulsomely extols the virtues of many soil conservation technologies and their capacity to reduce soil loss and in some cases increase productivity? As Bunch (1982:59), Chambers (1993), Douglas (1993) and Hallsworth (1987:148) point out, more often than not when farmers do not adopt recommended soil conservation technologies, they are accused of being ignorant, uncooperative, conservative and unwilling to change. This often strengthens the resolve of researchers, development practitioners and policy-makers to 'educate' farmers and to convince them of the virtue of adopting recommended soil conservation technologies.

In recent years, however, there has been a growing recognition and appreciation of the fact that resource-poor farmers are intimately dependent on the maintenance of the land's productivity. Farming is a gamble based on a farmer's personal assessment of the land's productivity in relation to the limited array of options and resources at a family's disposal (Dumanski et al. 1998:112; Shaxson et al. 1989:30). Farmers that survive do so because they have some expertise in managing their land and the resources at their disposal The "farmer-first" paradigm has encouraged a greater appreciation of the farmers' situation and has led some development practitioners to question whether farmers' unwillingness to follow recommendations stems more from the fact that soil conservation technologies devised by outsiders[2] do not accord with farmers' resources, needs and priorities (Shaxson, 1993).

Faced with this reality, it becomes clearer that while soil conservation has major on- and off-farm benefits, an alternative approach to achieving soil conservation is needed, one that better complements farmers' needs and priorities and one that better respects farmers' own expert assessment of those needs and priorities. The 'better land husbandry' approach is one such alternative. The better land husbandry approach aims to help farmers improve their husbandry of their lands. It stresses active farmer participation and

---

[2]In this book, the terms 'outsider' refers to anyone who is not a small farmer and who is deemed to be dominant or superior to or by farmers (Bunch, 1982:30). In this context, outsiders are researchers and development practitioners including extension agents.

targets the concerns that matter most to farmers while promoting integrated and synergistic resource management systems including crops, rainwater and soils. The approach seeks to combine farmers' concerns about productivity with conservationists' concerns about reducing soil erosion via practices that are both productivity-enhancing and conservation-effective.

1.3   BETTER LAND HUSBANDRY APPROACH

> *Good land husbandry is the active process of implementing and managing preferred systems of land use and production in such ways that there will be an increase—or, at worst, no loss—of productivity, of stability or of useful-ness for the chosen purpose* (Shaxson, 1997:11).

1.3.1   New Ways of Thinking

The obvious but oft-neglected reality is that farmers are primarily concerned with attaining economic and reliable production from their land (Hudson, 1983; Douglas, 1993; Shaxson, 1996). The conventional soil conservation argument is that erosion is a threat to farmers' livelihoods and should be controlled because of the link between soil loss and productivity (Hillel, 1991:161; Pagiola, 1994; Tengberg et al. 1998). As detailed in Chapter 3, there is much evidence, however, that the relationship between soil loss and productivity is elusive at best (Alegre and Rao, 1996; Herweg and Ludi, 1999; Lutz et al. 1994; Schiller et al. 1982; Stone et al. 1985). The reason is that productivity is governed more by the quality of soil remaining on the land than by the quantity lost through erosion (Shaxson et al. 1989:26).

In some cases, post-erosion yields will be lower because plants are growing in a poorer quality soil characterised by:

- Reduced depth for rooting and moisture retention
- Reduced quantities of nutrients and fewer available nutrients
- Less organic matter and reduced biological activity
- Poor soil architecture leading to reduced porosity, slower gas exchange rates and less plant-available water

The better the quality of soil as a rooting environment, in terms of its physical, biological and chemical status, the more productive it is, irrespective of how much has been eroded. While actual yields are, of course, determined by a complex interaction of a number of factors including soil quality, crop and land management system, and climate, the bottom line is that soil conservation technologies designed to increase productivity by controlling soil loss do no such thing. If the quality of the soil remaining, rather than the quantity lost, is a more important determinant of subsequent yields, then more attention needs to be directed at maintaining and improving soil quality.

A change in focus from the quantity of soil eroded to the quality of soil that remains in a farmer's field, also aids the understanding that soil erosion is a consequence rather than a cause of soil and land degradation. This is one of the founding principles or insights of the better land husbandry approach. The better the quality of a soil, the more organic matter it contains, the more open its texture, and the greater its capacity to absorb rainfall and restrict runoff. The onset of soil erosion is actually a consequence and not a cause of soil degradation. This fact is elaborated in Chapter 4 but in summary, decreased cover of the soil—in some cases caused by farmers burning their fields during land preparation—subsequently allows high-energy rainfall to impact the soil surface directly. The damage caused by raindrops leads to reduced porosity in the surface layers. This, in turn, causes more runoff (Shaxson et al. 1997). As the soil becomes more degraded, there is less infiltration and more runoff.

Conventional soil conservation programmes that focus on controlling soil erosion rather than maintaining or improving the quality of soil remaining in farmers' fields, address the symptoms of soil degradation rather than its causes. Conventional soil conservation strategies tackle neither the fundamental physical cause of soil erosion, which is soil degradation, nor the fundamental social and economic factors that prevent effective soil loss mitigation, i.e. the complex of constraints that affect the farmer's decision-making *vis-à-vis* land management. Most seriously of all, conventional soil conservation approaches tend to downplay the significance of the central purpose of the farmers' activities, which is to generate maximum livelihood benefit from the land through agricultural production.

Clearly, there is a need for a new way ahead. This new way must stress the central causes of land degradation, which are the quality of the soil and the means of production. Equally, it must shake off the negativity of the soil loss prevention ethos and focus, instead, on the positive aspects of land management. It must place the farmer first and must attend to soil quality and land productivity. It is precisely these ideas that provide the foundations and starting point for the better land husbandry approach to land management (Shaxson et al. 1989).

Better land husbandry strategies strive to address, simultaneously, farmers' concerns about productivity and, through a focus on soil quality, soil conservationists' concerns about minimising runoff and avoiding erosion. The better land husbandry approach signals a move away from looking at erosion in terms of 'what' is happening to questioning 'why' it is happening (Norman and Douglas, 1994:9). Examining the 'why' component facilitates identification of the causes of erosion and provides the basis for designing appropriate strategies to combat soil degradation by treating its causes, not just its symptoms (Norman and Douglas, 1994:45). It also shows where there are still gaps in our knowledge, particularly with respect to the beneficial roles played by the soil organisms.

This focus on 'why' erosion is happening requires a holistic analysis of the whole farming system. Embedded within this system may be found the agroecological, social and economic reasons why farmers' land management practices lead to soil degradation and why they are unable to change these circumstances. It aspires to discover an integrated framework within which land management issues can be analysed and practical, productivity-enhancing, strategies be formulated and implemented.

1.3.2   Agro-ecological Components of Better Land Husbandry

The focus of the better land husbandry approach is to improve soil quality, particularly soil architecture via soil protection, incorporation of organic matter, and the use of soil organisms. In this respect, it is in sharp contrast to the conventional soil conservation approach (see Box 1.5). The conventional soil conservation approach attempts to 'combat' soil erosion head-on by capturing soil once it has been eroded. The better land husbandry approach, on the other hand, aims to maintain optimum soil conditions—in physical, chemical, biological and hydrologic terms—for the acceptance, transmission and retention of water, and for root growth and crop production.

Box 1.5   Better Land Husbandry and soil conservation: altering some technical perspectives (from Shaxson et al. 1997b)

Better land husbandry perspective	*Conventional soil conservation perspective*
1. Chief causes for concern are (a) decline of *in-situ* productive potential of land, and (b) insufficiency of soil moisture.	1. The primary cause for concern is the quantities lost of soil particles and water.
2. Improving and managing soil to ensure optimum rainwater absorption and retention—to limit frequency and severity of growth-inhibiting water stress in plants and avoid excess of soil moisture—will have a greater beneficial impact on plant production than only constructing physical cross-slope works to catch or direct runoff water and soil already on the move.	2. It is commonly assumed that cross-slope physical conservation works will result in significant increases in yield, by holding back soil, water and nutrients in narrow bands across the slope.
3. Accelerated runoff and erosion (beyond that naturally	3. Accelerated runoff and erosion are visualised as primary

occurring in the landscape) are foreseeable ecological processes, and *consequences* of aspects of land degradation, especially of loss of cover and of soil porosity.

4. Post-erosion yields at any site after erosion are closely related to the quality of soil that remains *in situ*.

5. Rainfall's erosivity can be minimised by breaking the force of large raindrops by ensuring some form of cover over the soil surface.

6. Soil's erodibility is increased or diminished over time by effects of soil management

7. More intensive use of land at a particular site such that it (a) improves soil architectural conditions by favouring soil-organic transformations and minimising tillage-damage, and (b) increases density, duration and frequency of cover over the soil—can improve rather than diminish the conservation-effectiveness of the particular use.

8. Increased production of plant parts—including improvements in soil architectural (structural) conditions, and in the amounts of cover over the soil—is an effective way of *achieving* conservation of water and soil as a consequence of better husbandry within the farm production system.

active *causes* of land degradation.

4. It is generally assumed that decline in yields post-erosion can be related closely to quantities of water, soil particles and plant nutrients lost in the erosion process.

5. Erosivity of rainfall is usually implicitly assumed to be an unalterable feature of each rain event.

6. Erodibility of a soil series is assumed to be an inherent characteristic of that series.

7. If at a particular site the land use is "too intensive" for the Land Use Capability classification of that site, it is recommended to reduce the land use intensity until it matches that permitted for that class.

8. It is usually insisted that soil conservation be done/implemented before yields can rise.

9. Because the land system is dynamic, maintaining its capacity to continue producing what we want requires its active and conservation-effective management over time, at the same time as any reallocations of land uses and imposition of any physical works that are deemed necessary.	9. It is implied that land would be least subject to erosion when its uses are allocated across the landscape in accordance with the maps of "Land Use Capability Classification" and treated with types and layouts of physical and biological conservation measures.
10. Solving problems of low productivity and of erosion and runoff requires an inter-disciplinary approach to match the inter-relatedness of the problems' causes.	10. It has been assumed that soil conservation requires a mono-disciplinary specialist approach independent of other specialisms, and needs separate institutional arrangements.

Soil factors that favour root growth also favour better water relations in the soil and the conservation of soil and water *in situ* (Shaxson, 1993). Hence, improving soil structure and infiltration capacity will achieve both conservation and production (Shaxson, 1988). Farmers' concerns about productivity can be made more compatible with the conservationists' concerns about minimising runoff and avoiding erosion (Norman and Douglas, 1994:44). While conventional soil conservation programmes identify conservation as a precursor to increasing yields, the better land husbandry argument is that soil conservation can be achieved, more effectively, as a side effect of soil improvement. In such circumstances, improved production may go hand in hand with better erosion control (Hudson, 1988).

From the technical perspective, better land husbandry involves the active management of rainwater, vegetation, slopes and soils via the use of technologies and practices that, together or individually, are productivity-enhancing and conservation-effective. This may entail the use of agronomic, biological and mechanical practices. The most critical variables under control of the land user are cover and soil structure (Douglas, 1993; Shaxson, 1992; Lal, 1988). Soil surface cover, either living or dead, is the best single factor for reducing erosion (Foster et al. 1982; Kellman and Tackaberry, 1997:264).

Conventional soil conservation approaches do recognise the importance of soil cover and, in the last 15 years, there has been increasing interest in the use of cover crops and conservation tillage. However, soil cover alone may not be sufficient to maintain optimum soil structural conditions (Busscher et al. 1996) and more attention will often need to be directed at the activities of micro-, meso- and macro-organisms, both plants and animals, to enhance soil quality.

1.3.3   Social and Economic Components of Better Land Husbandry

In reality, the complexity of the factors that affect farmers' decision making is such that a successful agricultural development programme is unlikely to be built on a single component such as a soil conservation technology. The whole farming system must be considered (Bunch, 1982; Hudson, 1993b). In the past, when confronted with the tension between farmers' realities and practical action, professionals have often opted for 'simplicity', which is often manifested by mono-disciplinary recommendations that are not relevant to multi-faceted problems (Shaxson, et al. 1997). This is partly the outcome of an educational and training system that favours the production of narrowly specialised technical experts, who may find it hard to see the farmer's 'big picture' (Chambers, 1993; Doran et al. 1996). However, it is made worse by a tradition that views farmers as part of the problem rather than the core of the solution (Enters and Hagmann, 1996; Hudson, 1988; Shaxson et al. 1989:13).

Farmers should not be addressed as targets; they should to be treated as collaborators and controllers (Shaxson, 1997). Unless farmers are enabled to express their priorities, and through participation to direct research and extension into avenues that support their needs, the technology provided to them is liable to be inappropriate (Chambers, 1993). In sum, the farmers' problems are precisely those problems that need to be resolved. So, the new starting point is the farmers' reality and its impact on their decision-making (Scoones and Thompson, 1994). The differences in social and economic perspectives between the better land husbandry and conventional soil conservation approaches are detailed in Box 1.6.

Box 1.6 Better Land Husbandry and soil conservation: altering some social and economic perspectives (from Shaxson et al. 1997b)	
Better land husbandry perspective	*Conventional soil conservation perspective*
1. Farm families have their own observations and perceptions about land degradation, and other views of the reality than those of development specialists, they should be allowed to judge what is best in their situation.	1. Specialists' perceptions of land degradation problems and solutions have been presumed to be the correct ones.
2. The rural community and the development of its abilities to manage its own environment is the most appropriate focus of outside assistance.	2. Land conservation, production and economic efficiency have usually been proposed as the primary foci for development assistance.

3. Resource-poor small farmers have considerable knowledge about their environments and make rational decisions about allocation of their resources in the context of a number of constraints. The challenge is to reduce these constraints.

3. It has often been assumed that resource-poor farmers are by nature conservative, irrational and ignorant of good land use; the task is to change farmers' rationality.

4. Rural families ultimately decide what will be done on the land and whether it would be in their interests to change according to recommendations; resource-poor farmers are more vitally concerned than any outsider to maintain their lands' productivity in both the short- and long-term.

4. Governments have assumed that they should decide what is done on the land because they have a greater long-term concern to maintain productivity and halt land degradation than do small farmers with (supposedly) short-term horizons.

5. To get conservation-effective agriculture improved, it is important to start from where people are now, assist them to do better what they are already trying to do and remove constraints that inhibit their doing better.

5. Adoption of recommended changes and innovations have been promoted as being essential for getting agriculture moving.

6. A community, and the land it occupies and uses, is the optimum focus for planning and for integrating inputs of various "disciplines".

6. The topographic catchment/watershed with the people it contains, has generally been stated to be the logically optimum unit for programme planning, and for demonstrating the effects of technical recommendations.

7. "Participation" signifies technical advisers participating with farm families in helping people to identify and rank their most important problems, to decide what to do about them, to implement decided actions and to monitor the outcomes.

7. "Participation" has commonly been taken to mean "the people participating in implementing plans" devised by professionals, which are considered good for them.

8. Advisory workers should be promoters of dialogue and of two-way information-transfer, catalysts of interactions, and facilitators of interchange and of farmers' well-informed actions.	8. Extension workers have been trained as demonstrators and one-way transmitters of information to farm families in a process of "transfer of technology".
9. Until they have proved themselves to the satisfaction of individual farmers, technical advisers have very low credibility at the outset of their interactions with farm families.	9. Technical advisers armed with scientific knowledge assumed themselves to be 100 per cent credible from the outset.

In the context of better land husbandry, any farming practice is conservation-effective if it promotes and maintains soil in optimum condition for water acceptance and plant growth. The first step is to identify which of the farmers' existing practices are beneficial from a conservation point of view and which are not, along with the farmers' reasons for practising them (Douglas, 1993). Is the soil surface exposed to the full impact of rain drops because farmers regularly burn their fields, destroying vegetative cover and the soil organisms that play a critical role in maintaining soil fertility? Can this be prevented? Do farmers burn their fields because it is a labour-saving way to prepare their fields for planting? These are the sorts of questions that need to be answered prior to working with farmers to improve land management.

Understanding the causes of soil degradation necessitates finding out about the farmers' perspective. Suitable land management options can then be framed, strategies that are adapted to the goals and circumstances of farmers. In some cases, social and economic constraints may have to be tackled before the farmers are enabled to start on-farm conservation action. In some circumstances, these constraints may not be resolvable. However, much can still be achieved, for example, improvements in crop husbandry practices such as early planting, optimum density and leaving crop residues on the surface, can reduce splash erosion, water infiltration and lead to a rapid improvement in soil quality (Sillitoe, 1993; Gardner and Mawdesley, 1997; Stocking, 1994).

Better land management initiatives, that put the farmer first, encourage farmers to take charge of their lives and solve their own problems by involvement and innovation (Bunch, 1982; Edwards, 1989). The driving force behind participation is the enthusiasm that comes from programmes that address farmers' priority problems, work with farmers, and bring about early recognisable success. As farmers participate in programmes, they gain self-confidence, pride and the satisfaction of having made significant achievements (Bunch, 1982:28). The confidence that comes from this process of empowerment, increases farmers' ability to learn and experiment, skills that are vital in an increasingly globalised world that places a premium on adaptability and responsiveness (Ellis and Seeley, 2001).

1.4   WINNING THE DEBATE?

1.4.1   The Message is Not Getting Through Fast Enough

Soil and land degradation is a real threat to achieving a sustainable agriculture. The conventional soil conservation approach with its emphasis on transfer of technology and capturing soil once it has been eroded is, however, misdirected. By focusing on soil erosion rather than soil quality, we are potentially doing a disservice to farmers because any relationship between productivity and the soil depends much more on the quality of the soil remaining on the land rather than the amount of soil removed through erosion.

It is not too much of an exaggeration to claim that the conventional soil conservation approach has largely been a waste of money and effort. There are of course major off- and on-farm benefits to reduced soil loss. A growing body research suggests that these benefits can be attained via alternative approaches, such as better land husbandry, that promote strategies that seek to combine farmers' concerns about productivity with conservationists' concerns about reducing soil erosion, often via soil cover-management and an improvement in soil quality.

Over ten years ago, Hudson (1995:204) speculated that *"if the present trend continues 'soil conservation' may become a pejorative term"*. This has not happened yet. The conventional soil conservation paradigm, supported by some scientific research that shows that the technologies promoted do contribute to soil conservation and increased productivity, is still in the ascendancy (Scoones and Thompson, 1994; Segerros and Kerr, 1996). Despite evidence of low farmer adoption and adaptation rates of soil conservation technologies and a growing body of evidence that supports the rationale of the non-adopters, there is still a tendency to assume that soil conservation specialists know best. There is, therefore, a need to win over the sceptics and those who are either disinterested or uninterested. Informed debate backed up by rigorous research is an effective way to do this. This book is designed to contribute to this debate.

1.4.2   Structure of the Book

This book is made up of six chapters. Chapters 2, 3 and 4 lay out a compelling case for the paradigm shift from conventional soil conservation to better land husbandry. The narrative in these three chapters is supported by other people's research but throughout the author draws heavily on work that he carried out on degraded hillsides in Honduras over a three-year period (1996-1998). The research in Honduras was designed to shed light on farmers' reasons for non-adoption of soil conservation technologies and to identify a better, alternative strategy for achieving both soil conservation and a more sustainable agriculture in steeplands.

Chapter 2 details the changes in concept and practice between the conventional soil conservation and better land husbandry approaches with respect to the social and economic components of land management. The chapter looks at the importance of adopting a 'farmer-first approach', it identifies a number of research tools that can be used to capture the complexity of farmers' realities and details some of the key social and economic constraints that prevent farmers better managing their land.

Chapter 3 looks at the agro-ecological components of land management and the changes in concept and practice between the conventional soil conservation and better land husbandry approaches. The chapter details the wealth of research that has been carried out on measuring and assessing soil loss. Based largely on the results from the author's work in Honduras, the chapter goes on to question the effectiveness of the conventional soil conservation approach in reducing soil loss and, more importantly, contributing to increased production and soil productivity.

Chapter 4 details the importance of focusing on improving and maintaining soil quality. It looks at the physical, biological and chemical components of soil quality and demonstrates how, by improving soil quality researchers, extension agents and farmers are able to achieve soil conservation as well as simultaneously addressing farmers' main concern *vis-à-vis* improved and reliable agricultural production.

Chapter 5 provides a number of case studies of better land husbandry in practice. Each of the three sets of case studies is designed to illustrate one or more key components of the better land husbandry approach. The case studies from South and Central America look at the importance of soil cover. The case study from Honduras demonstrates the importance of active farmer participation and farmer empowerment. The case studies from parts of Africa clearly show the importance of conserving soil moisture. The chapter ends with a word of caution, namely that despite the advantages of the better land husbandry approach, it is not a panacea and cannot protect hillsides from extreme rainfall events.

Chapter 6 looks at the importance of the better land husbandry approach being supported by an enabling policy environment and a cadre of development professionals. It also outlines why better land husbandry needs to be put into a wider perspective that takes account of other rural-based activities, such as non-agricultural employment, that can contribute to a reduction in rural poverty.

Appendix 1 provides some basic approaches and tools for impact monitoring and assessment. Better land husbandry is a cross-disciplinary approach to improved land management that involves social, economic and agro-ecological aspects. Any impact monitoring and assessment system needs to measure or assess impact in all these areas. The appendix, therefore, covers environmental impact (covering the agro-ecological aspects of better land husbandry) and livelihood impact (covering the social and economic aspects of better land husbandry).

2

Social and Economic Components of Better Land Husbandry

"While it is certainly true that soil erosion is undermining agricultural production in certain places, the crisis may not be as prevalent as some commentators suggest. Indeed, our concentration on soil loss as the major issue has eclipsed other important production constraints, such as water quality available to plants, soil nutrient levels, labour availability, market incentives, and so on" (Scoones et al. 1996:4).

2.1 CONVENTIONAL SOIL CONSERVATION APPROACH

2.1.1 Promotion of Soil Conservation Technologies

As outlined in Chapter 1, pressures on the land are increasing and the challenge in the tropics and subtropics is to sustain the land's productivity without destroying its quality (Dumanski et al. 1998:108; Pieri et al. 1995:1). The front-line of this struggle is steeplands, where economically disadvantaged and often inexperienced farmers are cultivating the most fragile and erosive lands. The result is often accelerated soil and land degradation (El-Ashry, 1988; Hudson, 1988).

Faced with the danger that soil and land degradation will undermine efforts to increase agricultural productivity on a sustainable basis, one common response is investment in soil conservation technologies. In the last 40-50 years, soil conservation programmes have become commonplace in many parts of the developing (and developed) world (Hudson, 1995:354). The soil conservation approach identifies runoff and soil loss as the principle causes of land degradation. Based on this premise, the solution has been to provide farmers with technical advice and soil conservation technologies that can be used to restrict soil and water losses. These have included live barriers, terraces and, sometimes, the use of cover crops.

Government experts and non-governmental organisations have often implemented soil conservation programmes as part of planned national programmes designed to educate and involve the uninformed farming communities (Norman and Douglas, 1994:55; Young, 1989). There is one fundamental problem with this approach: farmers have not been particularly enthusiastic and active participants in these soil conservation programmes.

In any assessment of the success of a soil conservation programme both adoption and adaptation should be addressed. Adoption rates of particular technologies *per se* reveal little about the impact on the livelihoods of farming households (Hudson, 1992b:71) and little about farmers' capacity to experiment, learn and adapt technologies in the face of changing social, political, economic and technical circumstances (Bunch, 1999; Pretty, 1998). Adoption merely shows that farmers have accepted a technology but seldom sheds light on whether farmers are merely adopting the techniques as passive users. Of more importance and interest is the issues of adaptation whereby farmers incorporate the technique as a tool that they own and then subsequently adapt for their own purposes. Hinchcliffe *et al.* (1995:15), therefore, propose that criteria of success should be partly based on the degree of adaptation rather than adoption of technologies. Bunch (1982:190) argues that an assessment of the degree of innovation should be made about five to seven years after external funding finishes.

Three criteria can be used to judge whether soil conservation programmes are successful:

- Degree to which smallholder farmers (the target audience) are adopting and/or adapting the technologies.
- Degree to which the technologies reduce erosion and runoff.
- Degree to which the technologies have led to increased and/or sustained productivity.

The criteria are linked because farmers are more likely to adopt and sustain a soil conservation technology if it leads to an increase in productivity. Despite this, farmer-adoption is often quantified in terms of the number of farmers who have adopted a particular technology and/or the length of live barriers or terraces established. Much less attention has been directed at the impact of the technologies on productivity.

It is not always easy to assess the degree of adoption or adaptation of a soil conservation technology. For example, does a poorly maintained soil conservation structure represent adoption or abandonment? This is exemplified by Rwanda, where in a survey, farmers referred to terraces that they had constructed even though many of the terraces had been washed away and hardly served a conservation function (Ndiaye and Sofranko, 1994). It is also important to note that poorly maintained soil conservation technologies can even exacerbate the erosion problem (Pretty and Shah, 1997).

Based on the criteria of farmer adoption and adaptation of technologies and practices, the results of many soil conservation projects to date have been disappointing (Hudson, 1991; Pretty and Shah, 1997; Pagiola 1994; Fujisaka, 1991; Blaikie, 1989; Hinchcliffe et al. 1995:3; Stocking, 1994; Bewket and Sterk, 2002).

2.1.2   The Use of Incentives in Soil Conservation Programmes

Faced with the poor uptake of soil conservation recommendations, many researchers and practitioners have argued that incentives are needed in order to encourage farmers to adopt recommended soil conservation practices, at least during the first few years when there may not be tangible benefits (Napier et al. 1994; Southgate, 1994). Incentives are 'any inducement on the part of an external agency (government, NGO or other), meant to both allow or motivate the local population, be it collectively or on an individual basis, to adopt new techniques and methods aimed at improving natural resource management' (Laman et al. 1996).

Most authors (e.g. Sanders et al. 1999) distinguish between direct and indirect incentives (see Figure 2.1). The former includes cash payments for labour, grants, subsidies, loans, and also in kind payments such as the provision of food aid (food-for-work) and agricultural implements. Indirect incentives include fiscal and legislative measures such as tax concessions, secure access to land, and the removal of price distortions (Sanders and Cahill, 1999). The provision of indirect incentives is often dependent on policy decisions made at central government level. Soil conservation programmes run by private and public organisations are, therefore, unable to offer most indirect incentives and have tended to focus on the use of direct incentives.

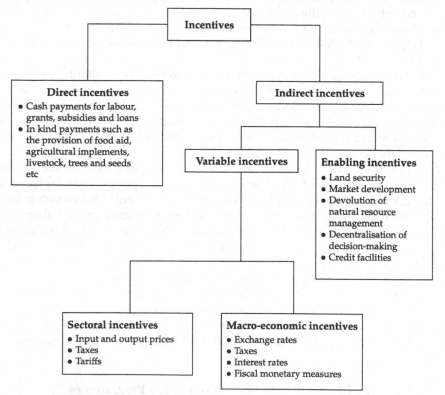

Fig. 2.1   Typology of incentives commonly used in soil conservation programmes (from Hellin and Schrader, 2003 and based on Enters, 1999 and Sanders and Cahill, 1999)

One of the justifications for offering incentives to farmers is that the incentives represent a legitimate payment for the off-site benefits of conservation that are enjoyed by society (Stocking and Tengberg, 1999). These benefits include reduced downstream siltation of reservoirs and impairment of aquatic ecosystems (Huszar, 1999). It is also argued that incentives at the beginning of a soil conservation programme are critical because farmers may not be able to afford investments in soil conservation. Furthermore, the economic benefits of soil conservation, in terms of improved yields can be delayed for several years (Perich, 1993; Heissenhuber et al. 1998).

In theory, once farmers are aware of the benefits of the soil conservation technologies, direct incentives can be phased out. The problem is that subsidies tend to buy short-term acquiescence and do not necessarily lead to long-term changes in attitudes and values (Pretty, 1998b:293). Subsidies generally only persuade farmers to modify their behaviour as long as they continue to be paid (Lutz et al. 1994; Sanders, 1988b). Hence, whilst farmer implementation rates worldwide have been enhanced by these temporary subsidies, more often than not farmers abandon the technologies once external support is withdrawn (see Box 2.1 and Figure 2.2).

Box 2.1 Worldwide examples of farmer abandonment of soil conservation technologies once subsidies have been withdrawn

Many soil conservation programmes have sought to overcome farmers' lack of interest by offering subsidies to adopt particular technologies (Pretty and Shah, 1999). There are many examples that demonstrate that the mid- and long-term implications of offering subsidies have not been encouraging in terms of farmer adoption of technologies (Sanders, 1988b).

Farmers abandoned alley-cropping with *Leucaena leucocephala* in Nigeria (Dvorak, 1991) and contour hedgerows in the Philippines (Fujisaka and Cenas, 1993) when subsidies finished. In Thailand, incentives such as free inputs of seeds and fertiliser, and cash incentives were offered for in return for the construction soil conservation technologies. When the subsidies were withdrawn, the number of farmers establishing and maintaining the soil conservation technologies dropped (Enters and Hagmann, 1996). A similar situation occurred in the Dominican Republic (Carrasco and Witter, 1993) and in tree-planting programmes in Honduras in the 1980s (Tschinkel, 1987).

Hellin and Schrader (2003) report on a soil conservation project in Honduras that ran from 1980-1991. Soil conservation technologies were promoted from 1980 onwards, but direct incentives were used only from 1984 to 1991. These incentives included cash payments for farmers' labour input; free seedlings, barbed wire and fertiliser; and subsidised

---

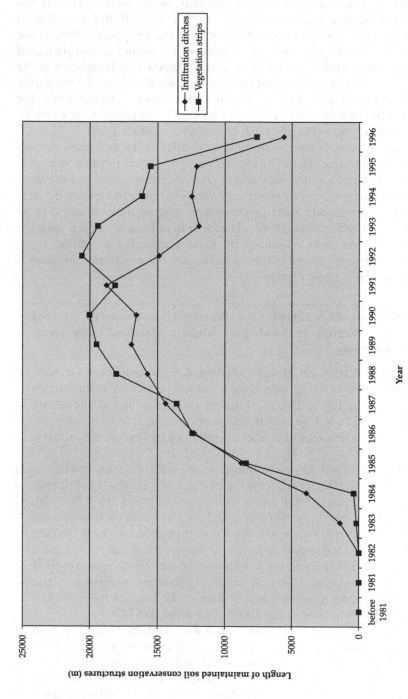

Fig. 2.2   Adoption and abandonment of soil conservation technologies in Honduras. Project duration 1980-1991. Direct incentives provided 1984-1991. Number of farms = 147. From Hellin and Schrader, 2003.

credit. Figure 2.2 shows the establishment and maintenance rates of the two most heavily promoted soil conservation technologies – infiltration ditches and live barriers - and their abandonment once incentives were withdrawn.

Figure 2.2 suggests that when farmers establish the technologies, their main objective is to gain access to the direct incentive being offered. This is demonstrated by the increase in the length of soil technologies maintained when direct incentives were introduced in 1984 and high abandonment rates following the removal of the incentives in 1991. Furthermore, most participating farmers established the technologies in a small plot in order to qualify for the incentives. Technologies were not established on the majority of farmers' cultivated land.

The adoption of soil conservation technologies may also reflect the effect of negative incentives, such as the threat of resettlement, which 'encourage' farmers to participate in soil conservation programmes (Enters, 1996). In the 1930s and 1940s colonial governments in Africa often sought to impose soil conservation programmes on local people (Critchley et al. 1994). These programmes that ignored indigenous conservation practices, were seldom seen by local people as relevant to their needs and often generated more resistance to change (Robinson, 1989). This led to what Blaikie (1993) refers to as 'fortress conservation', where conservation efforts were imposed on local people. Farmers resisted and in some cases sabotaged the soil conservation technologies.

Another danger is that the use of incentives can reinforce the unequal relationship between farmers and outsiders. Poverty is not only an economic state but also a psychological attitude; in this context, farmers may end up deceiving outsiders by feigning incapacity and incapacitate themselves by internalising these same outsiders' belief in the farmers' inabilities (Chambers, 1997:206; Altieri, 1990). Hence, the use of incentives can engender apathy and a degree of fatalism, whereby farmers eventually become convinced that they are incapable of making progress by themselves (Bunch, 1982:20; Pretty (1998b:209). For example in the Cape Verde Islands, government soil conservation programmes have resulted in farmer apathy and lack of initiative at the local level (Kloosterboer and Eppink, 1989).

2.1.3 Time to Ask Ourselves a Few Questions

The promotion of soil conservation technologies in the last few decades is testament to the growing recognition of the dangers of land degradation by researchers and policy-makers, and of the need to mitigate the problem. Inevitably, therefore, the search for reasons as to why soil conservation practices have or have not been adopted has therefore focused on smallholder farmers and the decision-making process at the farm level.

Chambers (1993) and Bunch (1982:59) have pointed out that if farmers do not adopt recommended soil conservation technologies they are often accused of being ignorant, uncooperative, conservative and unwilling to change. Many reach this conclusion because of the abundance of literature that extols the virtues of many soil conservation technologies in reducing soil loss and in some cases increasing productivity. Farmers, however, ultimately decide what happens on their land (Hallsworth, 1987:143; Shaxson et al. 1989:30; Dumanski et al. 1998:112). If resource-conserving technologies, such as soil conservation practices, are to contribute to a more sustainable agriculture, they have to be seen by farmers as appropriate, relevant or necessary.

On this basis, a more informed analysis is needed into why farmers do not adopt official soil conservation recommendations, one that seeks answers to the following questions:

- Do farmers recognise the need to control soil erosion?
- Is soil conservation a priority for farmers or are there other more pressing problems that affect their livelihoods and which they would like to see addressed first?

And, for those who wish to find answers to these questions, there is the key issue of how outsiders gain an understanding and appreciation of farmers' realities so that soil conservation approaches can be modified to complement these realities.

2.2   FARMERS' REALITIES

2.2.1   Livelihood Strategies

Resource-poor farmers are probably more desperate than their governments and any number of soil conservation experts about the land's quality because their livelihoods are intimately dependent on the maintenance of its productivity (Shaxson, 1993). Farmers' attitudes, observations, perceptions, suggestions, knowledge, insights, interpretations and comments are, therefore, absolutely fundamental to any initiative to improve land management. Recommendations for change, therefore, must be perceived to be beneficial by the supposed beneficiaries; they must be perceived to be contributing to the beneficiaries' livelihoods.

The term 'livelihood' is commonly used in development. A useful definition is that provided by Chambers and Conway (1992): *"a livelihood comprises the capabilities, assets (including both material and social resources) and activities required for a means of living. A livelihood is sustainable when it can cope with and recover from stresses and shocks and maintain or enhance its capabilities and assets both now and in the future, while not undermining the natural resource base"*. Building on this definition, a very useful conceptual tool for measuring the impact on peoples' livelihoods is the Sustainable Livelihoods Framework (SLF) (see Figure 2.3).

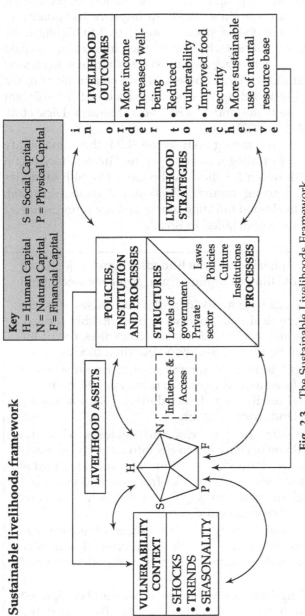

Fig. 2.3   The Sustainable Livelihoods Framework

The Sustainable Livelihoods approach, represented by the SLF is a way of looking at development in a way that is concerned principally with people. The approach seeks to understand people's strengths, including their skills and possessions, and how they use these assets to improve the quality of their lives. The sustainable livelihoods approach can be used to highlight the interactions between soil conservation technologies, the vulnerability context of farmers, their asset base, the intervening institutions, and farmers' livelihood strategies. Although the individual, farm household and community are the primary level of analysis, livelihood approaches also seek out the relevant interactions at micro-, intermediate- and macro-levels (Meinzen-Dick et al. 2004; Adato and Meinzen-Dick, 2002).

As we can see in Figure 2.3, the starting point in the SLF is the vulnerability context within which people, including farmers, operate. The next focus is on the assets that people can draw on for their livelihoods. The SLF identifies five classes of assets: human, social, natural, physical and financial capital. These assets can be seen as livelihood building blocks and are represented by the pentagon in Figure 2.3 and are detailed in Box 2.2.

Box 2.2   Assets and the Sustainable Livelihoods Framework. Based on http://www.livelihoods.org/info/info_guidancesheets.html

The sustainable livelihoods framework identifies five classes of assets: human, social, natural, physical and financial capital. In this context, capital does not mean capital stocks in the strict economic sense of the term. The five capitals can be seen as livelihood building blocks.

- **Natural capital**—is the term used for the natural resource stocks from which resource flows and services (e.g. nutrient cycling, erosion protection etc.) useful for livelihoods are derived. Natural capital includes vegetation, land, water and air.

- **Social capital**—reflects the patterns and systems of social organisation that facilitate or constrain co-operative enterprise, inter-household relations and individual entitlements. Includes formal and informal organisations and networks from community-based organisations to religious groups to neighbours who help each other out by sharing food, money and child-care etc.

- **Human capital**—equates broadly with levels of education, knowledge and health that enable people to experiment and solve their own problems and ultimately pursue different livelihood strategies.

- **Physical capital**—comprises the basic infrastructure and equipment and property needed to support livelihoods. The following components of infrastructure are usually essential for sustainable livelihoods: affordable transport; secure shelter and buildings; adequate water supply and sanitation; and clean, affordable energy.

> • **Financial capital**—is the financial resources that people use to achieve their livelihood objectives and include access to credit, loans, savings and remittances.

Sustainable livelihood development depends not on advances in access to just one of the classes of assets but on systematic cross-sectoral approaches to achieve an appropriate balance between these essential assets. The five classes of assets interact with policies, institutions and processes to shape the choice of livelihood strategies that farmers follow in pursuit of the livelihood outcomes such as more income, increased food security and more sustainable use of natural resources. For example, Mazzucato et al. (2001) report that farmers in Burkino Faso invest heavily in social networks (building up social capital) and that this gives them flexible access to resources necessary for agriculture and soil conservation (building up natural capital).

One of the important characteristics of the SLF is that it recognises that farmers may well pursue multiple strategies sequentially or simultaneously in pursuit of their preferred livelihood outcomes. In managing land to meet family livelihood requirements, farmers therefore have to consider a wide range of influences including risks and opportunities for profit. In this context, farmers are embedded in a range of social, economic cultural and political relationships that affect their access to property and labour, and their decision-making power within their communities and households with respect to livelihood options (Gardner and Lewis, 1996:87) (see Figure 2.4).

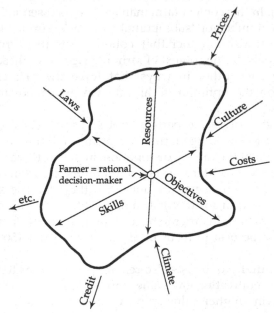

Fig. 2.4   A farmer makes rational decisions *vis-à-vis* livelihood options within an 'envelope' of constraints and potentials (from Shaxson et al. 1989:31)

Rural livelihoods, therefore, tend to be diverse and complex, with farmers reliant on non-agricultural and non-farm as well as agricultural and farm sources. Reflecting the complexity of their livelihoods, farmers' problems are often multiple and interrelated. Farmers are commonly faced with a range of adverse agro-ecological, social and economic conditions including erratic rain, low fertility soils, fluctuating market prices for agricultural products, and labour shortages (Douglas, 1993). In addition, farmers minimise risks and seldom take chances that might lead to hunger, starvation or loss of their land. Complex and diverse livelihood and farming systems reduce vulnerability and enhance security.

Granting legitimacy to people's beliefs and knowledge is likely to reveal that much farmer decision-making is actually rational (Kiome and Stocking, 1995; Unwin, 1985) (see Box 2.3). Harrison, (1992:84) gives the example of Madagascar, where farmers are aware that with 60 kg of fertiliser it is in theory possible to produce enough rice to last all the year. This represents a gain worth many more times the initial investment. However, farmers do not have sufficient money to purchase the fertiliser and are reluctant to borrow money in case the harvest fails.

Box 2.3   Rationale behind farmer decision-making is not always obvious to outsiders

Local people often know far better than development planners do how to strategize to get the best from difficult circumstances (Gardner and Lewis, 1996:15). In this context farm management is essentially a process of balancing a number of 'sub-optimal systems' (Suppe, 1987). Each system has optimisation foci that cohere with the farmer's own management goals. In the context of farm management, these strategies interfere with each other in ways that force the farmer to make compromises on the optimum technical management of most of the farming system.

Collinson (1981) gives the perhaps not so hypothetical example of outsiders trying to encourage farmers to plant a 180-day maize variety that can fully exploit the rainfall distribution, in place of farmers' practice of planting four weeks after the start of the rains. There may be sound reasons for farmers choosing to plant when they do. The soil is hard and farmers must wait for the first rains to soften the soil before they can prepare it by hand. In order to be relevant to farmers' circumstances, outsiders would be best to focus on productivity from 150-day maize varieties.

Some agricultural economists also remain perplexed as to why many farmers insist on cultivating low-value crops such as maize and sorghum instead of growing higher-value crops. What they fail to understand is that, besides seeking to reduce risk, aspects of subsistence agriculture are intimately related to culture (Pawluk et al. 1992; Durham, 1995).

With respect to indigenous farmers in Bolivia, Rist (1992) refers to a *cosmovisión andina* in which nature is intricately linked with society and culture. Any change is evaluated in relation to the *cosmovisión andina* and, as such, technical change promoted by outsiders is often judged by local people to be inappropriate.

Maize is indigenous to Mexico and has been cultivated there for approximately 6,000 years (see Plate 3). Many indigenous groups in the southern states of Mexico, such as Oaxaca, believe that the maize plant represents the origin of life itself (The Economist, 2004). Many farmers in Latin America, therefore, conceive growing maize as a cultural tradition, one that they are determined to continue irrespective of the outcome of the analysis of costs and benefits of alternative crops (O'Brien, 1998:92) (see Plate 4). Central American and Mexican farmers are also interested in more than just maize yields (Bunch, 1982:120). Maize stalks are used for fencing, husks for wrapping hot food and leaves for fodder. Furthermore, in Mexico, the desire to participate in religious festivals drives resource-poor farmers to harvest their maize early and to sell their grain before the price reaches its maximum (Meinzen-Dick et al. 2003; Bellon et al. 2003).

Wilken (1987:17) adds that in subsistence systems particular tasks may result in net losses but are justified by such non-quantifiable returns as security or prestige. There are also examples from the developed world e.g. Douglas *et al.* (1994) report an atavistic desire to maintain traditional farming methods in the Alpujarras mountains in southern Spain even though the main source of family income is non-agricultural.

Plate 3   Maize is indigenous to Mexico and has been cultivated in parts of Mexico and Central America for approximately 6,000 years. Guatemala. (Hellin, J.)

Plate 4   Many people in Latin America perceive growing maize as a cultural tradition.
Argentina. (Hellin, J.).

In the context of farmers' realities, decision-making *vis-à-vis* adoption of
soil conservation technologies is likely to be influenced by a number of factors.
As a result, soil conservation recommendations have been made that fail to
take into account the totality of the farming system and the envelope of
constraints and potentials within which the farmer makes decisions.
Unbeknown to the soil conservation expert, recommendations that lead to a
change in management of one component of the farming system may directly
or indirectly influence the feasibility, productivity and profitability of the
other components (Shaxson, 1999:69).

Many farmers depend upon production from their land and also upon off-
farm income-generating activities. Farmers' decision to pursue multiple
activities in order to achieve desired livelihood outcomes often, therefore, has
far-reaching implications for the availability of labour at different times of
year. This clearly affects their willingness to invest labour in the establishment
and maintenance of soil conservation structures such as terraces. Understanding
the true complexity of farmers' livelihoods can, therefore, help researchers
and development practitioners to develop soil conservation technologies and
approaches that better complement farmers' complex livelihood strategies.

When seeking to understand farmers' livelihoods and the rational of their decision-making it is also important to remember that within the village boundary, the water supplies, land types, vegetation, livestock and people are inter-linked and inter-acting in such a way that the village as a whole is a dynamic ecosystem (see Figure 2.5). In this case recommended changes to a farming system, such as the introduction of a new crop, can have unforeseen consequences for the functioning of the village ecosystem.

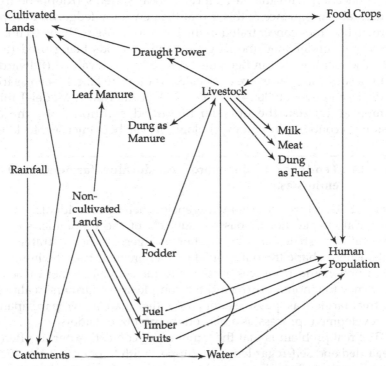

Fig. 2.5 Village as an ecosystem (from Shaxson, 1999:90)

2.2.2 Outsiders' Failure to Appreciate Farmers' Realities

One of the great problems is that sometimes outsiders have failed to understand and appreciate the complexity of farmers' realities and the impact that this complexity has on farmer decision-making. More often that not, when faced with a tension between farmers' realities (characterised by complex diversity) and practical action, the tendency amongst professionals has been to 'simplify' and as a consequence to make inappropriate recommendations (Chambers, 1997:49). All too often mono-disciplinary recommendations that reflect the technical focus of the expert adviser are made to deal with what are, in fact, multi-faceted problems (Shaxson, et al. 1997).

Chambers (1997:68) refers to this as the 'Model-T approach' where technology is uniform and mass-produced as a standard package. Technologies are deemed to be universal and are applied widely irrespective of their compatibility with local conditions (Warren and Cashman, 1988:4; Pretty and Shah, 1997; Hallsworth, 1987:145). This transfer-of-technology approach, which is characteristic of many soil conservation programmes, often involves a loss of social, political, physical and biological context in which the new practice loses connection with the variability of local systems (Kloppenburg, 1991). The natural consequence of this is that the new knowledge, and the power to control it, becomes concentrated in the hands of those with the technical skills necessary to understand the language and methods being used (Edwards, 1989). The result is that indigenous knowledge is devalued (Edwards, 1989) and local social structures are over-ruled as they are largely seen as irrelevant (Gardner and Lewis, 1996:94) (see Box 2.4). Furthermore, Model-T misfits are common and when the outsider-identified solution fails, the damage physically, economically and psychologically can be immense (Stocking, 1993).

Box 2.4   Tendency to disregard and devalue farmers' skills and enthusiasm

Long (1992) refers to the human agency, whereby individuals participate in social change not as passive subjects of the economic, social and institutional structures, but rather as agents whose strategies and interactions shape the outcome of development within the limits of the information and resources available to the actor. The root cause of the tendency to ignore or disregard the complexity of farmers' realities and to treat farmers as 'passive adopters' rather than 'active participants' in a development process, is a bias on the part of outsiders.

The real problem is that the 'ignorant of context' expert is often both regarded and self-regarded as an 'upper' with 'modern', 'scientific' and 'superior knowledge'. Even when the 'upper' is a local person, it is often still in their nature to show that they have risen above the 'traditional' or 'peasant' knowledge.

Paradigms of development, under-development and the environmental crisis, whether from the left or right of the political spectrum, still, therefore, tend to emphasis the predominant position of the 'core' (urban centres of power in the developed and developing world). These paradigms have little to say about social and political processes in the 'periphery' (rural hinterlands in the developing world) except that their dynamics are geared to meet the requirements of the 'core' (Long, 1992; Gardner and Lewis, 1996:18). There is still a tendency to stress the inevitability of environmental destruction (e.g. de Janvry et al. 1989; Loker, 1996) and to portray smallholder farmers as condemned to a life which in *"solitary, poor, nasty, brutish, and short"* (Hobbes, 1651:143).

> An emphasis on the predominant position of the 'core' serves to efface the heterogeneity among societies that make up the system (Greenberg and Park, 1994:7) and to underestimate the innovative nature and resourcefulness of local peoples and their ability to be pro-active in adverse conditions. All too often farmers' entrepreneurial spirit is under-appreciated and "*caricatured in their vulgar terms, both neo-classical and Marxist theory render the individual virtually powerless to change the course of human affairs*" (Chambers, 1997:12).

It is this tendency to devalue farmers' skills and enthusiasm that partly explains why soil conservation initiatives have not been particularly successful. Outsiders have judged what technologies are appropriate, often on technical grounds. The concept of transfer of technology has, by its very nature, precluded farmers' participation. However, a conventional soil conservation programme, with an emphasis on the Model-T approach, is unlikely to accord with farmers' resources, needs and priorities. The inevitable result of the transfer-of-technology approach is that farmers are seen as 'adopters' or 'non-adopters' of technologies, rather than originators of either technical knowledge or improved practice (Scoones and Thompson, 1994b).

2.3 PUTTING FARMERS FIRST

2.3.1 Changing paradigms

If local peoples' knowledge and capacities are valued and even granted legitimacy within scientific and development communities, researchers and extension agents will pay greater attention to the priorities, needs and capacities of rural peoples. This will enable outsiders to make more appropriate and sustainable recommendations (Scoones and Thompson, 1994b). Farmers' perspectives must be understood and taken into full account if programmes to assist them are to succeed. This is because it is impossible to understand real-life problems fully unless the multitude of constraints, imperfections, and emotions that shape the actions and decisions of real, living people are also understood (Edwards, 1989). The challenge is for agricultural and development workers to learn to be genuinely sensitive to the problems faced by farmers (Pawluk et al. 1992).

In recent years, research and extension agendas have sought to find out more about farmers' realities and to seek active farmer participation at all stages of the development process. The so-called farmer-first approach, and more recently the focus on sustainable livelihoods, represents a radical departure from conventional approaches that tended to undervalue farmers' resourcefulness. The farmer-first approach represents a paradigm shift whereby local people are encouraged to express their own realities rather than have outsiders' realities imposed on them (Chambers, 1997:103).

The farmer-first approach starts with the knowledge, problems, analysis and priorities of farmers and their household members (Douglas, 1993; Scoones and Thompson, 1994). In this context, if farmers are degrading land this does not necessarily mean that they do not care, but rather that they are unable to find appropriate solutions of any sort including any soil conservation recommendations. One such recommendation may, for example, involve an attempt to maximise some sort of outcome e.g. productivity or effect some reduction in soil loss. However, if the farmer decides not to adopt the technology, this could be because the benefits of adoption are insufficient to compensate the farmer for the costs of establishing and maintaining the technology in terms of land, labour and capital inputs. Alternatively, the risk could be seen to be too great, or the farmer may be more interested in stable yields as opposed to maximum yields, or the recommendation may be at odds with the farmer's community and cultural context.

2.3.2   Farmer participation

In order for research and extension work to be more relevant to resource-poor farmers, the same farmers have to be engaged to help construct outsiders' understandings of the way in which their world operates (Edwards, 1989; Bellon, 2001). This is largely because the route to solving problems at the whole-farm level runs not through specialist science but through thinking in terms of whole farms i.e. farmers (Kloppenburg, 1991). Adequate representation of farmers' views and perspectives can be brought about only by farmer participation. Participation is needed at all stages because any outside intervention, such as a soil conservation programme or any development initiative, is an ongoing, socially-constructed and negotiated process, and not simply the execution of a specified plan of action with expected outcomes.

The reality of farmer participation, however, has changed less than the rhetoric (Warren et al. 1995; Bunch, 1982:58). The identification of farmers' problems, whereby farmers define and rank their priority needs, is an increasingly common aspect of participation but too often participation does not go any further (Fujisaka, 1989). There is a need to engage active farmer participation at all stages of project development and implementation (Altieri, 2002). This will avoid local participation being little more than politically correct rhetoric (Sillitoe, 1998). It will also help the project's development remain 'tuned' to its impacts and effects on the farmer's reality, including those impacts that even the land users did not anticipate.

Pretty (1995:173) identifies a seven-level typology of farmer participation that ranges from manipulative and passive participation, where people are told what is to happen and act out predetermined roles, to self mobilisation, where people take initiatives largely independent of external institutions (see Box 2.5).

Box 2.5 A typology of participation: how people participate in development programmes and projects (based on Pretty, 1995:173)

Passive participation—people participate by being told what is going to happen or has already happened. It is a unilateral announcement by an administration or project management without listening to people's responses.

Participation in information giving—people participate by answering questions posed by extractive researchers. People do not have the opportunity to influence proceedings, as the findings are not shared.

Participation by consultation—People participate by being consulted and external agents listen to views. These external agents define both problems and solutions and may modify these in the light of people's responses.

Participation for material incentives—People participate by providing resources, for example labour, in return for food, cash or other material incentives. Much on-farm research falls in this category as farmers provide the fields but are not involved in experimentation or the process of learning.

Functional participation—People participate by forming groups to meet predetermined objectives related to a project. This can involve the development or promotion of externally initiated social organisation.

Interactive participation—People participate in joint analysis that leads to action plans and the formation of new local institutions of the strengthening of existing ones. It tends to involve interdisciplinary methodologies that seek multiple perspectives and make use of structured learning processes.

Self-mobilisation—People participate by taking initiatives to change systems independent of external institutions. They develop contacts with external institutions for resources and technical advice they need but retain control over how resources are used.

The success of a soil conservation programme in terms of farmer adoption rests partly on the credibility of the extension agents and their ability to communicate with farmers (Paudel and Thapa, 2004; Bunch, 1982:157). Nimlos and Savage (1991) attribute some of the success of soil conservation initiatives in Ecuador to the fact that extension agents are motivated and are able to earn the respect of local farmers. Unfulfilled promises and failed recommendations, lack of attention to farmers' priority problems and outsiders' paternalistic assumptions that they know what is best for farmers, may combine to produce feelings of scepticism and apathy in which farmers' receptivity to new ideas is understandably low (Shaxson, 1992).

Ruaysoongnern and Patanothai (1991) conclude that soil conservation technologies have not been readily adopted in Thailand because, due to the lack of farmer participation, farmers generally consider the new technologies as 'official recommendations' where the recommendation is treated with the same enthusiasm as a new tax or bureaucratic regulation. In this situation, extension agents (who may be seen by farmers as little more than the mouthpiece for the government) have little credibility. Once lost, it is particularly difficult to regain farmers' trust.

Farmers may offer deference to an outsider but credibility must still be earned. Outsiders' credibility can be raised if programmes are directed at farmers' priority problems and if allowance is made for the fact that these priorities and their rankings many change with time. When mutual trust and credibility increase between and among farmers and outsiders, information flows more freely and the merging of local priorities and externally perceived opportunities become possible and realistic (Shaxson, 1999, Pretty and Shaxson, 1998; Norman and Douglas, 1994:87; Uphoff, 1996:365). Cools et al. (2003) give the example of farmers and scientists in Syria working effectively together to plan and implement improved land management practices. One of the reasons for the success of the initiative was the effective communication between scientists and farmers.

At its best, active farmer participation stimulates a positive feedback system at all stages of the research and extension process. This merger between top-down and bottom-up approaches has been referred to as a multi-level stakeholder approach (Hurni et al. 1996:20). It is an approach that is characteristic of better land husbandry and it demonstrates the fundamental role that this approach to land management can play in contributing to sustainable rural livelihoods. Box 2.6 illustrates that initiatives, which encourage active farmer participation, also contribute to enhancing rural livelihoods by increasing human and social capital as outlined in Box 2.2 (McDonald and Brown, 2000; Pretty and Hine, 2001).

Box 2.6   Farmer participation and increased social and human capital

Farmer participation can form the basic building block of a successful development strategy. Development is a process of empowerment whereby people learn to take charge of their lives and learn to solve their own problems by way of participation and innovation (Edwards, 1989). It is the opposite to the type of paternalism that is characteristic of many conventional soil conservation programmes. Farmer participation in problem identification and subsequent project formulation and implementation is now recognised as one of the critical components of rural development (Pretty and Shah, 1997).

Rural and agricultural development is more a process of learning and action by farmers than the transference of technologies *per se* and the only way to avoid the growing dependency associated with subsidies and other forms of paternalism is to motivate farmers to do things for themselves (Bunch, 1982:23). The driving force behind participation is enthusiasm and this enthusiasm comes from programmes that address farmers' priority problems, work with farmers, and bring about early recognisable success (Bunch, 1982:24). As farmers participate in programmes, they gain self-confidence, pride and the satisfaction of having made significant achievements. In this context, success can be defined as *"the solution of a felt need with results that are both readily observable and desirable according to the culture's own value system"* (Bunch, 1982:25).

The confidence that comes from participation increases farmers' ability to learn and experiment (Edwards, 1989). Innovation is part of development and is vital because agro-ecological, social and economic conditions change and farmers need to be able to adapt to these changing circumstances. Bunch (1982:11) writes that the goal (of an agricultural programme) should not be to develop the people's agriculture, but to teach them a process by which they can develop their own agriculture. There are a growing number of examples of where active farmer participation has contributed to a process leading to greater human development. These examples include watershed management in Kenya (Thompson and Pretty, 1996; Thompson and Guijt, 1999); irrigation programmes in Sri Lanka (Uphoff, 1996) and agricultural programmes in Honduras (Bunch and López, 1999; Hellin, 1999).

2.3.3 Indigenous Knowledge

A rich source of information on farmers' realities is indigenous knowledge. Indigenous knowledge is the sum of experience and knowledge of a given ethnic group that forms the basis for decision-making in the face of familiar and unfamiliar problems and challenges (Warren and Cashman, 1988:3). For decades, anthropologists have acknowledged the importance of indigenous knowledge. In the last 20 years, there has been a growing appreciation of it by those involved in agricultural development largely because *"it is unproductive to seek to develop a package of improved practices without reference to the body of existing technological and farmers' indigenous knowledge"* (Norman and Douglas, 1994:139).

Soil conservation is a good example of the value of indigenous knowledge. There is often an assumption that soil erosion is a modern problem and that only modern techniques can be used to control soil loss. However, much can be gained by a greater appreciation of ancient techniques used to conserve soil and water and why in some cases their use has fallen into abeyance

(Hallsworth, 1987:146). Indigenous knowledge and the practices linked with this knowledge may provide the starting point for achieving more conservation-effective agricultural practices (Shaxson et al. 1997; Pawluk et al. 1992; Beach and Dunning, 1995; Sillitoe, 1998; Thurston, 1992:13; Kellman and Tackaberry, 1997:324; Hill and Woodland, 2003). This is especially the case because indigenous practices have often developed in marginal areas such as steep hillsides in response to sustained population pressure (Critchley et al. 1994). See Box 2.7 for examples from Africa and Latin America.

Box 2.7   Indigenous knowledge and soil and water conservation

A range of water management systems have evolved over the centuries for erosion control on steeplands (Lal, 1990b:515). Critchley et al. (1994) and Reij *et al.* (1996) describe a number of indigenous soil and water conservation technologies in Africa. These include planting-pits that are known as *zaï* in countries such as Mali (Wedum et al. 1996). Traditional *zaï* are about 20 cm in diameter and about 10 cm deep. They are dug out with a traditional hoe and organic matter is sometimes added in order to improve soil fertility.

Malley et al. (2004) and Temu and Bisanda (1996) describe an indigenous manual cultivation practice that is used on steep slopes in parts of southwest Tanzania. The ngoro conservation system consists of a matrix of pits with surrounding bund walls. Grass is cut prior to cultivation and is normally laid out in a 1.5 m x 1.5 m square. Soil is then dug from the middle of each square and is placed on top of the grass to form four bunds surrounding each pit. Crops are subsequently planted on these bund walls. The decomposing plant residues provide nutrients to the developing crops.

Wilken (1987) observes a variety of conservation-effective soil and crop management techniques used by farmers in Mexico and Central America. These techniques include irrigation systems and terraces. The main functions of sloping terraces are to control erosion and accumulate moisture. The central Mexican *metepantle* is where low embankments are paralleled immediately down slope by drains. Maize is planted in the fields and *maguey* (*Agave* spp.) on the embankments to stabilise them. Other common plants found on the embankments are *mesquite* [*Prosopis* spp.] and native fruit trees (Wilken, 1987:105). Meanwhile in South America, between AD 530 and the early 1500s, the people of the Andes built complex systems of irrigation, aqueducts and raised fields in the mountains of present-day Bolivia, Ecuador and Peru (Zimmerer, 1995; Hallsworth, 1987:69).

Altieri (2002) sums up the importance of indigenous knowledge in the context of land management: *"the stubborn persistence of millions of hectares under traditional agriculture in the form of raised fields, terraces,*

> *polycultures, agroforestry systems etc. are living proof of a successful indigenous agricultural strategy and comprises a tribute to the 'creativity' of small farmers throughout the developing world".*

Ancient practices have not always functioned effectively. Civilisations, such as the Maya in present-day Mexico and Central America, are believed to have declined as a result of soil degradation (Abrams and Rue, 1988) (see Plates 5 and CP 2). In the case of the Maya, this was despite the fact that they had learnt to reduce soil loss with a variety of conservation techniques including terraces (Hughes, 1999). Haug *et al* (2003), however, argue that the collapse of the Maya civilisation occurred during a century-long decline in rainfall that was punctuated by more intense droughts in 810, 860 and 910 A.D. The lack of rainfall, therefore, put a huge strain on a civilization that had expanded rapidly during a relatively wet period from 550 to 750 A.D.

Plate 5   Civilisations such as the Maya in present-day Mexico and Central America are believed to have declined as a result of soil degradation undermining agricultural systems. Guatamala. (Hellin, J.)

Based on an analyses of sediment cores from Lake Pátzcuaro in southern-central Mexico, O'Hara *et al.* (1993) conclude that extensive land clearance, leading to accelerated soil erosion, began before European contact. Hence, indigenous land management practices may not have been as conservation-effective as is sometimes portrayed. Pollard (1994) contends that the increase in soil erosion prior to the Conquest could be due as much to increased rainfall as to deforestation. This indicates that there may have been factors, other than soil loss, that were a limiting factor to sustained and increased agricultural productivity.

There is, however, a danger of romanticising indigenous knowledge (Sillitoe, 1998; Kellman and Tackaberry, 1997: 325; Marsden, 1994; Dickson, 2003; Bellon, 2001). As Stocking (1993:25-26) writes, *"local people may have a view of the world just as a person lost in the midst of trees has a view of the forest – desperate, not necessarily balanced and maybe not even sensible"*. Outsiders have potentially useful knowledge that farmers may not have. There is, therefore, a need to find a balance between the two extremes that claim that experts know best or local people know best (DeWalt, 1994; Scoones and Thompson, 1994b).

What is clear that greater appreciation and understanding of farmers' realities helps outsiders identify why farmers do not readily adopt recommended soil conservation technologies and why they continue to pursue land management practices that lead to soil and land degradation. Too often farmers' opinions have been ignored. It is precisely their view of the potential benefits and costs arising from the adoption of a particular soil conservation practice that will determine whether the practice is adopted. Furthermore, their decision is generally based on a careful assessment of the balance between risks, costs and benefits in the light of their objectives. But how do we find out about farmers' realities? This is the subject of the next section.

2.4   UNLOCKING THE COMPLEXITY OF FARMERS' REALITIES

2.4.1   Grappling with the Challenges of Validity and Reliability

Knowing that the traditional knowledge of land users is important is not, of course, the same as knowing what that knowledge is. Endeavours to find out about farmers' realities are plagued by difficulties. Inevitably, outsiders are dependent on information from the farmers themselves. They need to question continually whether the indicators developed are valid i.e. do they measure the concept they are designed to measure, and whether the information they collect is reliable i.e. a question is of little use if farmers answer it in one way one day and another the next. Ensuring a high degree of validity and reliability is one of the persistent concerns in any social science research strategy (Bleek, 1987:315). It can be particularly difficult in the context of smallholder agriculture.

As in all social surveys, problems arise because questions can be asked in a number of ways. Farmers' responses are conditioned by how the question is phrased, how it is asked and who asks it. As such, there are no guarantees that the correct answers will be given, or even that there are 'correct' answers (Gardner and Lewis, 1996:101). The ways in which the outsider questioner is perceived may also influence how farmers' answers a question (Christiansson et al. 1993; Pretty, 1995:165; Cornwall et al. 1994). Researchers associated with aid agencies may be seen as potential 'providers'. In these cases farmers may represent themselves more in terms of 'needs' than of self-sufficiency. If the questioner is associated with government, people may disguise information about income etc. (Gardner and Lewis, 1996:101). There again, if the questioner comes from the same community or peer group, then the respondent may emphasise the successes and achievements.

There is also a tendency on farmers' parts to romanticise the past and to paint a rosy picture which is not necessarily indicative of what that past was really like (Hudson, 1993:13). This tendency is not, of course, unique to smallholder farmers in the developing world. In the developed world, surveys regularly demonstrate people's perception that there was less violence in the past or that the threats to our children from sexual predators are greater today. Official figures do not always support these perceptions.

Farmers' concerns are also seasonal, for example in March, prior to sowing, potato farmers in Peru are most concerned with the cost and quality of seed, whilst in August just before the harvest, their main concern is insect pests (Rhoades, 1991:32). The timing of research can, therefore, bias a survey. Although in order to mitigate this problem, farmers can be questioned about problems they faced in earlier periods (Horton, 1990:7). Furthermore, any quest to collect valid and reliable data is hampered by the fact that what people say they are doing may not be the same as what they actually do. For example, farmers in western Honduras consistently denied, during semi-structured interviews, that they burnt their fields prior to sowing basic grains, despite substantial evidence to the contrary (Hellin, 1998).

One way to increase the reliability of data is to compare farmers' responses to field observations i.e. to 'triangulate' the information by using more than one research tool. Information gleaned from farmers can be confirmed or repudiated by comparing it against personal observations. A further advantage of triangulating information is that it may expose why there are discrepancies between what farmers say they do and what they do. Exploring why farmers are reluctant to share information or why they give misleading information can lead to "*important new insights into personal, social, and cultural aspects of their lives*" (Bleek, 1987:314).

There is also a real danger that researchers interpret what farmers are saying incorrectly. To an extent, social realities depend upon the subjective perspectives of those viewing the situation, we understand things in certain ways because we have adopted, either tacitly or explicitly, certain ways of seeing (Silverman, 1993:46). As Chambers writes (1997:76) "*believing becomes seeing*". Faced with all these dangers, the question arises as to how best to find out about farmers' realities? One of the most effective ways is to triangulate information by using a combination of qualitative and quantitative information sources.

2.4.2   Capturing Farmers' Realities: Qualitative and Quantitative Research Tools

Qualitative data can be defined as information that is not collected in numerical form and that is not easily quantifiable whilst quantitative data are any data that can readily be summarised in numerical form (Floyd, 1999). Quantitative data are generally seen as objective and qualitative data as subjective although this is not necessarily the case. However, there has been a tendency, especially among policy-makers, to value quantitative data more highly than qualitative (Chambers, 1997:39; Silverman, 1993:vii). The reality is that the boundary between qualitative and quantitative data in terms of objectivity and subjectivity is blurred. Scores and ranks assigned by farmers may well be subjective, but it is possible to quantify and statistically-analyse the data and hence transform it into quantitative data (Pound, 1999).

One of the most widely used quantitative research tools are questionnaires. The drawback of relying exclusively on a research tool such as a questionnaire is that there is no way in which increased rigour during analysis can compensate for the unknown and degree of inaccuracy involved in the measurement process (Gill, 1993:13). Furthermore, questionnaires may entail interpersonal relationships of power and distort farmers' realities by fitting them into centrally pre-set frameworks. Questionnaires may also suffer from the same degree of subjectivity as that normally attributed to qualitative research by reflecting the predisposition of the researcher (Guba, 1981). In summary, the extent to which results based on questionnaire data are socially and personally constructed remains under-researched, under-reported and under-recognised (Chambers, 1997:95).

Quantitative research tools, such as questionnaires, often also fail to capture many of the nuances of farmers' realities, the reason being that farmers' knowledge systems are often not verbally or numerically codified. In this case, qualitative data, such as that gained by participant observation techniques, may better represent their perceptions and realities (Sutherland, 1999, Norman and Douglas, 1994:98). Indeed, observation, interview and casual conversation also cause less suspicion and less guarded comment than research methods that involve outsiders writing down responses (Bunch, 1982:53).

In the context of smallholder agriculture and farmers' realities, the best way to proceed may involve a judicious combination of quantitative and qualitative research tools. When quantitative studies are combined with a credible understanding of complex real-world situations that characterise good qualitative studies, outsiders can gain a sound understanding of the problems and opportunities faced by farmers (Miles and Huberman, 1994:40-41; Horton, 1990:7; May, 1997:104).

Box 2.8 shows a range of qualitative and quantitative research tools that have been used in agricultural research worldwide. These include participant observation, questionnaires, structured and semi-structured interviews and focus group meetings. In recent years the growth of the use of Rapid Rural Appraisal and Participatory Rural Appraisal has highlighted a variety of additional qualitative research tools, which have been used in a variety of sectors ranging from natural resource management to health and nutrition (Chambers, 1997:120-121; Pretty et al. 1995).

Box 2.8   Agricultural research worldwide using a combination of qualitative and quantitative research tools

- A study of the impact of agricultural technologies on the rural poor in Bangladesh, Kenya, Zimbabwe and Mexico used a combination of qualitative and quantitative tools (Meinzen-Dick et al. 2004; Bellon et al. 2003).
- Structured interviews and participant observation were used in a study of the adoption of soil conservation technologies in southern Mali (Bodnar and De Graaff, 2003).
- Formal household surveys, focus group discussions and field observation were used in a study of farmer adoption of soil conservation technologies in Ethiopia (Bewket and Sterk, 2002).
- Eighty-five farmers were interviewed in north-central Bolivia. Open-ended, followed by close-ended questions were used (Sturm and Smith, 1993).
- A study of farmers' adoption of soil conservation technologies in Tanzania included the use of a household survey, group discussions and transect walks (Tenge et al. 2004).
- Survey of farmers' perceptions of an agroforestry project in Costa Rica was carried out using direct observation, unstructured and structured interviews (Marmillod, 1987).
- Study of farmers' perceptions of resource problems and adoption of soil conservation practices in Rwanda. Informal discussions and a formal survey were used (Ndiaye and Sofranko, 1994).
- Research on farming activities was carried out in southern Honduras using qualitative ethnographic fieldwork and quantitative survey methods (Stonich, 1991).
- Participant observation and structured and open-ended interviews were used to identify the amount of land under cultivation and crop yields in Peru (Painter, 1984).

- Study of indigenous soil conservation in India. Three different but consecutive phases: participant observation, unstructured interactions, and a survey were used (Rajasekaran and Warren, 1995).
- Questionnaire and informal survey used in study of adoption of soil conservation practices in El Salvador (Sain and Barreto, 1996).

2.4.3   Research on Better Land Management in Honduras

The problems, processes and pitfalls of farmer research (and agro-ecological research) are best illustrated by example. In this book, the author draws on his experience working in Honduras in the mid-, late-1990s. Honduras provided an excellent case study for an investigation into land degradation and the most appropriate means to mitigate the problem. Central America's mountains and heavy rainfall make much of the region particularly vulnerable to soil degradation, especially when more marginal lands are brought into production and fallow periods are shortened (Lutz et al. 1994; Stonich, 1995) (see Box 2.9). Leonard (1987:4) states that the hilly and highland zones in each of the five Central American countries make up between 73 percent (Costa Rica) and 95 percent (El Salvador) of the total area. Honduras, with a figure of 82 percent, lies in the middle of the range and typifies the worldwide problem of inequalities in land distribution and associated land degradation in steeplands.

Box 2.9   Reported erosion rates in Central America

- In El Salvador, in an area with annual rainfall of 1900 mm, Sheng (1982) reports an annual soil loss of 127 t ha^{-1} on a field of bare soil with a 30 percent slope, planted with maize running up and down the slope.
- Annual soil loss measurements from northwest El Salvador on clay loam to clay soils on 20 m long plots with 30 percent slopes under traditional corn and bean cultivation range from 13 to 137 t ha^{-1} (Wall, 1981).
- Wiggins (1981) reports annual soil loss rates of 15 to 150 t ha^{-1} on cultivated steeplands of the Acelhuate river basin in El Salvador.
- Research by Rivas (1993) in Nicaragua on bare slopes of 15 percent documents annual soil losses of 78 t ha^{-1}.
- The *Instituto Interamericano de Coorperación para la Agricultura* (IICA) (1995) reports that serious erosion is affecting 170,000 hectares in hillsides per year in Honduras with the annual rate of soil loss ranging from 22 to 46 t ha^{-1}.

Scherr and Yadav (1996:21) identify sub-humid Central American hillsides as examples of areas suffering from extreme land degradation (see Plate

CP 3). This assertion is borne out by the figures. Human-induced soil degradation worldwide has affected 1,966 million ha or 15 percent of the total land area. One of the most affected areas is Central America where almost a third of the total land area has been degraded. This figure includes 75 percent (28 million ha) of agricultural land (Oldeman, 1994). The main cause of this soil degradation is two-fold: much of Central America is steeplands and inequalities in land distribution have forced many resource-poor farmers to seek out some degree of livelihood security by farming steeplands.

The abandoned cities and agricultural terraces of Mexico and Central America are testament to the fact that land degradation in the region is not, uniquely, a recent phenomenon. There are, therefore, sound historic reasons for being concerned about the sustainability of current agricultural development, and to be wary about trying to reintroduce indigenous practices that have fallen into abeyance.

Although there are different agro-ecological zones within Honduras, there are similarities in climatic conditions in much of the country, especially in the central and southern zones. The climate in Honduras is rainy with an extremely dry season that lasts from November/December to April/May. The wet season is bimodal and lasts from April/May until October/November. During July and early August there is a marked reduction in rainfall known as the *canícula*. Based on rainfall data from 1966 to 1985, Zúniga (1990:26) reports that, throughout the country, annual average rainfall is approximately 1250 mm, although the range is from 746 mm to 3235 mm. In many parts of the country, rainfall distribution rather than rainfall total is one of the main limiting factors to productivity.

Farming practices are similar throughout the country albeit with minor adjustments to the timing of operations and sequence of events. Traditionally shifting cultivation was practised. Plots were managed on a cyclical basis, alternating between food crop production for three to four years, followed (in some cases) by cattle grazing. There was then a fallow period of generally five to six years. Many farmers, however, now cultivate plots on a continuous basis due to shortages of land. The land tenure situation in Honduras is complex, some farmers have title to their land (*dominio pleno*) and others have usufruct rights (*dominio útil*). Many farmers also rent from larger landowners or occupy land illegally.

As a result of government programmes and those run by non-governmental organisations (NGOs), fewer farmers now burn their fields prior to cultivation, or they burn their fields less frequently. Agricultural systems are therefore evolving (albeit slowly) from slash and burn agriculture to slash and mulch agriculture, where the crops are sown into the mulch of cut vegetation (DeWalt, 1985) (see Box 2.10). Wealthier farmers use oxen to plough their land, but the traditional practice is to use a planting stick.

Box 2.10   Steepland agriculture in Honduras

During the rainy season the main crops grown by resource-poor farmers are maize, sorghum and beans. Approximately 73 percent of the maize, beans and sorghum grown in Honduras are produced on cleared hillsides (IICA, 1995). Steeplands are, therefore, the foundation of food production in Honduras (Toness et al. 1998:35). There is some variation in the sequence of crops that are cultivated, the degree of intercropping and the timing of agricultural activities.

There are normally two harvests per annum although some farmers sometimes avoid producing two successive maize crops on the same land to avoid impoverishing the land further. Crops are sown at the beginning of the rains and harvested in August/September. This is known as the *primera*. The *postrera* is sown immediately afterwards and is harvested in December/January. The land can be bare and more vulnerable to erosion at the beginning of the *primera* (especially if fields have been burnt) and at the beginning *postrera*. Farmers with access to irrigation may grow an assortment of vegetables.

The issue at stake for many farmers is that soil degradation exacerbates an already precarious struggle for food security. For example, Bunch (1988) reports maize yields as low as 400 kg ha^{-1} in areas suffering from soil degradation in central Honduras. By 1989, the area in southern Honduras cultivated in maize had declined to 49 percent of its 1952 level, while per capita production fell to 28 percent. The area cultivated in beans fell to 15 percent of the levels in 1952 and to 5 percent of per capita production (Stonich, 1993:73). Conroy et al. (1996:30) report that throughout Central America in the 1980s, maize and bean production per capita fell by 14 percent and 25 percent respectively. Faced with declining food security, smallholder farmers are increasingly working off-farm (Stonich, 1993:134).

Within Honduras, there have been a series of soil conservation initiatives, which mirror worldwide efforts to reduce soil erosion (Tracy, 1988). Issues of soil quality and active farmer participation are often ignored and the emphasis has been on the transfer of cross-slope technologies and other soil erosion control measures. Many of the soil conservation initiatives have been disappointing in terms of farmer-adoption and adaptation of the technologies promoted.

With its combination of steeplands, resource-poor farmers, land degradation, failed soil conservation initiatives and declining food security situation, Honduras provided the author with a challenging environment in which to conduct cross-disciplinary research on better land management. The author carried out research in two areas of the country: Güinope and Choluteca (see Map 2.1).

Map 2.1 Honduras and location of Güinope and Choluteca research areas

As outlined in Figure 2.6, the author's own farmer research in Honduras used a combination of quantitative and qualitative research tools in a research process recommended by Horton (1990). Horton (1990:11) advises that a profile of a farming system can best be captured if an informal survey is followed by a sharply focused questionnaire. The need for some real understanding of the farming system prior to the use of a questionnaire is critical because farmers judge outsiders on the basis of their behaviour, attitudes and questions, hence, irrelevant and culturally insensitive questions can result in scepticism, distrust and lack of co-operation (Norman and Douglas, 1994:112).

A summary of the key themes surrounding farmers' realities and their livelihood strategies that were explored is outlined in Table 2.1. Social and economic surveys were conducted in three communities in each of two regions in Honduras – Choluteca and Güinope. In the case of the semi-structured interviews and focus group meetings, in each community, one adopter and one non-adopter of soil conservation technologies was interviewed and a focus group meeting was organised for approximately 20 farmers. There was scope for a diversity of issues to be discussed under each of the themes outlined in Table 2.1 but the discussion was guided so as to avoid the danger of saying 'a little about a lot' rather than 'a lot about a little' (Silverman, 1993:3). The themes were not discussed in any particular order and all the meetings were tape-recorded with the farmers' consent. The recordings were subsequently transcribed.

Participant observation
- Fundamental to much qualitative research especially anthropological research (Silverman, 1993:9).
- Leads the inquirer to a greater understanding of the characteristics of the situation being researched (Guba, 1981).
- Many key issues and themes surrounding farmers' livelihood strategies were identified through participant observation (Hellin and Haigh, 2002).

Semi-structured interviews and focus group meetings
- Guided conversations in which topics are predetermined and during which new questions and insights arise as a result of the discussion and visualised analyses (Pretty et al. 1995:73).
- They are more an art than a set of fixed procedures and the interview process is dynamic and iterative (Norman and Douglas, 1994:91; Rhoades, 1991:10).
- One-to-one conversations and group meetings are needed because a frequent bias in agricultural development is to think in terms of 'the farmer' despite the fact that decisions about farming are not made by the farmer in isolation and decision-making is influenced by social pressures and beliefs (Rhoades, 1991:17).
- Furthermore interviews with groups of farmers may be more instructive than those with individual farmers because group members have an overlapping spread of knowledge, which may cover a wider field than any single person (Chambers, 1997:148; Pretty, 1995:132).

Questionnaire
- Quantitative data permit a more objective assessment (Floyd, 1999) and facilitate an assessment of larger-scale patterns, trends and relationships including, in this context, farmers' priority problems and needs, and their understanding of land degradation.
- Questionnaires focused on *what* farmers are doing, qualitative research tools not only provided a means to check the reliability of data from questionnaires, but can also gave more insight into *why* farmers manage the land the way that they do and *how* they formulate land management decisions.

Fig. 2.6   Selecting the research tools used in farmer research (Hellin and Haigh, 2002)

The questionnaire was administered to a total of 213 farmers – 104 in Choluteca and 109 in Güinope – See Figure 2.7. Farmers questioned largely came from the same communities where the qualitative research was conducted. The questions covered the same themes as those in the semi-structured interviews and focus group meetings. Since good questionnaires

Table 2.1 Themes discussed during the semi-structured interviews and focus group meetings and then explored during the questionnaire

---

OBJECTIVES
- **Determine whether farmers identify soil erosion as a threat to their livelihoods**
- **Determine whether soil conservation is a priority, and to identify farmers' priority problems**
- **Identify farmers' understanding of land degradation processes**
- **Identify the obstacles to the adoption and adaptation of soil conservation technologies**

THEMES DISCUSSED
Theme 1   Farmers' objectives
What are the farmers' livelihood objectives?
How does the farm contribute to meeting these objectives?

Theme 2   Land
How does the land contribute to production on the farm?
How is land impoverished?

Theme 3   Problems faced by farmers
What are/were the problems faced by farmers? (e.g. pests and diseases, water shortages etc.)
Do farmers understand the cause(s) of the problem? (e.g. are water shortages due to reduced rainfall and/or reduced infiltration and increased runoff? If so why)?

Theme 4   Mitigating the problems
How do/did farmers mitigate the problem(s)? (e.g. migrate, leave land in fallow, irrigate etc.)
Why have they sought to mitigate the problem the way they have?

Theme 5   The farm and farmer
Name and age of farmers; rents or owns land; number of years in the community; does the farmer periodically work off-farm etc.
Number of plots; quality of plots; access to irrigation; crops grown etc.

Theme 6   Farmer-identified obstacles
If a technology is shown to work, what are the obstacles farmers feel would prevent then from adopting the technology?
Can farmers overcome these obstacles?

Theme 7   Soil quality
Which characteristics of the land govern production (e.g. colour, soil depth etc.)
How do farmers determine what is a good quality?
Do farmers see land/soil as 'dead' or 'alive'?
Do farmers believe that land/soil can be 'improved'?

Theme 8   Relationship with extension agents
What is your impression of the extension programmes that have worked in this area?
What has your relationship been like with the extension agents?
What have been the strengths and weaknesses of different extension programmes?

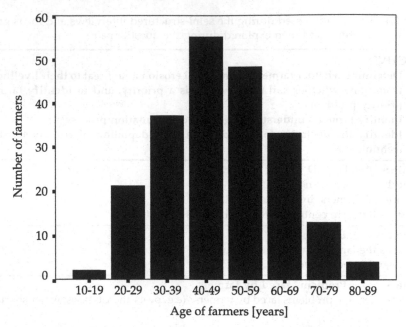

Fig. 2.7   Age of farmers interviewed (total 213 farmers)

do not just happen, but involve careful thinking, numerous drafts, thorough evaluation and extensive testing (de Vaus, 1996:105), much attention was directed at the following areas: question content; wording of the question; use of open and closed questions; logical flow of the questionnaire; and the process of pilot testing. Box 2.11 gives more detail about each of these aspects of questionnaire design in the context of the research in Honduras.

Box 2.11   Issue addressed in the design of the questionnaire used in Honduras

Question content can include behaviour, beliefs and attributes. Questions that are formulated to establish what a person does (behaviour) are different from those designed to measure beliefs (de Vaus, 1996:82). Farmers are not always able to behave as they might like and the failure to distinguish between the different types of question content can lead to the collection of the wrong types of information. Different questions have to be asked to differentiate what people believe from what people are actually doing. Attribute questions are more straightforward and are designed to obtain information about the respondent's characteristics.

The wording of questions is also fundamental because of the dangers of misunderstanding that questions can provoke in the respondent.

Questioning has been compared to trying to catch a particularly elusive fish, in terms of casting different types of bait at different depths, without knowing what goes on beneath the surface (Oppenheim, 1986:49). Key issues include whether the language used in the questionnaire is simple, whether the question can be shortened and whether the question is a leading one.

Questions can also be closed or opened. The former is where a number of alternative answers are provided from which respondents select one of more. Opened questions are where respondents formulate their own answers. There is no right or wrong approach (de Vaus, 1996:87). A good questionnaire is also one in which there is a good logical flow and which is interesting to the respondent (Oppenheim, 1986:37).

In the questionnaire, questions were designed to elicit farmers' priority problems (beliefs) and what farmers are doing to mitigate the problems (behaviour). Attribute questions covered the characteristics of the farm and farmer. Suitable wording of questions was assured by designing the questionnaire after some of the nuances surrounding the vocabulary used by farmers had also been identified during the qualitative research and by working with two local farmers. Closed questions were used to ascertain characteristics of the farm and the farmer. Open questions were used to explore the themes of farmers' priority problems and food security.

The questionnaire was pilot tested with 20 farmers on an individual basis. According to de Vaus (1996:99), pilot testing can involve *declared* and *undeclared* testing. During the pilot testing phase in Honduras, ten farmers were told that the questions were being developed during the questionnaire. These farmers were asked to comment on each question after it had been asked i.e. to identify questions, or aspects of the question, that they did not understand etc. In the case of the other 10 farmers, the complete questionnaire was administered without farmers being asked to comment. On the basis of the pilot testing, several individual questions were rephrased and reordered.

Interviews and focus group work had demonstrated that Honduran farmers are often eager to talk about their farms. Hence, the questionnaire started with questions about the characteristics of the farm before moving onto the other themes.

Farmers' responses to each of the questions were scored (Miles and Huberman, 1994:55) and analysed (via the Statistical Package for Social Scientists (SPSS)). Most of the data collected during the questionnaire was nominal and hence a non-parametric statistical test, the *Chi*-square test, was used in the analysis. In this context, the null hypothesis is that no difference exists between the data from farmers from two groups with respect to the relative frequency with which group members fall into the various categories of the variable of interest. Data on area of farm, average number of plots and average plot size, were collected and measured on an interval scale. A parametric test, an independent *t*-test, was used to compare the mean response between two groups and to test whether the observed difference in the means of the two groups is 'real' or due to random variation. The null hypothesis tested by the *t*-test is that the means of two populations are equal.

2.5   FARMERS' PRIORITY NEEDS

2.5.1   Making Sense of the Complexity

An enhanced understanding and appreciation of farmers' realities, may enable outsiders to understand better why farmers do not readily adopt soil conservation technologies. The search for this understanding may also encourage outsiders to work with farmers in their quest to identify and implement suitable practices that mitigate the problems of soil and land degradation. One of the challenges in any analysis of farmers' realities and the appropriateness (or otherwise) of a soil conservation initiative is to avoid getting overwhelmed by the plethora of reasons which account for the lack of farmer enthusiasm for outsiders' recommendations. These are summarised in Box 2.12 and some of these factors are considered in more detail below and in Chapter 3.

Box 2.12 Reasons for non-adoption and adaptation of soil conservation technologies (based on Hellin and Haigh, 2002)

- Farmers feel that they are unlikely to reap expected benefits because of a lack of secure access to land (Bunch, 1982:45; Comia et al. 1994; Sanders, 1988; Wachter, 1994).
- Labour costs involved in establishment and maintenance of technologies are too high, especially if farmers periodically work off-farm (Critchley et al. 1994; de Janvry et al. 1989; Douglas, 1993; Garrity et al. 1997; Lal, 1989; Stocking, 1995; Zimmerer, 1993).
- Technologies do not reliably reduce soil loss and, even if they do, the reduction does not, immediately, raise yields (Pretty, 1995:51).
- Farmers believe that the economic contribution of their plots to their livelihoods is so small that it is not worth investing time and money in 'improving' the plots (Scherr, 1999:11).
- Technologies of physical earthworks and cross-slope barriers do not, of themselves, lead to improvements in productivity and even if they do farmers expect low economic returns from the technologies available (Bunch, 1982:139; Critchley et al. 1994; Herweg and Ludi, 1999).

- Technologies often require farmers to take land out of production and they are unwilling to do this (Douglas, 1993; Fujisaka, 1991b; Pretty and Shaxson, 1998).
- Farmers do not rate soil erosion as a key problem that needs to be addressed and so soil conservation recommendations are seen as a waste of time and effort (Ashby, 1985; Blaikie, 1989; Fujisaka, 1989; Rhoades, 1991:40; Shaxson et al. 1997).
- Resistance by local peoples to 'top-down' soil conservation programmes (Blaikie, 1993; Gardner and Lewis, 1996:65; Kloosterboer and Eppink, 1989; Robinson, 1989).
- Technologies exacerbate other problems such as water-logging, weeds, pests and diseases (Herweg and Ludi, 1999; Pretty, 1995:51).
- Farmers do not feel that they own the technologies due to 'transfer-of-technology' extension practices (Dvorak, 1991; Enters and Hagmann, 1996; Fujisaka, 1991; Pretty and Shah, 1997).
- Technologies do not address, or may even increase, the risks inherent in agricultural production, especially if their implementation involves investment and additional debt (Napier and Sommers, 1993).
- Farmers lack access to the capital necessary to establish and maintain soil conservation technologies (Scherr, 1999:11).
- Soil conservation practices require changes in farming systems that do not suit the economic or cultural realities of that system (Lutz et al. 1994).

2.5.2 Farmers and Soil Loss: Perceptions and Coping Strategies

Of course, a deeper understanding of farmers' realities is not, in itself, always sufficient to appreciate the rational behind farmer decision-making. Box 2.12 above demonstrates that one of the reasons that farmers do not adopt soil conservation recommendations is that they seldom identify soil erosion as a problem that needs addressing. Box 2.13 demonstrates that farmers' non-identification of soil erosion as a problem is a worldwide phenomenon. For those outsiders who are convinced of the threat of soil erosion and the virtues of soil conservation technologies, this can be particularly galling and may reinforce their belief that farmers are conservative and unwilling to change.

Box 2.13 Farmers worldwide seldom identify erosion as a problem

A worldwide study in the 1980s, in which more than 10,000 farms were visited, demonstrated that whilst almost all the farmers knew that their lands were declining in fertility, few of them associated this decline with erosion (Hallsworth, 1987:84). Likewise, Howeler (1994) reports that interviews with 240 farmers in northern Thailand indicated that farmers were not seriously concerned about soil erosion even though they were aware of a decline in soil fertility and productivity due to continuous farming of the same land.

Cartier van Dissel and de Graff (1999) observe that farmers in Zululand in South Africa only consider erosion a problem when gullies were formed. In fact in the same study, when asked to interpret the word 'erosion', scientists refer to all forms of soil loss whilst farmers only refer to gullies and rills. Similarly, Kiome and Stocking (1995) report that when discussing erosion with 38 farm households in Kenya, the majority spoke of gullies but none referred to sheet erosion.

Farmers' tendency not to identify soil erosion as a problem is not confined to the developing world. Osterman and Hicks (1988) report that in the Missouri Flat Creek area in the United States, farmers' perception of highly erodible land represented only 28 percent of what the United States Department of Agriculture defined as highly erodible. Hallsworth (1987:150) cites an example from New South Wales, Australia where surveyors assigned 197 farms to three erosion classes. Almost 90 percent of farmers did not consider erosion a problem even though the surveyors placed 70 percent of the farms in the moderate- and severe-erosion categories.

Ericksen (1998) likewise observes that Honduran farmers do not identify soil erosion as a cause of reduced productivity even though they are aware of the factors that favour erosion. Toness *et al.* (1998:5) report that farmers in southern Honduras cite rapid depletion of soil moisture and low fertility as the main constraints to crop growth.

Deeper analysis can reveal why farmers seldom identify soil erosion as a problem and why they are often disinterested in soil conservation recommendations. The following discussion is based around three themes:

- Farmers do not recognise that soil is actually being eroded.
- Farmers regard soil loss as a problem but a manageable one.
- Farmers often identify other factors that are perceived as greater threats to productivity e.g. water availability.

Farmers' lack of concern about soil loss is partly explained by the fact that many of them simply do not recognise that soil erosion is occurring. Annual soil loss rates of 20 to 40 t ha^{-1} are difficult to notice even by trained observers (Mutchler et al. 1994). In addition, recognition of erosion is made more difficult by the fact that changes in yield can very seldom be attributed to soil loss.

In addition, as outlined in Chapter 1 and detailed in Chapter 3, actual yields are determined more by the quality of soil remaining than the amount of soil lost. Furthermore, climatic variability, particularly rainfall can mask any impact of soil loss on subsequent yields. Higher and better-distributed rainfall one year may lead to greater yields irrespective of the amount of soil eroded. This may largely explain observations made in El Salvador whereby in a survey of 302 farm households many farmers reported greater erosion on steep slopes compared to shallow slopes, but far fewer suggested that the erosion would have any effect on yields (Pagiola, 1997).

The Honduran farmers interviewed as part of the author's study also recognised that more soil is washed off steep slopes than on shallow slopes. They know that the losses are greater at the beginning of the rains, when some fields have been burnt and the soil surface is unprotected. However, the farmers see soil erosion as an unavoidable nuisance that has no immediate impact on productivity. By contrast, Honduran farmers readily regret that their *abono* is being washed away or carried away by landslides. *Abono* is a generic term used to describe topsoil, rotting vegetation on the soil surface and also fertiliser (organic and inorganic). According to farmers organic matter makes the soil more *suelto* or open-structured, improves infiltration and reduces the risk of erosion and waterlogging. Farmers also regretted the destruction of organic matter through burning their fields as part of agricultural preparation prior to the onset of the rains – although the benefits of burning were, once again, thought to outweigh the costs.

Some of the discrepancies between the dangers of erosion perceived by outsiders and farmers could be due to the inaccuracy of much research directed at calculating erosion rates i.e. researchers may be using data that exaggerate soil loss (see Chapter 3). Exceptional, dramatic or interesting erosion results are more likely to be published and cited than those that are modest, unexceptional or mundane. Even if the data are representative, another critical factor explaining farmers' lack of concern about soil loss is that they are masters in survival and have learnt how to cope with the problem. What is clear is that many farmers have developed practices that alleviate the detrimental consequences of soil erosion (see Box 2.14). Farmers can mask the effects of erosion by growing crops that are less demanding and/or by increasing levels of inputs such as irrigation and fertilisers (Scherr and Yadav, 1996:5; Enters, 1998:9). By so doing, soil erosion becomes less of a threat to their livelihoods.

Box 2.14 Farmers' practices which alleviate the detrimental consequences of soil erosion or declining soil fertility on agricultural production

Ashby (1985) documents an example in Colombia where farmers' local soil classification system identified three types of soil: black, red and mixed. Farmers considered black soil as the most fertile. Red soils are eroded soils from which the black A horizon has disappeared. As the black soil became more degraded and turned to red, farmers recognised that it was no longer possible to grow maize and beans. They started growing a variety of cassava that was better suited to the less fertile red soil and for which there was a demand from a local starch-processing industry. Hence, erosion of the black soil was not perceived to be a constraint to production.

Sillitoe (1993) argues that soil erosion in the highlands of Papua New Guinea is not seen as problem by the Wola-speaking people despite the fact that the climate is wet all year and farmers cultivate steep slopes semi-continuously as part of a system of shifting cultivation. One of the reasons is that when fertility declines due to continuous cultivation and soil erosion, farmers switch from a diversity of crops to sweet potato (*Ipomoea batatas*). Sweet potato is tolerant of very low concentrations of phosphorous and can be grown for many years on degraded land (Sillitoe and Shiel, 1999).

The difficulties farmers face in recognising that soil erosion is occurring, together with their use of strategies to alleviate any of the detrimental impacts of erosion, helps explain why for many farmers there are other factors that are seen to be a far greater threat to productivity and ultimately farmers' livelihoods. It may well be that soil erosion is undermining agricultural production in certain places, but by focusing on soil loss as the major issue, other production constraints may be being ignored. These constraints may include water availability to plants, soil nutrient levels, labour availability, and market incentives (Scoones et al. 1996:4). Farmers may prioritise these constraints ahead of the issue of soil erosion. Box 2.15 shows this to be the case for many farmers from Bolivia to Kenya to Ethiopia.

Box 2.15 Farmer-identified threats to production

- In Bolivia, group problem ranking exercises by farmers in the lowlands demonstrated that loss of productivity and decreasing water and soil humidity are higher priorities than the control of soil erosion (Lawrence, 1997).
- In Peru, farmers did not rank soil problems very highly and were more concerned with insect and irrigation problems and the costs of inputs (Rhoades, 1991:40).
- Farmers in the Machakos district in Kenya consider low soil fertility and erosion as secondary constraints to crop production. Primary constraints include low rainfall, labour and lack of oxen for land production (David, 1995).
- Farmers in the rift valley in Ethiopia identified weeds as a major cause of yield loss and cited the additional problem of the high amount of labour required for crop production (Fujisaka, 1997).
- In Ecuador, farmers do not identify erosion as a major problem. They are more concerned with erratic rainfall, shortages of agricultural labour, cost of insecticides and difficulties of securing loans (Stadel, 1989).

Despite growing evidence that farmers do not see soil erosion as a problem, some outsiders readily point to indigenous soil conservation technologies (c.f. Box 2.7) as evidence that farmers are concerned about soil loss and in some circumstances are prepared to do something about it. There is a degree of self-delusion in all this: in many cases these indigenous soil conservation practices are designed to conserve moisture more than soil. Terraces used by the ancient Maya, for example, were probably not really conserving soil *per se* but were used to make different types of water-conserving soil platforms by damning eroding soil and perhaps adding organic soil (Beach and Dunning, 1995). The same is true of the ancient terrace systems of South East Asia (Pereira, 1989). This is in sharp contrast to most conventional externally driven soil conservation initiatives where in general the priority is to control soil loss (Critchley et al. 1994).

Furthermore, where farmers have adopted soil conservation recommended technologies, their motivation has been more on water conservation rather than controlling soil loss. In Kenya, trash-lines established by farmers are seen as a way to increase water infiltration rather than control soil erosion (Kiome and Stocking, 1995). Whilst in Bolivia, Lawrence (1997) argues that live barriers are of more interest to farmers because the barriers maintain soil moisture. Wenner (1989) concludes that one of the major reasons that farmers have maintained terraces imposed by colonial governments in Lesotho is because of soil moisture conservation.

The situation is made more complicated by examples where erosion is seen in a favourable light because it can provide a cost-less means of transporting soil to places where it will be more productive (Bocco, 1991, Chambero, 1993). Lewis (1992) observes a situation in Rwanda where, in response to exposure of poor sub-soils (and subsequent reductions in yield) along the back portions of terrace benches, farmers remove a portion of the terrace berm. The soil from the berm is then spread over the lower bench, especially along the back portions. Hence, these farmers in Rwanda consciously accelerate the down-slope movement of large amounts of soil on terraced fields. Meanwhile, Hallsworth (1987:106) cites the example of farmers in south-eastern Nigeria who believe that gullying is a punishment of the Gods and that no attempts should be made to mitigate the problem for fear of offending the Gods.

By focusing on the control of soil erosion, outsiders may be *"scratching where there is no itching"* (Bunch, 1982:99). Such soil conservation programmes are unlikely to succeed. For example, a soil conservation project in northern Laos failed partly because the technologies did not address the main farmer problem of weeds (Fujisaka, 1994). The question then arises: where should we be scratching? The author's work in Honduras provides some answers to this question.

2.5.3   Where Should we be Scratching?

The author's research in Honduras initially focused on farmers' priority problems. There are several advantages to the active participation of farm families in defining, ranking and analysing their priority problems. Firstly, it is relatively straight forward to identify the reasons for and the causes of particular problems, and farmers can contribute much to characterising the time dimensions of the problems: history, rates of change and sequences. Secondly, by collecting together, sorting, analysing, interpreting and clarifying relationships of data relevant to problems and potentials, appropriate ways of resolving or mitigating a problem can be identified.

This research confirmed that for many farmers, soil erosion is, indeed, not a priority problem. There is a host of other more pressing threats to farmers' livelihoods and issues that they would like addressed (see Plate 6). One key open question in the questionnaire asked the farmers to identify up to four major problems. The 213 farmers surveyed gave 448 responses and identified 28 problems, although in the case of 18 problems, there were fewer than 10 responses per problem (Hellin and Haigh, 2002). These 28 farmer-identified problems identified seemed to fall naturally into seven classes (Table 2.2).

Plate 6   Farmers' priority needs may have little to do with land management. In this case, farmers are calling for the provision of water and electricity. Bolivia. (Hellin, J.)

Table 2.2   Problems faced by farmers in Honduras. Based on data from a questionnaire (213 farmers and 448 responses). From Hellin and Haigh, 2002.

Problem	Responses	Percentage of responses
Pests and diseases	172	38
Drought and/or irregular rain	136	30
Low productivity	60	13
Quality of land (eroded, waterlogged etc.)	40	9
Availability of land	29	7
Few economic resources	7	2
Others	4	1

The results from the questionnaire in Honduras suggest that the major concerns of the farmers are pest and disease problems together with drought and/or irregular rains. However, as Rhoades (1991:32) cautions, farmers' concerns are seasonal. The author administered the questionnaire at the end of 1997, a year in which rainfall had been much reduced due to the weather phenomenon known as *El Niño*. Many farmers mentioned irregular and reduced rainfall as their priority concern (Box 2.15 shows that farmer in other parts of the world have likewise identified these problems as priorities).

By contrast, relatively few farmers identify the quality of the land, which includes eroded land, as a threat to their livelihoods, even though they understand the causes of land degradation and that their agricultural practices can contribute to the degradation process. In fact, out of a total of farmers' 448 responses, only three were about soil loss. This is less than 1 percent of responses!

When the semi-structured interviews and focus group meetings took place, at the beginning of the rainy season in 1997, the farmers main concerns were pests and diseases, lack of rural credit and the land tenure situation. The last two problems are often linked. In Honduras (and in many other developing countries) few farmers hold any official title to land, so it is very difficult for them to secure credit from banks and often the only source of capital is private money-lenders who often charge exorbitant interest rates (see Box 2.16). Farmers worldwide face a similar problem vis-à-vis securing credit from banks (Hallsworth, 1987:141).

Box 2.16 Importance of credit

Land tenure and secure access to land affect both investment incentives and the availability of resources to finance investment. Southgate (1994) points out that the supply of credit, especially from institutional sources, frequently depends on the borrower's ownership security. Provision of collateral is a common prerequisite for commercial bank loans and land is useful as collateral only if ownership by the borrower can be proven. Pagiola (1994) reports that in Kenya few farmers were prepared to use their land as collateral for fear of losing it.

Institutions also prefer borrowers with secure ownership even in the absence of the provision of collateral. Hence, resource-poor farmers are often forced to borrow from informal lenders. These loans are often at higher interest rates and it becomes increasingly difficult to justify investing in improved land management because farmers can ill-afford the cost of covering short-term capital costs associated with soil conservation.

In Honduras, the semi-structured interviews and focus group meetings explored the farmers' problems and their perceptions of their causes and effects. This work highlighted how easy it can be to misinterpret the results of a questionnaire. Table 2.2 suggests that the farmers' most worrying problems are pests and diseases (38 percent of responses) followed by irregular rains (30 percent of responses). However, deeper questioning discovered that farmers are not worried about pests and diseases or reduced rainfall *per se*. In the questionnaire, when asked about the effects of each of the problems they had identified, reduced productivity (360 responses) and hunger (46 responses) made up 91 percent of the total of 448 responses. Farmers real concern is the consequence of these problems expressed in reduced productivity. The results support the obvious, but oft-neglected, observation that farmers are primarily concerned with attaining, economic and reliable production from their land (Hudson, 1983; Douglas, 1993; Shaxson, 1996).

It is also important to point out that there are additional dangers of misinterpreting farmers' responses in the questionnaire. During one interview, a farmer in Güinope, Honduras explained:

"Often we see that the plant is not growing and we say that it is being attacked by a pest or disease and that we need to add pesticides, but the truth is that we do not always know what the problem is, it may not really be pests and diseases" (farmer, semi-structured interview carried out by the author, Güinope, Honduras).

This is another example of the danger of relying exclusively on indigenous knowledge and of the advantage to be gained by combining scientific and indigenous knowledge. It also raises the question whether pests and diseases are a real problem or a figment of farmers' imagination? Debate will continue on the severity of insect pests and diseases, but what is less equivocal is that for many farmers, the biggest challenge is how to maintain and increase production throughout the farm. From their perspective, soil conservation is just another component of the farm enterprise. Research results presented below show land and labour shortages have often led to farmers deciding not to adopt soil conservation technologies or to abandon particular measures. Land and labour shortages are critical factors in farmer decision-making and cannot be ignored in any initiative designed to improve land management.

2.6 OBSTACLES TO BETTER LAND MANAGEMENT

2.6.1 Land and Labour Shortages and Land Management

Soil degradation in steeplands is exacerbated by land and labour shortages, which in turn determine farmers' receptivity to soil conservation programmes. A number of theories exist to explain the link between land shortages, labour shortages and land degradation (see Box 2.17). Irrespective of the validity of some of these theories, what is clear is that faced with increasing costs of production, many farmers seek off-farm income-generating activities. This often has detrimental consequences for land management (see below) and farmers' receptivity to soil conservation initiatives.

Box 2.17 Labour shortages and land degradation

According to some theorists, land and labour shortages are explained by the concepts of '**functional dualism**' and the '**reproduction squeeze**'. In the case of **functional dualism**, farmers are obliged to work as part-time wage labourers to make up shortfalls of staples and cash requirements for household goods, as well as to pay for inputs for the production process itself on their farms (Blaikie, 1989). Employers accumulate super-profits by paying less than the cost of reproducing the farm household, since the smallholder's farm makes up the shortfall to the level of the full economic wage, which would otherwise have to pay for the full costs of living for the worker and his/her dependants.

Farm household production, therefore, acts as a subsidy to wages since part of the subsistence cost of farm households is borne by household labour (de Janvry and Helfand, 1990). Farmers, although increasingly dependent on non-farm sources of income, are unable to find sufficient employment opportunities either to migrate and abandon the agricultural sectors or to depend fully on wage earnings for their subsistence (de Janvry et al. 1989; Stonich, 1991). This predicament has implications for the amount of labour that farmers are able and willing to invest in soil conservation technologies and better land management practices.

The **reproduction squeeze**, on the other hand, is a process of impoverishment of smallholder farmers through the development of unfavourable terms of trade. Typically resource-poor farmers have to purchase commodities for consumption and production by producing primary products for sale. Faced with a deterioration of the terms of trade, for example governments ensuring artificially low prices for agricultural products, farmers have to reduce levels of consumption and/or intensify commodity production (Blaikie, 1989; Loker, 1996). The latter often involves more labour input on poorer or more distant soils, many of which are found in steep areas. This increases the costs of production and reduces the returns to labour (Bernstein, 1977; Zimmerer, 1992).

One the other hand, Boserup (1965) contends that in some situations it is the very shortage of land itself that is a stimulus to conservation. Tiffin *et al.* (1994) document the case of the Machakos district in Kenya where farmers were goaded into action to reverse the trend of increasing soil degradation on ever-decreasing areas of farmland. The study looked at the relationship between increasing population density, productivity and environmental degradation between 1930 and 1990. In the 1930s,

Machakos District was considered an environmental disaster. In 1990, however, the environment was in a better condition than in the 1930s despite the fact that the population of the district had increased more than five-fold. In response to land shortages and increasing soil degradation (as well as better contacts with markets), farmers had constructed terraces and had planted trees for fuel wood. In Mali, Bodnar and De Graaff (2003) have also noted that farmers in high land pressure areas more readily adopt soil fertility practices.

The phenomenon of farmers working off-farm is not confined to Latin America, for example in parts of Rwanda, 40 per cent of farm households do not consider agriculture to be their main source of income (Ndiaye and Sofranko, 1994). In the Himalaya, Ashish (1979) talked of a "money order economy", where rural villages survive only because of money orders sent back by family members working in the cities in the Gangetic plains. Worldwide, the demands of the off-farm jobs, which are often located far from the area, mean that farm labour can be in short supply at critical times of the year. These include prior to the beginning of the rainy season when farmers need to prepare their land for sowing (see below).

The situation is exacerbated in areas where farmers have insecure access to land. In these cases a higher percentage of farmers seek off-farm work opportunities in order to try and secure some degree of livelihood security. Naturally, such farmers are even more reluctant to invest time (and money) in better land management practices and/or the establishment of soil conservation technologies (Ruben, 2001). Table 2.3, again based on the author's research in Honduras, clearly demonstrates the importance of land ownership when it comes to farmers' decision to work off-farm. In this case, significantly more farmers who rent land are involved in a combination of agriculture and salaried (off-farm) activities (*Chi*-Square = 24.082 (Yates corrected), df = 1, p < 0.000).

Table 2.3 Land tenure and income-generating activities (dry season) in Honduras

Tenure arrangement	Income-generating activities during the dry season		Number of farmers
	Agriculture	Agriculture and salaried	
Own land and do not rent	51.4%	48.6%	142
Rent at least one plot	15.5%	84.5%	71
Total	39.4%	60.6%	213

The figures in Table 2.3 concur with Stonich (1991) who observes that in southern Honduras, 80-90 percent of male households considered themselves farmers, 50 percent report that they had secondary occupations and 70-85 percent have migrated to find work at some time in the past. Further research by Stonich (1993:134) shows that in southern Honduras male householders allotted 38 percent of their time to production of basic grains, while wage labour, and petty commodity manufacture and vending took up 23 percent and 21 percent respectively of their time.

The author's research in Honduras also reveals that farmers, in general, understand the causes of land degradation and acknowledge that their agricultural practices can contribute to the degradation process. Farmers are known to romanticise the past (c.f. Hudson, 1993:13), however, almost unanimously, the Honduran farmers interviewed agree that agricultural productivity was greater in the past. They attribute this to the land being more fertile (stronger), rainfall being more plentiful and pest and disease problems less severe.

This may not simply be a case of the 'skies being much more blue in the old days'. In the past, farmers were able to leave more land in fallow for longer and they know that yields from land previously left in fallow are higher than from continuously cultivated land. Today, because of land shortages, there is less opportunity to leave land in fallow. The Honduran farmers interviewed attribute land shortages to population increase, the growth of agricultural cash crops for export in the fertile valleys and the fact that wealthier farmers have, over one or two generations, managed to monopolise land ownership within the hillside communities.

In the questionnaire, next to pests and diseases, Honduran farmers cited drought as their main concern. Many blamed the reduced and irregular rainfall on deforestation, which farmers agreed, was largely caused by the practice of burning fields to clear unwanted vegetation before the beginning of the rainy season. Incomplete records of deforestation and rainfall make it very difficult to compare scientific evidence with farmers' perceptions of changes in rainfall and to determine if there is less rainfall now than in the past. However, the problem may be due to reduced water availability rather than reduced rainfall *per se*. Farming practices may have lead to a deterioration in soil quality, a loss of water-holding organic matter for example, which in turn has led to a reduction in water–holding capacity.

The main reasons for burning is that it is the easiest way to clear land and control weeds as well as pests and diseases. The practice saves enormous amounts of labour in land preparation and sowing. Thurow and Smith (1998:7) calculate that in southern Honduras, preparing the land for planting using the slash and mulch method requires 12 man-days ha^{-1} compared to six man-days ha^{-1} for the traditional slash and burn method. As Table 2.3 demonstrates, many of the land-poor Honduran farmers work off-farm in the dry season to supplement their income. Labour is at a premium, so burning is commonly supposed to be the only rational way to prepare land for planting prior to the rains. In addition, for those farmers renting land, short-term rental agreements mean that they are highly unlikely to benefit from the potential medium- and long-term increases in productivity associated with a slash and mulch system.

Table 2.4 clearly shows that, although there are other advantages associated with burning fields, lower labour costs are by far the most important. The next most important advantage is pest and disease control. Kellman and Tackaberry (1997:212) stress that soil sterilisation brought about by burning is especially important when land is left in fallow for a short duration because there may be a carry-over of pests and seeds of various weeds from the previous site usage. In this context, burning is a very attractive proposition for farmers who do not have secure access to land, who face labour shortages and who cannot afford pesticides to control increased pest problems.

Table 2.4   Advantages of burning plots prior to sowing identified by farmers in Honduras (209 farmers and 352 responses)

Advantages of burning	Responses	% Responses
Saves labour (cleaning fields)	204	58
Reduces pest and disease problems	69	20
Better productivity	64	18
No advantages	15	4

Burning can also be very destructive. Fire destroys the vegetative cover, which is important for protecting the soil surface from high-intensity raindrops, and also soil organisms and organic matter, both of which play a critical role in maintaining soil quality (see Chapter 3). The Honduran farmers interviewed acknowledge that burning fields prior to sowing destroys the organic matter that, they believe, makes the land 'stronger' and more productive.

While yields increase for the first 1-2 years after burning, farmers worry that regular burning impoverishes land in the medium- and long-term. Despite this, they carry on burning because in a situation where the return on the labour invested in basic grain production is low and unpredictable, farmers strive to minimise the costs of production and/or maximise the short-term marginal return on their investment.

Within the context of the constraints and opportunities faced by Honduras farmers (and resource-poor farmers worldwide), decisions as to farming techniques used and the intensity of land use are not based on the maximisation of total production but on the marginal returns to household labour (Stocking and Abel, 1992). Fujisaka (1991b) cites the example of farmers in Mindanao in the Philippines who plant maize even though the market price is lower than other crops such as rice. Farmers show a preference for maize because labour and material costs are low, the crop faces few risks if planted early, and, because of the low labour demands, farmers are able to engage in off-farm activities. The intricate links between land and labour shortages have substantial implications for farmers' response to soil conservation technologies.

2.6.2 Soil Conservation Technologies and Competing Demands for Land and Labour

Establishment and maintenance costs of soil conservation technologies can be high (Critchley et al. 1994; Wiersum, 1994). Resource-poor farmers often find that labour, essential for investment in soil improvement or maintenance of conservation structures, needs to be diverted to the immediate goal of primary production or off-farm activities (Stonich and DeWalt, 1989; Collins, 1987; Zimmerer, 1993; de Janvry et al. 1989, Stocking, 1995). Faced with these labour constraints and given the importance of cash in fulfilling household requirements, it is not surprising that farmers often chose not to engage in labour intensive conservation measures but decide to invest their labour in off-farm activities that lead to short-term increases in income.

Soil conservation programmes often try to encourage farmers to abandon the practice of burning their fields. A technology such as the construction of live barriers, whose adoption requires the abandonment of the practice of burning, would have to offer sufficient benefits (in terms of increased productivity) to compensate farmers for increased labour input associated with forgoing the practice of burning. There are numerous examples of where soil technologies have proved to be inappropriate because of high labour demands (see Box 2.18).

Box 2.18 Soil conservation technologies and competing demands for labour

Stocking and Abel (1992) cite the example of southern Africa, where out-migration had led to the absence of male labour on the farm, which in turn prevents activities such as soil conservation. Blaut *et al.* (1959) and Blustain (1982) conclude that in mountainous regions of Jamaica, labour shortages limit the potential for implementing soil conservation technologies. Meanwhile, acute seasonal shortages of labour in the Bolivian highlands, especially in the period 1953-1991, led to the abandonment of soil conservation techniques, and a worsening soil erosion problem. This was largely due to the fact that labour-scarce households were unable to recruit labour for labour-intensive conservation tasks (Zimmerer, 1993).

Carter (1995) cites additional labour demands and the inflexibility of timing of these demands as one of the largest disincentives to the establishment of live barriers. Specifically, Ya (1998), Fujisaka *et al.* (1994) and David (1995) report that the labour demands of live barriers are a disincentive to adoption in the Hindu-Kush-Himalayan region, the Philippines and Kenya respectively. In Honduras, farmers abandoned live barriers of *Pennisetum purpureum* (Elephant grass) partly because of the extra labour costs of stopping the grass from invading neighbouring

---

fields (Hellin and Larrea, 1998). Enters and Hagmann (1996) report a similar situation in Thailand where farmers complained that the grass live barriers (*Brachiaria ruziensis*) invaded their fields and increased labour demands.

The problem is not confined to the developing world. Douglas *et al.* (1994) report that labour shortages over the last 40 years are largely responsible for the abandonment of some of the 500-year-old system of terraces found in the Alpujarras mountains in southern Spain. A similar process is occurring in Yemen where ancient terraces and canals are being abandoned as labour is drawn towards jobs in the capital city Sana'a and to the oil-rich Gulf States (Vogel, 1988; Hudson, 1995:209; Hallsworth, 1987:146).

Other studies have demonstrated a link between the adoption of soil conservation technologies, farm size and labour. In Southern Honduras, for example, the mean number of conservation practices varied from 0.8 practices on farms of less than 1 ha to 6.3 practices on farms from 5-20 ha (Stonich, 1993:147). The latter group invested the most household labour in on-farm agricultural activities while the smallholder farmers had to divide their labour between subsistence activities and a variety of other economic tasks.

A cost-benefit analysis of soil conservation technologies in Kenya reveals vastly different returns from different technologies with a large part of these differences due to the varying labour demands that each technology requires (Kiome and Stocking, 1995). In terms of soil conservation techniques, the lowest labour requirements are for biological systems and the highest requirements tend to be for terraces.

In addition to the social and economic reasons detailed above, there may be sound technical reasons for promoting the least labour-intensive technology. There is evidence that technologies with varying establishment and maintenance labour costs do not differ in their ability to control erosion. Siebert and Lassoie (1991), for example, report that the use of bench terraces, grass bunds, and grass plus *Gliricidia sepium* bunds with mulch, resulted in significant reductions ($p<0.05$) in soil loss and water runoff in comparison with conventional cultivation methods on hillside farms in Sumatra. Interestingly, there were no significant differences in soil erosion rates between conservation treatments.

In some cases, however, terraces or stone walls have proved popular with farmers although soil conservation has not always been the primary motivation for building them. One of the reasons why farmers in central Honduras readily adopted stone walls as part of a World Neighbors' soil conservation programme is that they had traditionally cleared stones from fields but had piled them up in mounds dotted around their fields. Constructing walls did not therefore represent an enormous amount of extra labour. A similar situation has been reported in Kenya and elsewhere (Kiome and Stocking, 1995).

Despite the importance of labour to farm household decision-making, labour requirements are often forgotten or ignored in the design of soil conservation schemes. Tacio (1993) refers to a cost-benefit study of contour hedgerows in the Philippines where labour costs were ignored on the basis that farmers use their own labour! Meanwhile, some researchers and extension workers argue that soil conservation is an activity for the dry season, when there is a lull in agricultural activities and a slack in labour demand. This is often a misconception because in many parts of the world the dry season is precisely that period when many farmers work off-farm to supplement incomes (Gill, 1993:8) (see also Table 2.3).

The issue of labour availability and the establishment and maintenance of soil conservation technologies is even more important in the context of the human immunodeficiency virus/acquired immunodeficiency syndrome (HIV/AIDS) pandemic. There are an estimated 40 million cases of HIV/AIDS worldwide and 95 percent of these are in developing countries. Southern Africa is the worst affected region (UNAIDS, 2004). The impact of HIV/AIDS in terms of morbidity and mortality is particularly severe in the agricultural sector in Africa. According to the FAO some 7 million farmers and farm workers in 25 African countries had died of AIDS by 2000 and that 16 million more will die by 2020 (FAO, 2001). The FAO also anticipates that in the most affected countries, HIV/AIDS is going to reduce the agricultural labour force by 10-26 percent between 2000 and 2020. The impact that HIV/AIDS will have in Africa is being likened to the bubonic plague in Europe in the 14th Century when populations in villages and regions were reduced by two-thirds (Runge et al. 2003:24).

There is now a danger that Southern Africa has crossed a threshold and that, irrespective of climatic conditions, food insecurity will last for years because of the debilitating impact of HIV/AIDS. In some cases agricultural output can decline by 50 percent in farm households affected by HIV/AIDS. Remote fields tend to be left in fallow and, in general, yields decline as a result of delays in essential farming operations, lack of resources to purchase agricultural inputs and the abandonment of soil conservation measures. Also as people fall ill, agricultural educational and extension services may be disrupted. In addition, children may be taken out of school to fill labour and income gaps created when adults fall ill and/or start caring for family members with HIV/AIDS (Meinzen-Dick et al, 2004). In this context, farmers are likely to be even more reluctant to establish soil conservation technologies.

The issue of labour availability is also very germane when it comes to recommendations to re-introduce (or restore) indigenous soil conservation technologies, such as the terraces found in parts of Central and South America. A realistic assessment demonstrates that the labour demands of such an enterprise are enormous and quite unsuited to present-day labour availability. This is likely to be the main reason why farmers view ancient Maya terraces, found in Guatemala, as no more than 'interesting relics' (Beach and Dunning,

1995). Altieri (1999) reporting on a project to restore terraces in Peru, cites that it would require 2,000 worker days to reconstruct terraces on only 1 ha. Without substantial inducements, it is highly unlikely that farmers would either reconstruct or maintain such high labour-intensive structures.

Aside from the labour demands associated with the establishment and maintenance of soil conservation technologies, it has long been recognised that these technologies also tend to remove land from production. From the farmer's perspective this reduction in land area for growing crops may be seen as too great a sacrifice (see Box 2.19). Douglas (1988) points out that farmers with only 1 or 2 ha of land, who are struggling to produce sufficient food, cannot afford to take land out of food crop production to put it under physical conservation structures. This is particularly the case if, as described above, farmers seldom see soil erosion as a problem, rarely see soil conservation as a priority, and have a shortage of labour.

Box 2.19 Soil conservation technologies and land taken out of production

In the case of cross-slope soil conservation technologies, extrapolation of the slope:horizontal spacing relationships from flatter lands to steeplands, often gives unacceptably close-spacing between the technologies and the loss of about 20 percent of the cultivable area (Shaxson, 1999:87). This is a major disincentive to adoption. Pellek (1992) describes the prescribed spacing for contour hedgerows that were used by an extension programme in Haiti. Farmers were loath to follow the recommended close spacing because of the amount of cropland taken up by the conservation technology.

Hwang *et al.* (1994) also describe soil conservation work in the Bao watershed in the Dominican Republic where extension agents promoted a number of technologies such as live barriers and hillside ditches. Live barriers were considered the most cost-effective because of lower establishment and maintenance costs, however, there was a corresponding disastrous reduction in net farm income of 54 percent because of the amount of productive land taken up by the barriers. Hellin and Larrea (1998) have documented the case in Honduras of recommended live barrier species being so aggressive that, if not properly managed, they invade farmers' remaining agricultural land and compete with food crops such as maize and beans.

Clark *et al* (1999) report that in Bolivia, the use of stone boundary bunds can lead to loss in crop area of 19-28 percent. Furthermore, in many cases the increase in yields in the area of sediment accumulation above a bund is unlikely to compensate farmers for the land taken out of production. In some cases, however, non-yield benefits of boundary bunds such as protection of crops from livestock, compensate farm households for the loss in crop area.

Much has been written about the link between land tenure (and tree tenure) and the adoption of soil conservation technologies or use of better land management practices (e.g. Bunch, 1982:45; Sanders, 1988; Pellek, 1992; Carter, 1995; Napier, 1989; Wachter, 1994; Murray and Bannister, 2004). The bulk of this research suggests that lack of tenure is a disincentive to the adoption of these technologies. Evidence comes from the Philippines (Fujisaka, 1991), Vietnam (Fahlen, 2002), Jamaica (Blustain, 1982), Benin (Versteeg and Koudokpon, 1993), Thailand (Feder and Onchan, 1987) and Niger (Gruhn et al. 2000:9). The Honduran farmers interviewed as part of the author's research confirmed this link (see Box 2.20).

Box 2.20 Link between secure access to land and the adoption of soil conservation technologies in Honduras

As part of the author's research, Honduran farmers pointed out that now that land is in short supply, the costs of buying and renting land have escalated out of the reach of most farmers. Rental agreements are often made on a year-to-year or harvest-to-harvest basis at the discretion of the landowner. Other researchers have pointed out that landowners in southern Honduras fear that tenants who use the same plot for any length of time will be less willing to leave than is the case if they farm the plot for one or two harvests only (Stonich, 1993:149).

Honduran farmers stress that, if they do not own or have secure access to land, there is little incentive to farm it wisely. The majority of the farmers that the author interviewed acknowledge that over the medium- and long-term, yields are generally higher in plots where farmers use conservation practices such as minimum tillage but the incentive to do so is non-existent if they are only going to be farming the plot for a short period. They also explain that often it is the poorer quality land that is rented out (the landowners keep the better quality land for themselves) and that if they did anything to 'improve' their plots, such as establishing soil conservation technologies, landowners would be more likely to take the land back. This is a general problem, not confined to Honduras (Pretty, 1995:270).

Farmers added that even if land were more readily available to rent or buy, few could afford the rents charged or the price asked. In addition, the area that they could farm would be limited by the amount of fertiliser and pesticides they could buy (as well as the labour available). A lack of economic resources also prevents farmers from buying hose pipes for the establishment of a basic irrigation system. It remains very difficult to secure formal rural credit without legal title to land.

What is less clear is whether individual land ownership is essential for the implementation of soil conservation measures. There are conflicting reports, for example, soil conservation measures in El Salvador are located on both privately owned and rented land (Pagiola, 1997; Sain and Barreto, 1996). However, in the case of El Salvador, the soil conservation technologies were promoted as part of a package that ensured rapid increases in productivity. Hence, unlike many soil conservation programmes, farmers with insecure tenure could still expect to benefit from the programme.

Farmers without legal title to their land have also adopted soil conservation technologies in Honduras (Lutz et al. 1994) and Indonesia (Wiersum, 1994). Although, in some cases, it may be the case that farmers renting land adopted soil conservation technologies because of inducements such as subsidies and/ or because the landowners asked the farmers to establish the technologies.

The key issue may well be one of secure access to land rather than land tenure *per se* (Bonnard, 1995). Sufficient security, for example, may be provided by traditional tenure systems (Tobisson, 1993). In Kenya, land tenure constraints do not play an important role in the adoption of soil conservation measures because, even without title deeds, most farmers feel secure about their tenure (Pagiola, 1994). Similarly many of the farmers in western Honduras who have adopted an indigenous agroforestry practice, known as the Quezungual System, have customary and not legally recognised titles to their land (Hellin et al. 1999). Indeed, worldwide there are reports that trees are planted by individual households in a wide variety of tenurial contexts other than private property (Arnold and Dewees, 1997).

2.7   FARMERS' RELATIONSHIP WITH OUTSIDERS

If a soil technology is shown to be technically effective and to complement the farming system, farmer adoption can still be inhibited by the way that the technology is presented. This includes the behaviour of extension agents, the content of the extension message, and the language used to convey the message (Bellon et al. 2003). The importance of these factors can be demonstrated by the results of the author's work in Honduras (see also the case study from Güinope, Honduras in Chapter 5) together with that of other development practitioners (also see section 5.4.3).

Many farmers interviewed in Honduras complained that extension agents from different organisations were condescending and arrogant, and that this was a disincentive to meaningful collaboration. Bunch (1982:156) stresses the importance of extension agents visiting farmers' fields and showing interest in their farming situation. Farmers in Honduras commented that many extension agents only work with farmers living by the side of the road, the more important people in the community and/or farmers who belong to the ruling political party. Hallsworth (1987:113) has also reported that despite large number of extension workers worldwide, many farmers in more remote areas have no contact with the extension services.

The language used by farmers and extension agents, and hence the extent to which they are communicating, can also determine the outcome of a programme irrespective of the effectiveness of a soil conservation technology, the behaviour of extension agents, or the content of the extension message (see Box 2.21).

Box 2.21 Importance of communication between farmers and extension workers

Increased communication engendered by a common language is likely to strengthen farmers' confidence in outsiders' recommendations and increase their receptivity to new ideas especially when these ideas focus on farmers' needs and concerns. There is evidence from Honduras that demonstrates that communication between extension agents and farmers can be greatly enhanced by the use of a shared terminology.

Bentley (1993) gives the example the term 'ice' that is used by Honduran farmers to describe various plant diseases, especially plant leaf pathologies. Several extension agents believed that farmers genuinely thought that their crops froze and hence ridiculed the farmers. As soon as the outsiders understood the true meaning of the word 'ice', they began to respect the farmer more, and started to use the term when discussing plant diseases, i.e. they began to share a common language and to communicate better with farmers.

The definitions of land and soil quality used by scientists, extension agents and farmers, indicates the degree to which they share a common language. It is important for scientists to understand farmers' systems for classifying soils in order to communicate more effectively with these same farmers (Bellon, 2001; Barrios and Trejo, 2003). Kerven *et al.* (1995) point out that in northern Zambia, researchers and extension agents were unable to communicate with farmers about their soils because they lacked a common terminology. A greater appreciation of the terms used by farmers to define soil quality, and use of these terms, is likely to facilitate communication and understanding between extension agents, researchers and farmers (Sikana, 1994; Romig et al. 1995). This, in turn, is likely to enhance farmers' ability to monitor the health of their plots and the effect of management practices (Harris and Bezdicek, 1994). There has been a tendency for researchers to use analytical definitions and for farmers to use descriptive ones (See Box 2.22).

Box 2.22 Analytical and descriptive definitions of soil quality

Diagnostic properties characterising soil quality, divide into two components: analytical and descriptive. Properties under the analytical component are more quantitative and have been favoured by researchers.

These include chemical characteristics such as pH, nutrient status of the soil, particle composition and bulk density. Since ancient times farmers, however, have tended to use descriptive terms, based on observation, to assess the condition of the soil and its performance as a sustainable crop production system (Harris and Bezdicek, 1994).

The tendency of farmers in Mexico, Central America and South America to use classification systems that are utilitarian rather than taxonomic and analytical has been documented by Bellon (2001), Oberthur et al. (2004), Ericksen and Ardon (2003), Thiele and Terrazas (1998), Williams and Ortiz-Solorio (1981) and Hecht (1990). Wilken (1987:29), for example, reports that in the area around La Malinche, in Mexico, traditional soil identification and evaluation are based on a limited number of characteristics, primarily colour and texture. Other factors such as structure, salinity, depth and parent material are included if important. Soils are rated superior or inferior primarily on the basis of anticipated crop yields. Yet, even with this almost limitless array of descriptive possibilities, farmers are reluctant to pass final judgement on a soil from inspection only. A soil's true nature can be known only by working with it over several seasons. Guillet *et al.* (1995) also report that folk soil classification in Peru is governed by terms linked to texture, water-holding properties and productivity.

Meanwhile Sikana (1994) observes that in northern Zambia, farmers' main criteria for classifying soils were colour of the top horizon, consistency and organic matter content. Bellon (2001) observes that in Chihota, Zimbabwe, the underlying soil properties of farmers' classification of soils are texture, colour, water-holding capacity, ease of work, inherent fertility, response to fertilizers and proneness to water logging.

The descriptive terms that tend to be used by farmers worldwide are often directly related to production, are inherently qualitative and subjective and, consequently, have been undervalued by natural scientists. As a result, the potential contributions of farmer indigenous knowledge to the concept of soil quality has, however, yet to be fully explored (Pawluk, et al. 1992) despite the fact that farmers can contribute experience-based descriptive knowledge of soil quality and scientists can provide analytical expertise (Garlynd et al. 1994).

The research carried out by the author in Honduras confirmed that resource-poor farmers' assessment of soil quality is largely based upon descriptive and easily observable characteristics that directly relate to production, such as texture (often related to the ease with which land could be prepared), available soil moisture and yields. Relatively few farmers in Honduras refer to soil (*suelo*) and more often use the term land (*tierra* or *terreno*). Honduran farmers interviewed often differentiated soils by:

- Colour (largely organic matter);
- Workability of the soil (structure and compaction);
- Wetness or dryness of soil (drainage, storage and infiltration capacity);
- Topsoil texture and depth (indicators of soil erosion and production potential).

Farmers also placed great emphasis on the advantages of farming land close to their homes and/or a road (see Table 2.5). In these cases, it is easier to deal with animals that enter the plots and also farmers do not have to get up in the early hours of the morning to start walking to distant plots. Another reason is that land in the mountains is often fertile and weeds grow fast. Hence, if the plots are not cleaned on a regular basis, there is a danger that yields will be reduced due to competition from the weeds.

Table 2.5 Summary of farmers' definition of good soil or land based on results from the questionnaire (213 farmers and 527 responses). From Hellin, 1999b.

Definitions of good land	Responses	Percentage of responses
Gives a good harvest (land is strong)	93	17.6
High organic matter content	52	9.9
Land that has a dark colour	118	22.4
Presence of earthworms	17	3.2
Woody vegetation as opposed to grasses	47	8.9
Easily worked[1]	60	11.4
Good structure	40	7.6
Does not need much fertiliser	14	2.7
Thick fertile cap[2]	14	2.7
Land with a water source	15	2.8
Land with good infiltration	11	2.1
Land that you own	5	1
Land that has been left in fallow	2	0.4
Flat land (less erosion)	6	1.1
Does not erode	5	0.9
Disease-free	4	0.8
Land that has not been burnt[3]	5	0.9
Land that retains soil moisture[4]	7	1.3
Land that does not get water-logged	7	1.3
Easy to pull out vegetation	1	0.2
Land that is close to home and/or road	2	0.4
Land with SC technologies	2	0.4

[1]Farmers favoured land that was not 'hard' and that was 'light', and easy to plough.
[2]Farmers used the term *abono* rather than the term *capa fertíl*. *Abono* also refers to the rotting vegetation on the soil surface and fertiliser (organic and chemical).
[3]Farmers said that 1-2 years after a plot had been burnt, it became 'weak' and production fell.
[4]Retention of soil moisture was attributed to the different proportions of sand and clay.

One of the most cited criteria of good land used by the Honduran farmers is that the land gives a good harvest. Farmers refer to such land as 'strong' and land that does not produce is seen as 'weak'. Weak land needs more fertiliser than strong land. Interestingly, Pretty (1998b:119) reports that farmers in the Alpujarras mountains in southern Spain plant a green manure when they consider the land to be 'weak'.

The criteria used by Honduran farmers can be further divided into two connected categories: ease of management (*suave* i.e. easily worked, does not flood and easy to plough) and productivity (*dark* and high organic matter content). Table 2.15 is very similar to the descriptive terms used by farmers in Wisconsin in the United States to describe healthy soil (Romig et al. 1995).

Honduran farmers also have their own terminology for the erosion process. Farmers in Choluteca do not use the term erosion but discuss the fertiliser/ organic matter (*abono*) being washed away. In the Dominican Republic, Ryder (1994) observes that farmers were more comfortable with the term scrape (*arrastre*) than with erosion. Blaut *et al.* (1959) observe that farmers in the Blue Mountains in Jamaica refer to the soil as either 'strong' or 'weak' and the 'strength' of the soil lies in the 'juices' which are the remains of the decayed grass. The 'juices' are lost in runoff and/or are used by crops.

Farmers' lukewarm response to soil conservation programmes becomes more understandable when the issue of a common language is taken into account. Farmers in Honduras tend to refer to land rather than soil, they seldom recognise erosion as a problem and if they do talk about soil loss they refer to losses of *abono* or the juices of the earth being lost. On the other hand, many soil conservation programmes stress the need to control erosion. These same programmes use terms such as soil fertility instead of terms that are more comprehensible to farmers such as the quantity of earthworms in a farmer's field, the ease with which a *machete* can be pushed into the ground or the dark colour of the soil.

2.8   TAPPING INTO FARMERS' RESOURCEFULNESS

The analysis of farmers' realities sheds some light on the obstacles to better land management and a greater understanding and appreciation of why farmers do not readily adopt recommended soil conservation technologies. Besides the more obvious issues such as labour and land shortages, there are a host of more subtle reasons for non-adoption. These include the issue of the rate of change that potential adopters can tolerate at any one time.

The rate of change can become so accelerated by the adoption of certain soil conservation technologies that some farmers cannot accommodate the disruptions to their life styles and livelihood systems (Napier, 1991; Pretty and Shah, 1999). Hence, it may be better to introduce relatively small changes over a longer time period. Bunch (1982:141) also points out that the social consequences of a crop failure are high. When a crop is lost, a farmer is often seen as having failed the entire extended family. The farmer's pride can be severely damaged. Hence, the farmer's cautious approach to making wholesale changes to their farming systems by, for example, adopting a specific soil conservation technology.

The rule in many soil conservation programmes has been to plan from the top down. Normally, the *"conservation expert identifies the problem in the field (usually perceived as loss of soil, gullying or downstream sedimentation), arrives at a solution with the aid of pre-determined technical guidelines, and only involves the farmers through an extension package at the implementation stage"* (Douglas, 1993:12). The underlying assumption is that the outsiders know best. From this perspective stems the view that farmers ought to be concerned about soil loss, they ought to adopt soil conservation recommendations, and their failure to co-operate is the major obstacle to better land management.

Finally, after many years of complaints and coercion, the reality is dawning that it is the outsiders who may actually be wrong. Changing outsiders' understandings, insights, attitudes and approaches may be much more valuable and relevant than trying to change the farmers' attitude. A farmer's priority is normally increased and sustained agricultural production rather than the control of soil erosion or soil conservation *per se*. Many soil conservation programmes, however, focus almost exclusively on controlling soil erosion, while sustained agricultural productivity is often seen as a by-product. In fact several soil conservation technologies may not be effective at controlling soil loss or increasing yields.

This chapter has considered some of the difficulties that affect smallholder farmers. It has also shown farmers' reluctance to adopt some of the soil conservation technologies promoted can be a sound and rational decision. Most farmers are faced with a plethora of problems that they feel ought to be addressed before tackling relatively non urgent concerns such as soil erosion. Some of the problems they face such as insecure access to land, may encourage them to pursue activities that are detrimental to better land management in the long-term e.g. burning their fields, because the long-term condition of the land is not their priority concern.

Sometimes, of course, farmers' lukewarm response to soil conservation initiatives, however appropriate and relevant they might be, may be linked to the outsiders' use of an alien terminology that does little to enhance communication and trust. Simply, the outsiders do not seem to be talking about or addressing aspects of land management or the wider livelihood context that really matter to farmers. Faced with a plethora of problems and outsiders spouting incomprehensible jargon, is it any surprise that farmers show little enthusiasm for adopting the soil erosion control and soil conservation technologies promoted by outside organisations? Table 2.6 examines the issues at stake, with respect to one heavily promoted soil conservation technology – live barriers of vetiver grass.

Table 2.6   Vetiver grass live barriers: outsiders' versus farmers' realities (From Hellin, 1999c)

Outsiders' reality	Farmers' reality
Vetiver grass live barriers are an effective way to control soil erosion.	Farmers seldom recognise soil erosion as a priority problem. They are more interested in sustained and enhanced agricultural production.
Recommendation spacing of live barriers is determined by slope angle.	Prescribed spacing of barriers can lead to a substantial reduction in area for cultivation.
In the mid-, long-term terraces will form behind the barriers (although there is evidence that the effectiveness of barriers to retain soil diminishes as angle of the slope increases).	Farmers may not have secure access to land and are largely interested in changes in the short-term. The most important change is in agricultural production rather than the amount of soil accumulating behind a barrier.
Maintenance of the barriers is not too time-consuming.	Barriers have few useful by-products and maintenance can be an onerous activity.
Extension agents refer to soil erosion and the need to combat it and use largely analytical terms to describe changes in soil structure.	Farmers seldom use the term 'soil' and tend to use descriptive terms when talking about the characteristics of their plots.

So where do we go from here? Firstly, a purely technical fix is not the answer. Many factors leading to degradation especially on steep lands have their root cause outside agriculture (Hudson, 1988). In the last 15 to 20 years, there has been a growing recognition that land degradation is a social, economic and political, as well as technical issue (Blaikie, 1989). In general, explanations of soil erosion and land degradation involve the wider political economy and a number of non-technical issues. In the context of farmers' realities, a successful agricultural development programme cannot be built on a single component such as a new technology. All other relevant aspects of a farming system have to be considered and any new measure must be fitted into and tuned to the system (Hudson, 1993b).

Secondly, outsiders must learn to respect and tap into farmers' resourcefulness. Despite the all too common caricature of smallholder farmers as being powerless to change, there is a wealth of evidence that many of these farmers are skilled entrepreneurs and will strive to invest their labour for future benefits. Admittedly, some farmers are resigned to fate and see their suffering as the will of God. Often this fatalism is manifested in docility and belief that the situation is inevitable. Many farmers, though, are not passive in the face of adversity and their enthusiasm and affinity with the land is a strong foundation on which to base efforts to improve land management. This is why World Neigbors and other successful NGOs emphasise strengthening the farmers' own sense of self-belief in their projects and promote wider aspects of empowerment as strongly as any agricultural technology (Bunch, 1982).

There are growing numbers of examples of farmers practising sustainable agriculture despite problems in access to resources (Pretty, 1995). There is also a wealth of evidence that peripheral rural areas can be characterised by a dynamic farmer and small-scale entrepreneurial sector. These examples demonstrate that farmers' abilities, skills and innovative nature are a vast and, to date, largely untapped resource. Furthermore, they suggest that this is exactly the resource that we, the outsiders, ought to be utilising in our endeavours to bring about better land management.

Farmer non-adoption is often seen as the principle obstacle to soil conservation. Traditionally farmers have been seen as part of the problem. Their input in problem identification, analysis and solution has seldom been sought (Enters and Hagmann, 1996). This approach, like so much in traditional soil and water conservation, is an inherently negative one. It undervalues farmers' skills, resourcefulness and innovative nature. It also deprives outsiders of the information and insights they need to put their work on a sound footing, one that makes rational sense to the farming communities where they work. Norman and Douglas, (1994:8) contend that land has been studied too much in soil conservation programmes, whereas the farmer and farming community has been studied too little.

The problem posed by farmers' reluctance to adopt recommended soil conservation technologies can be resolved. What is required is new ways of thinking on the part of outsiders. The foundation for much of this new thinking is the farmer-first paradigm with its implicit acknowledgement and appreciation of farmers' realities. Rather than seeing farmers as uncooperative and unwilling to change, an alternative interpretation is that farmers' unwillingness to follow official soil conservation recommendations stems more from the fact that technologies devised by outsiders do not, in general, accord with farmers' resources, needs and priorities (Shaxson, 1993). Put simply, soil conservation technologies and the entire soil conservation approach do not complement farmers' realities.

A farmer-first approach, which by considering the totality of the farming systems is inherently cross disciplinary, can reveal much about farmers' realities and suggest alternative approaches to achieving better land management. Researchers and extension agents can start thinking of farmers as part of the solution, and not as part of the problem (Hudson, 1988; Shaxson et al. 1989:13). In order to be successful, soil conservation has to be supported enthusiastically by farmers and so practices need to be farmer-friendly in terms of being easily adopted and offering tangible benefits (Sanders, 1988). This requires a bottom-up and farmer-first approach that matches conservation activities to the needs and wishes of the farming community (Douglas, 1993; Hudson, 1988).

In place of direct incentives, policy changes may be needed to facilitate the adoption of better land management practices (Pretty, 1995:243). The aim should be to make it easier for farmers to adopt conservation-effective farming practices, perceived as beneficial to farmers, rather than using direct incentives to 'encourage' farmers to conserve soil. Breakthroughs are unlikely to come through 'add-on' soil conservation technologies. They are more likely to come through building on farmers' initiatives and making changes to the farming system that also improve soil quality.

Recognition that the conventional soil conservation approach is unlikely to mesh with farm families' social and economic reality is an important step. However, what is also required on the part of outsiders is an additional and fundamental mental flip. Outsiders have to recognise that the conventional soil conservation approach, with its focus on soil loss and erosion control, may also be technically flawed. This means that farmers may be acting even more rationally when they reject soil conservation recommendations. This issue is the focus of the next chapter.

3

Agro-ecological Components of Better Land Husbandry

"Conservation was built on an engineering approach. If the soil should start to move, then the response was to build a barrier to arrest further movement. It sounded sensible but it ignored a fundamental factor—namely, that the conservation measures were tackling the symptoms and not the disease" (Stocking, 1991:7).

3.1   QUESTIONING THE SOIL CONSERVATION APPROACH

Chapter 2 showed how farmers are primarily concerned with stable and economic production (Shaxson, 1996; Hudson, 1983, Thurston, 1992:9). They readily identify a number of obstacles to attaining higher and more stable yields, including pests and diseases and a lack of economic resources. Few farmers identify soil erosion as an obstacle to productivity, despite the fact that much of the research and extension agenda is driven by the goal of erosion control. This may partly explain why farmers have not readily adopted soil conservation recommendations although, as Chapter 2 clearly illustrates, there are also a host of compounding reasons for farmer non-adoption including labour and land shortages.

The research and extension agendas are largely based on an assumption that there is a direct relationship between soil loss and productivity. This assumption has been an article of conventional soil conservation faith for many years and underpins the belief that, in the absence of obstacles to better land management, farmers really ought to be concerned about soil loss and would benefit were they to adopt soil conservation recommendations. However, the long assumed link between soil loss and productivity is now disputed. This raises several fundamental questions:

- If soil loss and the impact on productivity is as severe as researchers have claimed, why is it that so few farmers identify soil erosion as a process in the degradation of their land?
- Are farmers acting rationally when they reject recommended soil conservation technologies?
- Is the conventional soil conservation approach, with its focus on preventing soil (and water) loss and its emphasis on the technologies of erosion control, misdirected?

- If farmers are acting rationally, and even taking into account the extraneous factors of land and labour shortages, what is a more appropriate approach to mitigating the problems of soil and land degradation?

Chapter 2 addressed some of these questions from a social and economic perspective. This chapter examines these issues by analysing the agro-ecological components of the conventional soil conservation approach. It critically examines outsiders' concern with soil loss, demonstrates that efforts to tackle soil erosion are highly unlikely to address farmers' concerns about stable and reliable agricultural yields, and suggests that a new approach is needed, one that focuses on soil quality.

This analysis is based in part on secondary sources and in part on detailed formal scientific field experiments in Honduras that gave equal weight to the concerns of soil conservation researchers and resource-poor farmers. Specifically, this research sought to clarify whether, on steep slopes cultivated by resource-poor farmers, a typical soil conservation technique – live barriers of *Vetiveria zizanioides* (vetiver grass) – made any tangible difference either to soil and water loss or to agricultural production. Conventional wisdom might regard this as a ridiculous question – of course the conservation technique both reduces soil and water loss and helps increase agricultural production! In fact, the truth was not so straightforward.

Today there is an on-going debate about the future direction of conservation efforts. In the hilly tropics, smallholder agriculture continues to encroach on marginal hillsides. The result of this encroachment is soil and land degradation, a degradation that is threatening to undermine the livelihoods of millions and millions of farmers and their families. Clearly, the problem must be resolved. The question remains how best to shift this agriculture from self-destruction to self-sustaining. This chapter focuses on the problems inherent in the current soil conservation approach, the next focuses on a solution to the soil and land degradation problem.

This chapter further sets the scene for a discussion in Chapter 4 of the better land husbandry approach, an approach to improved land management that, by focusing on soil quality, simultaneously addresses farmers' priority concerns about agricultural production and productivity and conservationists' concerns about soil loss and land degradation.

3.2   SOIL EROSION PROCESSES

3.2.1   The Research Agenda

Researchers have been obsessed with soil erosion for several decades. Following patterns of work in the United States, the bulk of experimental research on soil and land degradation has been directed at measuring or estimating soil loss, particularly that caused by water erosion. By contrast, there are relatively few studies of aspects of degradation, such as soil degradation *in-situ*, that affect agricultural productivity.

The figures presented in Table 3.1 give some idea of the apparent soil erosion problem. Once the current rate of erosion has been calculated (or estimated) the research community has sought to find out how the rate can be reduced by modifying the factors responsible so as to bring the calculated rate of soil loss to within 'acceptable levels'.

Table 3.1 Global extent of soil degradation (millions of hectares) due to erosion
Source: Oldeman et al. 1991 (quoted in Scherr, 1999:19)

Region	Area eroded by water erosion				Area eroded by wind erosion				Total area eroded
	Light	Moderate	Strong & extreme	Total	Light	Moderate	Strong & extreme	Total	
Africa	58	67	102	227	88	89	9	186	413
Asia	124	242	73	441	132	75	15	222	663
South America	46	65	12	123	26	16	Negligible	42	165
Central America	1	22	23	46	246	4	1	5	51
North America	14	46	Negligible	60	3	31	1	35	95
Europe	21	81	12	114	3	38	1	42	156
Oceania	79	3	222	83	16	Negligible	27	16	99
World	343	526	223	1,094	269	254	26	548	1,642

There is no clear definition of what constitutes acceptable levels. One school of thought is that these are levels of soil loss at which soil fertility can be maintained, with realistic inputs of technology, during a time frame of approximately 30 years (Stocking, 1995). Alternatively, Wischmeier and Smith (1965) define the *soil loss tolerance* as the maximum rate of soil erosion that will economically and indefinitely permit sustained crop productivity.

Irrespective of the lack of consensus over what are acceptable or unacceptable levels of soil loss, results from one soil erosion experiment indicating high soil loss can be seized upon as indicative of the situation at a larger scale. Often these data are cited to justify spending millions of dollars on the promotion of soil conservation technologies. As we saw in Chapter 2, in the face of research data that demonstrate the huge threat that soil erosion poses, farmers' indifference to the problem of soil erosion and their subsequent lack of interest in adopting soil conservation technologies are seen as an obduracy that must be overcome. It is only in recent years that there has begun to grow doubts about the scientific rigour and representativeness of soil erosion data and, still more recently, a questioning of researchers' obsession with soil loss.

3.2.2   Erosivity and Erodibility

Soil erosion is one of two categories of soil degradation (see Box 1.1). The physical processes of soil erosion are well known, as they should be considering the amount the money that has been spent on research. Soil erosion involves the expenditure of energy in all phases from breaking down soil aggregates to runoff (Hudson, 1995:73). The amount of erosion, however, depends upon a combination of erosivity, which is the power of the rain to cause erosion, and erodibility, which is the ability of the soil to resist the rain.

The first phase of the erosion process is called inter-rill erosion. It involves movement by rain splash and transport of raindrop-detached soil by a thin surface flow. The kinetic energy of raindrops (and possibly runoff) separates detachable particles from the soil mass and these are subsequently transported by runoff (and possibly rainfall in the form of rain splash) (El-Swaify et al. 1982:75). Splash or the impact effect of raindrops is the first and most important stage in the phase of inter-rill erosion. Soil aggregates are broken down and soil particles are detached and thrown into the air. Detachment is caused by the locally intense shear stresses generated at the soil surface (Rose, 1993). The detached particles can be moved further by runoff.

The erosive capacity of this flow is increased by turbulence generated by raindrop impact (Lal, 1990b:141). This increases the capacity of runoff to scour and to transport soil particles. As a result of the surface tension properties of water, runoff concentrates as it flows downhill causing rills (this is the second phase of the erosion process) and possible gullies (the third phase). There is no clear dividing line between the second and third phases of the erosion process (Hudson, 1995:43). Rills and gullies can form, even on a relatively flat slope (Mutchler et al. 1994) although mass movements such as landslides are associated with saturated soil on steep slopes (Perotto-Bladiviezo et al. 2004).

The falling raindrop is an erosive agent within itself (Hudson, 1995:27) and the intensity of rain is a potential parameter of erosivity. The kinetic energy of rainfall is dependent on rainfall intensity, which is influenced by the median raindrop size and the terminal velocity of the free-falling raindrops (Mikhailova et al. 1997; Lal, 1990b:45). High rainfall intensity is related to the relatively big drop size and to the number of drops falling per unit area per unit time (Lal, 1990b:33). Hudson (1995:73) points out that the kinetic energy available in falling rain is substantially higher than that of runoff. Hence, the detaching action of raindrop splash is the most important part of the erosion process (Hudson, 1995:74; El-Swaify and Dangler, 1982)

Erosion is more of a problem in the tropics than in temperate zones because the energy load of tropical rains is more than temperate rains due to larger drops and a greater number of falling drops per unit of time (El-Swaify et al. 1982:86). In temperate climates, the rainfall rate rarely exceeds 75 mm hr^{-1} while in many tropical countries intensities of 150 mm hr^{-1}, albeit sustained over short periods, are common (Hudson, 1995:49). According to Hudson (1995:79), the practical threshold separating erosive from non-erosive rain is 25 mm hr^{-1}. In temperate areas, approximately 95 percent of the rain falls at non-erosive intensities. (i.e. < 25 mm hr^{-1}). In tropical countries only about 60 percent of rain falls at less than 25 mm hr^{-1} (Hudson, 1995:88).

The primary functions of soil, relative to erosion by water, are to enhance infiltration; to facilitate the water-holding capacity of the soil; and to resist physical and chemical degradation forces associated with raindrop impact (Karlen and Stott, 1994). Erodibility is governed by a number of factors including texture (sandy soil is more easily eroded than a hard clay); soil structure (a better structured soil will allow more water to enter the soil matrix); aggregate stability; slope (a steep slope will generally erode more rapidly to a non-productive state than a shallow slope) and the amount of cover provided by a crop and/or plant residues (Hudson, 1995:71; Lal, 1990b:60).

The factors that determine the erodibility of a soil mean that farmers' management practices can dramatically alter a soil's erodibility. For example, practices leading to reduced soil cover allow high-energy raindrops to reach the soil surface. As a consequence of reduced cover, soil structure, particularly in the surface layers is degraded, leading to surface sealing and a reduction in porosity (Moldenhauer et al. 1994; Hillel, 1991:38). This, in turn, leads to more erosion and runoff. As we will see later, erosion can, therefore, be much reduced by protecting the soil surface from raindrop impact.

Losses of soil nutrients are also accentuated by splash erosion and runoff. This is due to the differential removal of the lighter and smaller, more nutrient-rich, clay particles and organic matter. Eroded soil materials may be 4-5 times richer in plant nutrients than the soil left behind (Shaxson, 1999:23).

3.2.3 Vulnerability of Steeplands

Erosion occurs naturally but accelerated soil erosion, caused by man's activities, is a severe hazard on most lands with undulating to sloping terrain. Outsiders' concerns about soil erosion, and the need to combat soil loss, grow as increasing numbers of smallholder farmers cultivate marginal steeplands. The threat of accelerated soil erosion on steeplands to farmers' livelihoods is that the effective soil depth may well be less than the necessary rooting depth of the farmers' annual and perennial crops. In these situations, reduced rainfall is likely to have more severe effects on yields than would be the case if the soil depth exceeds that of the plant's roots. Soil conservation interventions are often justified on the basis that cultivating slopes of more than 30 percent in the humid tropics, without conservation methods, can cause annual soil losses of 100-200 t ha^{-1} (Sheng, 1990).

Although the rainfall on steeplands may be no more erosive that that on lowlands, the erosive forces – splash, scour and transport – all have a greater effect on steep slopes and can provoke severe rill and gully erosion (Lal,

1990; Cassel and Lal, 1992). Conventional wisdom is that increases in soil surface slope tend to increase erosion rates by promoting greater down-slope displacement of soil particles in the splash process, and by increasing the velocity of overland flow and hence its capacity to detach and transport soil particles (Kellman and Tackaberry, 1997:264). A surface-flow of about 25 cm/s is considered erosive for bare soils (Shaxson, 1999:17).

Soil splash is also affected by slope steepness and the direction of the rainfall in relation to the direction of slope. Soil particles that are thrown into the air by the impact of raindrops may move horizontally by as much as 1.5 m from the original location (Lal, 1990b:136). During the landing, under the force of gravity, the amount of material splashed down-slope is often more than that splashed up-slope. Hence soil can move down-slope even if there is no runoff (Lal, 1990b:136).

If other factors are the same, velocity will be slower on a shallow slope than a steep one (Shaxson et al. 1989:47). Length of slope is also important because it allows a progressive build up of both runoff volume and velocity (Hudson, 1995:98). In addition, when slope exceeds a critical steepness, rill erosion begins and this can cause soil loss to increase rapidly with increasing steepness (Hudson, 1992b:145; Sheng, 1982). Due to the steep terrain involved, gully formation is a common mode of erosion on steeplands as are mass movements such as landslides and debris flows (Hudson, 1995:46). This is especially the case during periods of high rainfall intensity, for example during hurricanes and topical storms.

Steeper slopes are also more affected by the creep of soil. In nature, the swelling and contraction of the soil due to wetting, drying, freeze-thaw, bioturbation, trampling and rain-splash sequentially open voids in the soil and close them again. In agriculture, ploughing effects a far greater loosening of the soil. Inevitably, loosening causes the soil to expand normal to the slope, while compaction always involves a vertical component due to gravity. As slope angle increases, soil creep can rival surface soil erosion as a source of sediment.

Steeplands are also vulnerable to another type of soil loss - mass movements such as landslides and debris flows. Haigh (1999) suggests that many steep hill slopes have evolved to a condition that lies close to their margins of stability. In this condition, the forces that preserve the slope and resist failure (strength of the rock, tree roots and soil etc.) are in balance with the forces that encourage the slope's failure (gravity and weight of trees etc.). As such they exhibit behaviour associated with 'self organised criticality'; slopes become vulnerable to very small disturbances and fail unpredictably in individual cases but predictably in statistical terms.

3.3 ARE THE DATA RELIABLE?

3.3.1 Challenging Accepted Wisdom

The scientific justification for the promotion of soil conservation technologies has been based on soil erosion research that demonstrates high soil loss. Estimates of soil loss can be based on predictive models or on measurements in the field, which are then extrapolated to estimate soil loss from the entire land unit (Roels, 1985). Laboratory studies can also be carried out, but they are a precursor and not a substitute for field experiments (Bryan and Luk, 1981).

There are now growing doubts about the scientific validity of many soil erosion experiments. There is also a growing, albeit at times begrudging, acceptance that it is actually very difficult to draw conclusions from data available in the literature, much of which is compounded by emotional connotations and lack of objectivity (Stocking, 1995b). *"The literature is full of horror stories. From the level of technical information presented, however, it is often difficult to judge whether an author is 'crying wolf' or a threat to natural resources and the environment is genuine"* (Lal, 1990:132). More information is needed on the conditions under which the experiment was conducted. For example, quoted erosion rates often relate to plots kept bare of vegetation despite the fact that this is untypical of farmers' fields (Evans, 1995). Frequently, experiments are not contextualised to the farmers' cropping cycle. Commonly, they deal with interesting extreme conditions rather than the mundane or run-of-the-mill.

This raises the questions of whether the data demonstrating high soil loss, essentially the *raison d'etre* of soil conservation initiatives worldwide, are in fact either inaccurate or not truly representative of field conditions? It is worth spending some time considering these questions because if the answer to both is 'yes', it would then force us to question whether efforts to alleviate the problems of land degradation have been appropriately directed and, furthermore, whether farmers are or are not justified in being unconcerned about soil loss.

3.3.2 Predictive Models

There are many predictive models to choose from to determine soil loss. The best-known example is the Universal Soil Loss Equation (USLE). The USLE has been modified and superseded, firstly, by the Revised Universal Soil Loss Equation (RUSLE), and more recently by the Water Erosion Prediction Project (WEPP) (Larose et al. 2004). Other models include the Soil Loss Estimation Model for Southern Africa (SLEMSA) (Elwell and Stocking, 1982), the European Soil Erosion Model (EUROSEM) and the Areal Non-Point Source Watershed Environment Response Simulation model (ANSWERS)

(Moehansyah et al. 2004). However, there are many problems in using predictive models to estimate soil loss (see Box 3.1). As a result, these models have often been misused (Wischmeier, 1976; El-Swaify et al. 1982:76).

Box 3.1   Disadvantages associated with the use of predictive models to estimate soil loss

Firstly, it is all too common for predictions from models to be treated with great respect because the computations are complex and beyond the understanding of the layperson. Without belittling the positive role that models can play in natural resource management, it is important to remember that a model is also only as good as the database available for input (Clark, 1996:17). These data may not be available or may be poor quality. For example, Van Zyl et al. (1998) tried to predict sediment yields in catchment areas of the Lesotho Highlands using the Agricultural Catchment Research Unit (ACRU) model. They concluded that although many models can predict accurately sediment yield, vast sets of data are needed and these data sets are not readily available in developing countries. In some cases, therefore, predictions from models may be a case of 'garbage in, gospel out'.

Secondly, many erosion predictions only consider the site factors and rarely take into account the social, economic, cultural and political contexts (Stocking, 1995). Predictions are also often based on long-term averages from areas under continuous cultivation, and are not appropriate for agricultural situations where farmers deal with soil loss and yields on a season-by-season basis, and where farmers practice short-term agriculture, for example, in a long rotation forest-fallow system. Hence, while models such as USLE may recognise variables that are directly or indirectly of relevance to farmers' livelihoods, they may be irrelevant by virtue of the time-scale used by farmers.

Thirdly, the magnitude of soil erosion at any location is determined by the interaction of five major factors: climate; topography; vegetation; soils; and time (Bryan and Luk, 1981). There is a wide range of values within each of these factors and soil erosion can vary considerably from location to location. Within each factor, a number of sub-factors have been identified. These include rainfall intensity and duration, antecedent moisture content, slope angle and length, vegetation type and density, and soil type. Models, such as the USLE, attempt to combine these data into predictive equations for soil loss. The problem according to Herweg and Ostrowski (1997:3) is that descriptive models such as the USLE reproduce the essential characteristics of the soil-plant-water system but without reproducing the system itself.

The design of models, such as USLE, actually involves a substantial and subjective reduction process whereby the model demands that each value for the factors in the equation are independent even though in reality they are interrelated (Stocking, 1995b; Valentin, 1989; Hudson, 1992b:146). Nearing (1998) argues that the implications of this random variation in soil loss essentially invalidate the basic assumption of deterministic models such as the USLE, the assumption being that a given combination of sub-factors will always produce approximately the same amount of soil loss.

Researchers have sought to validate the USLE model in field situations and the model has tended to over predict soil loss. This follows the good engineering tradition of 'over-design'. Thurow and Smith (1998:16), for example, tested the applicability of the USLE and the RUSLE at a trial site in Southern Honduras where soil loss was also measured. The models over-predicted soil loss. Meanwhile, in Taiwan, Wu and Wang (1998) found that on a 60 percent bare slope the use of the USLE, with the local rainfall erosivity index and estimated soil erodibility index, gave an estimated annual soil loss 73 times higher than the average actual field measurements.

Fourthly, Hudson (1982) points out the dangers of extrapolating an empirical relationship beyond the measured range. The USLE, for example, was developed in the United States and is based on conditions pertaining to the United States, in particular the corn-belt and Great Plains. Within the conditions for which it was devised, the USLE provides an estimate of the long-term average soil loss of arable land under different cropping regimes (Hudson, 1995:129).

The USLE is used to calculate average annual soil loss in t ha^{-1}, based on erosivity of rainfall, erodibility of the soil, length and steepness of slope, a crop management factor and a conservation practice factor. The original length x slope factor of the USLE computed by Wischmeier and Smith (1965) was not valid for slopes less than 3 percent, greater than 18 percent, or longer then 122 m (El-Swaify, 1997). Despite this, researchers have used the USLE in conditions where slopes are far steeper than 18 percent, for example in Papua New Guinea (Sillitoe, 1993) and Taiwan (Wu and Wang, 1998). Meanwhile, Foster et al. (1982) claim that that the USLE slope steepness relationship should apply well to the tropics for slopes up to 25 percent.

Hudson (1983) also notes that the nomograph for soil erodibility used in the USLE has the correct ingredients of easily measured parameters, but the problem is that many tropical soils have little silt and very fine sand and, hence, are squashed into the left-hand side of the nomograph. Soils are subsequently assigned similar erodibility values. Sillitoe (1993) also reports that in Papua New Guinea, the extremely high organic matter content of the soils is beyond the nomograph range used in the USLE. The USLE was designed for use in the USA cornbelt and prairie grasslands, not the tropics. USLE logic would dictate that a separate monograph should be constructed to resolve this problem. This, however, would require major research.

Fifthly, many of the models, including USLE, are designed to provide an estimate of long-term average annual soil loss. Hallsworth (1987:12) notes that the infrequency of large annual soil loss, coupled with the importance of large individual losses, makes the concept of an average soil loss almost meaningless. Models such as the USLE basically ignore soil loss from gullies and mass movements because the features are too large to be recorded at the plot-scale. However, it is the less frequent large events that erode most soil (Evans, 1995) and which are likely to have a far more severe impact on farmers' livelihoods than surface erosion especially on steep slopes. El-Swaify (1994) points out that gully erosion and mass movements such as landslides are common in steeplands and are less predictable than surface erosion. Also the USLE and some of the other models do not take into account deposition of eroded material within the catchment, or snow-melt erosion.

Models developed since USLE are more sophisticated, for example WEPP is a process-based erosion model that is applied at the same level and for the same uses as the USLE, but is applicable to a wider range of scales and land use possibilities (Hudson, 1995:146). However, these newer models still suffer from some of the disadvantages detailed in Box 3.1 and require large amounts of data. EUROSEM simulates erosion on a single slope plane or segment and can therefore be used to predict soil loss from individual fields. It is a more improved and sophisticated model than the USLE and does, for example, deal with detachment and deposition of soil particles by runoff. The main disadvantage of the model is the need for considerable data entry. Quinton (1997) also points out that, in the case of EUROSEM, there is uncertainty over the simulation results because of uncertainty over the value of the model's input parameters. In summary, many of the predictive models require good data and if these are not to hand, researchers are reduced to using guesstimates. This reduces any confidence in the output.

3.3.3   Field Experiments and Extrapolation

Pieri et al. (1995:37) argue that because of the dangers, detailed above, in the uncritical application of models, comparisons with indicators derived from direct measurements is always desirable. Four field measurement techniques are generally used for estimating sediment loss:
- Sediment collections from runoff plots or small catchments (Mutchler et al. 1994)
- Recording of ground loss by means of erosion pins or other point measures (Hudson, 1995:158)

- Use of a quantitative record of soil-surface features such as eroding clods, flow paths, rills and gullies (Herweg, 1996)
- Use of radioactive tracers, notably caesium-137 (Van Oost et al. 2003)

Each field measurement has its advantages and disadvantages (Bergsma, 1997; Haigh, 1977; Herweg, 1996; Stocking 1995). Runoff plots, for example, have been used for several decades in soil erosion research. One of the problems is that the data generated by field experiments are seldom accurate because the methods for measuring soil loss directly are essentially primitive. Results from erosion experiments, such as research plots, can be notoriously inaccurate because of differences in experimental design, problems with how the soil loss is estimated, and the complications due to year-on-year climatic variability.

The source of many of these inaccuracies is the non-comparability of research methods. This non-comparability is due to issues of time-scale, replicability and the fact that experimental plots based on short linear slopes are not representative of the landscape as a whole nor of conditions faced by smallholder farmers (Evans, 1995; Pretty, 1995:205; Gardner and Mawdesley, 1997; Stocking, 1987). See Box 3.2.

Box 3.2   Disadvantages of basing soil loss estimates on data from field experiments

Soil and land degradation processes such as erosion are also not always generally distributed. They tend to be confined to particular 'hot-spots' where economic and social factors have encouraged the use of the physical resource base beyond its sustainable productive capacity (Posner and McPherson, 1982). When the inherent variability in the bio-physical factors controlling soil erosion, for example, are taken into account, it is clear that there may be substantial differences in soil loss rates over small areas which may or may not be 'captured' by soil loss experiments (Roels, 1985). Furthermore, field devices have very low resolution and accuracy. Hence, researchers tend to seek out sites where they are sure of getting something to measure!

Data from one set of plots are, therefore, only a sample of the climate and land conditions at that particular site (Mutchler et al. 1994; Rüttiman et al. 1995). Hence, there is a danger of extrapolating plot results to all areas with a similar physiography (slope, aspect, soil conditions, cover and erosion features). Data from a single location must be combined with data from other locations to permit researchers to make erosion predictions over a large area.

Secondly, soil loss data come from plots or areas of differing size. The problem is that soil erosion processes are scale-dependent and the scale of the experiment will affect the results. Research plots may, therefore, not reflect the nature of the erosion process found at the field level

(Evans, 1995). It is customary to compute soil loss on a 'per unit area' basis, such as t ha^{-1}. This assumes that the complete area contributes to total soil loss. On large plots, where most erosion may take place in rills, this is demonstrably incorrect (Bryan and Luk, 1981; Dehn, 1995).

Meanwhile, at the micro-scale, the spatial variability of soil erosion is a function of exposure of bare soil, localised changes in soil texture and compaction, slope steepness, and opportunities for deposition such as above field boundaries (Harden, 1993). Hence, runoff plots do not deal with the questions of how much eroded soil is deposited, how much leaves the hillside, and how much is transported downstream (Dunne and Dietrich, 1982). As the assessment area increases, so does the possibility that eroded material will be stored within the bounded area and not recorded. In real field conditions, as much as 90-95 percent of eroded soil is re-deposited elsewhere in the landscape (Stocking, 1995).

Results from research plots can, therefore, overestimate regional soil loss because the plots often give the highest measured soil losses per unit area. This is because each soil particle detached by erosion is often captured and weighed (Hudson, 1993:39). Quite the opposite can also occur. Thurow and Smith (1998:14) point out that research plots are often too small for landslide processes to occur, and that soil loss from landslides can account for most of the erosion on farmers' fields. In this case, estimates of soil loss from field-scale catchment areas are likely to be higher than plot-scale estimates.

Another source of error is that it is not always clear if data come from on-station or on-farm research. Critics of on-station experiments point out that they are often conducted on soils that, for years, have been cultivated, fertilised and chemically treated in ways that bear little resemblance to farmers' practices. Because the research takes place in conditions that are not representative of those faced by smallholder farmers, results can not meaningfully be extrapolated to variable biological and social farming systems (Bunch, 1982:120; Suppe, 1988). This reveals a tension in research between those trying to understand soil erosion processes in ideal conditions and those who are more interested in the reality; soil loss experiments conducted under conditions faced by resource-poor farmers.

In agriculture, an additional complicating factor when it comes to predicting (or reducing soil loss on steep slopes), is that the amount of soil moved across the terrace or the inter-row area, as a result of agricultural practices, may be more important that the amount moved by erosion (Hudson, 1992b:160). Lewis (1992) and Lewis and Nyamulinda (1996) demonstrate that on steep slopes in Rwanda, field preparation, weeding and harvesting are important activities that contribute to soil loss. Turkleboom et al. (1997) also report that in Thailand, manual soil tillage carried out in order to control weeds, decreased the soil's resistance to detachment by raindrop impact and runoff and led to increased erosion especially on slopes above 60 percent.

Quine et al. (1999) find that gross rates of tillage erosion on fields with terraces and contour-strips in China, Zimbabwe and Lesotho, which were cultivated by animal traction, were comparable to, or greater than, gross rates of water erosion. Govers et al. (1994), based on research in Belgium, conclude that erosion and sedimentation rates associated with tillage, carried out with a mouldboard and a chisel plough, may be more important than those associated with water erosion on arable land in temperate areas. Nehmdahl (1999) observes that some of the terrace formation on research plots planted with barriers of vetiver grass in Tanzania was due to down-slope cultivation of the plots.

3.3.4 Shenanigans with the Data?

While soil loss data alone has been used to justify soil conservation programmes, the argument for such programme is strengthened by research that demonstrates that agricultural yields decline as erosion increases. The research literature is peppered with data on changes in soil loss and yields before and after the establishment of soil conservation technologies. Seldom, however, is there any indication whether the changes are due to the technologies *per se* or to changes in extraneous factors such as pests and diseases or natural variability through time particularly with respect to seasonal differences in rainfall. For example, with reference to a soil conservation project in Thailand, Schiller et al. (1982) observe that much of the variability in yield data between years can be attributed to differences in rainfall distribution rather than the effect of soil conservation practices.

Much more attention also needs to be given to statistical analysis of the data used by researchers because scientific data about erosion are used for particular ends and different actors can manipulate the data to suit their different agendas. Stocking (1995b) cites the example of 22 erosion studies in the Upper Mahaweli catchment in Sri Lanka. There is an almost 8000 fold difference in soil loss estimates from these studies, with the figures ranging from 0.13 t ha^{-1} to 1026 t ha^{-1}. Natural differences in soil erosion rates and measurement errors could possible account for these differences but Stocking (1995b) contends that the results may indicate systematic bias on the part of researchers. In the Sri Lanka example, the high estimates of soil loss appear in a report to demonstrate the seriousness of the erosion problem. The low estimate comes from a study carried out in a tea plantation, which aims to show that tea growing is conservation-effective.

The Sri Lanka example is not, of course, an isolated case. There are numerous examples of where science—well beyond research on soil loss— has become 'politicised'. In these cases, data are generated (or seen to be generated) for a purpose. The United States' government, for example, disputes the data used by Intergovernmental Panel on Climate Change to predict the severity of global warming.

3.3.5   Steeplands: The Justification for More Research Using Runoff Plots

The above discussion about the dubious accuracy of some of the field data on soil loss and the impact of soil conservation technologies means that soil erosion research can be more art, possibly political art, than science, and that figures on erosion rates have to be treated with some caution. Data from field measurements can, however, be very useful and there is ample justification for continuing to carry out good quality field research. Their usefulness is more apparent when research takes place on the steeper slopes now being cultivated by smallholder farmers.

As outlined in Chapter 1, steeplands are the front line in the struggle to achieve a more sustainable agriculture and while the accuracy of soil erosion data in general is questionable, that from steeplands (where increasing numbers of smallholder farmers are now found) is even less reliable. Research on steeplands, with slopes exceeding 20 percent, has been sketchy or neglected because cultivation of these sites has traditionally been considered inappropriate and unsustainable (Lal, 1988; McDonald et al. 2002). Also research on steeplands is difficult because of remoteness, difficult access and extremely variable landscapes and natural events (Hudson, 1992b:145).

As detailed above there are dangers in extrapolating the results from field research. This is particularly the case with soil research conducted on shallow slopes (Hudson, 1992b:113) largely because an understanding of the physical processes of soil erosion on steep slopes is limited (El-Swaify, 1997). The interactions of rainfall and soil characteristics, which control infiltration and flow initiation and which, together with hydraulic variables, determine soil detachment, are extremely complex. Any small variation in a contributory factor can disturb relationships and significantly alter results and the erosive forces of splash, scour and transport all have a greater effect on steep slopes than on shallow slopes (see section 3.2.3)

Slope shape and complexity alone are important modifiers of steepness and length influences on soil erosion (El-Swaify, 1997) and as Lal (1988) points out, regular slopes are the exception rather than the rule. On slopes steeper than 20 percent, the possibility of greater interaction between slope and other factors such as cover or length of slope is entirely plausible and would further complicate the process of predicting soil loss (Hudson, 1992b:146,149).

There is also field and laboratory research that confounds our expectations. Research has demonstrated that an increase in slope does not always lead to an increase in soil loss. Ahmed and Breckner (1973) demonstrate, on three Tobago soils, that erosion was less on steep slopes than on gentle ones. Sheng (1982) and Sillitoe (1993) posit that one explanation for this is that the kinetic energy of raindrops on steep slopes is smaller because the rain splashes on a larger unit than on gentle slopes. Sillitoe (1993) also reports that there is evidence in Papua New Guinea that soil loss peaks at about 70 percent slope and then decreases. Valentin (1989) argues that runoff is often stabilised and may even be reduced as slope increases as the surface roughness (and infiltration area), mainly caused by linear incisions, increases with gradient.

Another explanation is that the soils in steeplands have already been exposed to severe erosion and the erodible material e.g. sand has already been eroded leaving behind erosion resistant soil material. In general though, erosion processes can be expected to be more severe on steep slopes and, therefore, a technology that is effective at controlling soil loss on a 20 percent slope may not be as effective on a 50 percent slope. Much of the research on soil loss and the effectiveness of soil conservation technologies has taken place on shallow slopes. Could this be one of the reasons why farmers are not readily adopting some soil conservation recommendations, namely they do not work at the slope angles now being cultivated? Research in Honduras was designed to help answer this question.

3.4   TESTING THE SOIL CONSERVATION APPROACH

3.4.1   The Interest in Live Barriers

Throughout the world, researchers and development practitioners who are concerned about soil and land degradation have worked hard to encourage resource-poor farmers to adopt soil conservation technologies. In the context of smallholder farming systems on steeper slopes, many of these soil conservation initiatives are directed at the use of biological control measures such as live barriers of woody and grass species. These are considered more suitable for smallholder agriculture because of lower establishment and maintenance costs than mechanical protection measures such as stone walls.

Live barriers, therefore, merit special attention. Their effectiveness in reducing soil loss has been demonstrated in laboratory conditions (Boubakari and Morgan, 1999; Tadesse and Morgan, 1996) but of more interest are the results from field experiments. One of the most widely promoted live barrier species is vetiver grass (*Vetiveria zizanioides*) and there are also field data that demonstrate that vetiver grass barriers are effective at reducing soil and water loss and, in some cases, increasing productivity (e.g. Subudhi et al. (1998). See Table 3.2.

Much of the research summarised in Table 3.2, however, has been conducted on shallow slopes, often less than 20 percent. There is evidence that the effectiveness of live barriers to reduce soil loss is inversely related to slope-angle. Based on the results from 15 independent studies of runoff and erosion under live barriers, Young (1997b:70) concludes that the technology has been proved effective for erosion control only on slopes up to 20 percent.

Table 3.2 Effectiveness of vetiver grass live barriers in reducing soil and water loss and increasing productivity

Source	Country	Data	Spacing	Slope	Runoff	Soil loss	Yield
Bharad and Krishnappa (1990)	India	27 plot years of data 1987-1990	Vertical interval of 1.0 m	5 %	Vetiver grass barriers reduced runoff by 30% (± 23 %) and 47% (± 9 %) compared to graded bunds and across slope cultivation respectively	Vetiver grass barriers reduced soil loss by 43% (± 19%) and 74% (± 5 %) compared to graded bunds and across slope cultivation respectively	Vetiver grass barriers increased crop yields by 6% (± 10 %) and 26 % (± 20%) compared to graded bunds and across slope cultivation respectively. Maize yield reductions noted in two rows closest to vetiver grass barriers
Kon and Lim (1991)	Malaysia	1990-1991	Distance of 4.0 m	4 %	Vetiver grass barriers reduced runoff by 73% compared to bare soil plots	Vetiver grass barriers reduced soil loss by 93 % compared to bare soil plots	Vetiver grass barriers reduced maize yields by 10%. Vetiver grass competed with maize especially in two rows closest to barriers
Laing (1992)	Colombia	1990-1991 growing season	Unknown	Unknown	Vetiver grass barriers reduced runoff by 69 % compared to bare soil plots	Vetiver grass barriers reduced soil loss by 91% compared to bare soil plots	Cassava yields on plots with vetiver grass were 5% less than those from cassava grown on flat plots. Growth of cassava was not affected by vetiver grass barriers whereas elephant grass barriers had severe competitive effects on cassava. Hellin and Larrea (1998) reported on competitive effect of elephant grass barriers in Honduras

(Contd.)

(Contd.)

Source	Country	Data	Spacing	Slope	Runoff	Soil loss	Yield
Laing (1992)	Colombia	1991-1992 growing season	Unknown	Unknown	Vetiver grass barriers reduced runoff by 58 % compared to bare soil plots	Vetiver grass barriers reduced soil loss by 99 % compared to bare soil plots	Cassava yields on plots with vetiver grass were 4 % greater than those from cassava grown on flat plots. Vetiver grass had competitive effects on cassava yield in the row closest to the barrier. This was compensated for by higher yields in intermediate rows
Rodriguez (1995)	Venezuela	Simulated rainfall on research plots	Unknown	15 % and 26 %	Not available	Reduced to less than 1 Mg ha^{-1} by combining mulch with vetiver grass barriers compared to 8 Mg ha^{-1} on the bare soil plot	Not available
Bharad (1995)	India	Research plots	Unknown	Unknown	Vetiver grass barriers reduced runoff by 51% and 36% compared to control on shallow and deep soil respectively	Vetiver grass barriers reduced soil loss by 72 % and 70 % compared to control on shallow and deep soil respectively	Yields of pearl millet and sorghoum with vetiver grass increased 29 % & 14 % respectively compared to control
Sukmana (1995)	Indonesia	Various experiments	Vertical interval of 0.2-1.25 m	6-35%	Vetiver grass barriers reduced runoff by 57% compared to control plots	Vetiver grass barriers reduced soil loss by 82 % compared to bare soil plots	Not available

(Contd.)

(Contd.)

Source	Country	Data	Spacing	Slope	Runoff	Soil loss	Yield
Bhardwaj (1996)	India	Research plots	Vertical interval of 1 m.	4 %	Vetiver grass barriers reduced runoff by 27 % compared to cultivated fallow plots	Vetiver grass barriers reduced soil loss by 90% compared to cultivated fallow plots	No difference between treatments
Rao and Rao (1996)	India	Unknown	Unknown	2.5 %	Vetiver grass barriers reduced runoff by 66% compared to contour cultivation	Vetiver grass barriers reduced soil loss by 76% compared to contour cultivation	Vetiver grass barriers increased crop yields by 7–22% compared to contour cultivation
Shengluan and Jiayou (1998)	China	Research plots	Unknown	5 %	Vetiver grass barriers reduced runoff by 33 % compared to control plots	Vetiver grass barriers reduced soil loss by 64% compared to bare soil plots	Not available
Subudhi *et al.* (1998)	India	Unknown	Horizontal interval of 8.0 m	2 %	Vetiver grass barriers reduced runoff by 35% compared to control plots	Vetiver grass barriers reduced soil loss by 60% compared to bare soil plots	Vetiver grass barriers increased yields of rice by 93 % compared to control
Pawar (1998)	India	Observations at watershed level	Unknown	Unknown	Vetiver grass barriers reduced runoff by 8–47%	Vetiver grass barriers reduced soil loss by 13–43%	Vetiver grass barriers increased yields of sorghum by 29%.
Nehmdahl (1999)	Tanzania	Research plots	Vertical interval of 2.0 m	29 %	Not available	Average of 110 t ha^{-1} of soil were lost from control plots over three growing seasons compared to less than 2 t ha^{-1} from vetiver grass plots	Reduced yield on the first line of maize below a vetiver grass barrier

3.4.2 Rationale Behind the Design of the Field Experiments

As outlined in section 2.4.3, Honduras is a mountainous country that suffers from high rates of soil and land degradation. Much of this is associated with smallholder farmers' encroachment onto steeplands with slopes greater than 20 percent. As in many parts of the world, development organisations in Honduras have promoted the use of vetiver live barriers. Farmers, however, have not readily adopted the live barriers. Do they know something that we do not? Are the barriers effective on the steep slopes increasingly being cultivated? A formal scientific trial was designed to provide answers to these questions. Box 3.3 summarises the reasons why it was decided to establish a field experiment using runoff plots in Honduras.

Box 3.3   Field experiment to address conservationists' concern with soil loss and farmers' concern with agricultural production

Resource-poor farmers are primarily concerned with agricultural yields (see chapter 2). In order to gain more insight as to why farmers were reluctant to establish vetiver grass barriers as recommended by institutions such as the World Bank (World Bank, 1990), it is necessary to test the impact of this soil conservation technology on agricultural yields. There are several indirect ways of assessing yields but they have their drawbacks. Erosion rates can be used as proxies for changes in yield but they are poor proxies as yields are determined by other factors including rainfall (Clark, 1996). Furthermore, productivity data based on farmers' calculations is notoriously inaccurate. It seemed to be sensible to measure changes in agricultural yields under experimental conditions. Further justification for this project was the need to determine how effective live barriers of vetiver grass actually are at controlling soil loss.

In the context of land degradation and from the perspective of a soil conservationist, the amount of soil lost is the most important factor to assess. As outlined in Section 3.3.3 there are several ways to measure or assess soil loss. These include the use of runoff plots and the use of erosion pins to record ground loss. Despite the inaccuracies associated with the use of runoff plots, they are very useful in comparative studies to test the effect of soil conservation treatments on erosion and runoff (Hudson,1993:27). It was, therefore, decided to establish an experiment to examine, simultaneously, soil and water loss and changes in agricultural production.

The next question was whether to establish on-farm or on-station experiments. One of the concerns with on-station trials is that they are seldom representative of conditions faced by farmers. However, there are several advantages to on-station trials (Hudson, 1993:2,11,28). Firstly, it is easier to separate and control the variables. Second; there are fewer problems of interference with the equipment by the local population. Thirdly, there is greater reliability, which is assured by professional staff being able to get to the site at all times.

It was, therefore, decided to establish a series of on-station runoff plots. However, it was necessary to avoid the dangers of the plots not being representative of local farming condition. The best way of avoiding this problem was to have local farmers conduct the agricultural processes. This project, therefore, used what Norman and Douglas (1994:10) refer to as a superimposed trial. This type of trial includes elements of management from both farmers and researchers, with implementation primarily the responsibility of farmers. In this case, the research plots were located on slopes that are representative of those cultivated by local farmers. The plots were managed with active co-operation and guidance from local farmers. Traditional farming practices were followed and the only conditions were those required by the demands of the experiment, such as ensuring that activities such as weeding, sowing and harvesting of maize took place at the same time in all plots.

3.4.3   Location of Santa Rosa Experimental Station

The Santa Rosa Experimental station in southern Honduras, hereafter referred to as Santa Rosa, lies in the department of Choluteca (see Map 2.1 on page 49). Many smallholder farmers in the region cultivate small hillside plots while the better quality land in the valleys belongs to large landowners and international fruit companies. The challenge facing smallholder farmers is to maintain or increase productivity on a sustainable basis from these marginal steeplands. Denuded hillsides are evidence that the challenge is not being met.

Santa Rosa offered the ideal location for an experiment to test whether live barriers of vetiver grass can assist farmers in meeting this challenge. The experimental station is in the department of Choluteca in southern Honduras at 87° 04' W and 13° 17' N. The station is managed by the *Corporación Hondureña de Desarrollo Forestal* (COHDEFOR). The land where the trials were established had been used for cattle grazing until 1982 when ownership of Santa Rosa was taken over by a sugar company prior to it passing to COHDEFOR in 1990. Details of the climate at Santa Rosa are provided in Box 3.4 and a summary of the soil characteristics are provided in Table 3.3.

Box 3.4   Climate at Santa Rosa

The climate is similar to the rest of central and southern Honduras; rainy with an extremely dry season. The dry period lasts from November/December to April/May. During this period moisture is deficient for plant production. The wet season is bimodal, lasts from April/May until October/November, and accounts for approximately

90 percent of the annual precipitation, which exceeds 2,000 mm. During July and early August there is a marked reduction in rainfall known locally as the *canícula*. September and October are often the wettest two months.

Rainfall at Santa Rosa is higher than the rest of the lowland Pacific littoral because it is strongly influenced by the local orographic effects of nearby mountains. Smith (1997:104) reports that the average rainfall erosivity factor (R) in the foothills of Cerro Guanacaure for the three-year period 1993-1995 was 13,889 MJ mm ha^{-1} h^{-1} yr^{-1}. Average daily temperature is 27°C with a maximum temperature of 35°C and a minimum of 21°C. Relative humidity varies between 60 percent and 80 percent. The low humidity readings coincide with the dry season. The prevailing winds are the north-east trades (Zúñiga, 1990).

Table 3.3 Soil characteristics at the trial site in Choluteca (from Hellin and Haigh, 2002)

Soil type	Fine-loamy to coarse-loamy gypsic, isohyperthermic Ustropepts (Inceptisol) and Haplustolls
Soil texture	Sandy loam with an average particle size distribution of sand (62 percent), Silt (21 percent) and clay (17 percent)
Topsoil	Depths ranged from 18-45 cm
Bulk density	Ranged from 1.06-1.37 g cm^3 at 0-8 cm depths and 1.08-1.42 g cm^3 at 8-16 cm depths
Infiltration	Rates varied from 17-783 mm hr^{-1} with an average of 398 mm hr^{-1}
Organic carbon	Ranged from 1.62–2.26 percent in topsoil
pH values	Ranged from 5.1-6.5 in topsoil
CEC values	Varied between 11.49–15.36 cmol$_c$ kg^{-1} in topsoil

3.4.4 Experimental Design

Having decided to establish a field experiment and identified a suitable location, the next task was to decide on the experimental design. The theory behind experimental design was not the problem rather difficulties arose when turning to previous research for guidance.

The primary goal of any experiment should be to obtain good data that can be statistically analysed, interpreted and used to infer or test scientific principles. In order to obtain this data, there is a need to control and quantify the local variation (Pierce and Lal, 1994). Hence, the art of experimental design is to keep all variables constant between tests and their replication, with the exception of the variables that are being examined. In field experiments this is made more difficult because of site variability, a variability that can render even the most carefully collected data statistically non-significant (Frye and Thomas, 1991; Pushparajah, 1989).

Randomisation and replication, the two basic requirements for a good experiment, are critical to successful statistical analysis of field data (Mead et al. 1993:41). The difficulty for any would be researcher is that previous research offers few guidelines on the optimum design of field experiments, especially when it comes to decisions about plot size and the numbers of treatments and replications to use (see Table 3.4).

Table 3.4 shows that researchers have used a bewildering variety of different experimental designs. At the end of the day, any decision about experimental design is partly government by practical considerations. There is no absolute mathematical answer to the number of replications that should be used, although Hudson (1993:6) suggests that there should always be a minimum of three. Despite the use of the Universal Soil Loss Equation (USLE) plot size (4 × 22 m), Table 3.4 clearly shows that there is no universally accepted plot size. Furthermore, Hudson (1993:35) suggests that there is little justification for following the precise measurements, in metric units, of the USLE plot size and advocates a plot size of about 5 × 20 m. The reality is that there is often a compromise between size of the research plots and the number of treatments and replications. This is precisely because the size and design of an experiment has limitations such as cost, labour and area.

Eventually, the author decided to establish a trial consisting of 24 research plots each measuring 24 m × 5 m. This decision was based on previous research detailed in Table 3.4 and the resources available. The plots were replicated on two slope angles, 35-45 percent and 65-75 percent, which are those most representative of those cultivated by smallholder farmers in the area. There were two agricultural treatments, each replicated six times on each of the two slope angles: one set with live barriers of vetiver grass and a control set without barriers (see Table 3.5). The treatments were assigned randomly to each plot.

3.4.5   Establishment of the Experimental Plots

The creation of the experimental plots and their subsequent management sought to replicate local agricultural practices. Rather than burning, the hillsides were manually cleared of secondary vegetation. Although burning plots remains a part of traditional agriculture, it is a less common practice than in the past. The procedures used in constructing the plots were based on Hudson (1993). They are detailed in Box 3.5.

Toness et al. (1998:36) report that, because of the amount of land taken out of production, a spacing of approximately 6.0 m between soil conservation barriers is about the minimum distance that is acceptable to farmers in Honduras. At Santa Rosa, live barriers were established at 6.0 m intervals i.e. there were three barriers per plot. Single row barriers were used and vetiver

Table 3.4 Design of a number of agronomic experiments showing the number of replications and plot size

Type of experiment	Trial design	Source
Effects of mulching with multipurpose tree prunings on soil and water runoff under semi-arid conditions in Kenya.	Randomised complete block with three replications and seven treatments. Plots 10 × 5 m.	Omoro and Nair (1993)
Effects of perennial mulches on moisture conservation and soil-building properties through agroforestry.	Randomised block design with eight replications and six treatments. Plots 5 × 4 m.	Tomar *et al.* (1992)
Regional maize grain yield response to applied phosphorous in Central America.	Thirty-three experiments. At each location, randomised complete block with three replications and six treatments. Plots consisted of four to six maize rows, 5.5 m in length.	Raun and Barreto (1995)
Effect of intercropping maize and closely spaced *Leucaena* spp. hedgerows on soil conservation and maize yield on a steep slope at Ntcheu, Malawi.	No replications. Four treatments. Four plots in total each 4 × 24 m.	Banda, *et al.* (1994)
Effect on maize growth of the interaction between increased nitrogen availability and competition with trees in alley cropping.	Three replications and no treatments. No information on plot size.	Haggar and Beer (1993)
Nitrogen fixation in the component species of contour hedgerows.	Split-plot design with four replications. Main plots were contour hedgerows of three species. Each main plot randomly sub-divided into two subplots.	Garrity and Mercado (1994)
Productivity of annual cropping and agroforestry systems on a shallow Alfisol in semi-arid India.	Randomised block with three replications and eight treatments. No information on plot size.	Rao *et al.* (1991)
Alley cropping of maize with nine leguminous trees.	Randomised complete block design with 14 trees per treatment in one row. Four replications of nine treatments.	Rosecrance *et al.* (1992)
Productivity of hedgerow shrubs and maize under alley cropping and block planting systems in semi-arid Kenya.	Split-plot experiment with three replications. Each sub-plot 5 × 20 m.	Jama *et al.* (1995)

(Contd.)

(Contd.)

Type of experiment	Trial design	Source
C. cajan (L.) Millsp. as a potential agroforestry component in the Eastern Province of Zambia.	Unreplicated trial with 33 lines of *C. cajan*. Each plot had three rows, each 1.5 m in length.	Boehringer and Caldwell (1989)
Pattern of soil moisture depletion in alley cropping under semi-arid conditions in Zambia.	No replications. Plot size 10 × 11 m.	Chirwa *et al.* (1994)
The effects of between-row (alley widths) and within-row spacings of *G. sepium* on alley-cropped maize in Sierra Leone.	Three blocks replicated. No information on plot size.	Karim et al. (1993)
Productivity of alley farming with *L. leucocephala* and *P. purpureum* in coastal lowland Kenya.	Randomised block design with three replications. Plot size 16 × 8.5 m.	Mureithi et al. (1995)
Erosion hazard evaluation from soil microtopographic features in Thailand.	22 plots each 10 m × 36 m. Five treatments replicated four times. Two treatments not replicated.	Bergsma (1997)
Soil loss from the Ardèche drainage basin in France.	Sixty modified adjoining Gerlach troughs on each of two slope segments.	Roels (1985)
Effectiveness of conservation technologies to reduce soil loss in Taiwan.	Seven plots each 21.13 × 4 m on 60% slope. No replications.	Wu and Wang (1998)
Rainfall erosivity and erodibility in Colombia.	Plots 22.1 × 11 m established at two locations. Slope 8% to 17%. Three replications at one site and two at the other.	Ruppenthal et al. (1996)
Reduced soil loss with litter-mulch as a protective cover on maize plots in Mexico.	20-40 m² plots on 31-51% slopes. No information on number of replications.	Maass (1992)
Effectiveness of vegetative filter strips for removal of sediment and phosphorous from feedlot runoff.	Nine plots with a 5.5 × 18.3 m bare source area (simulated feedlot) and either a 0, 4.6 m or 9.1 m filter located at the lower end of each plot. Field plots were constructed on three different slopes (5, 11 and 16 %) *i.e.* no replications on each slope angle.	Dillaha et al. (1986)

(Contd.)

(Contd.)

Type of experiment	Trial design	Source
Effectiveness of conservation technologies to reduce soil loss in Thailand.	Two 5 × 71 m plots on 40 % slope *i.e.* no replications.	Sombatpanit et al. (1992)
Effectiveness of conservation technologies to reduce soil loss in Indonesia.	Randomised complete block design with three replications. Plots measured 1.5 × 10 m.	Siebert and Lassoie (1991)
Effect of *Cassia siamea* hedgerow barriers in Kenya.	Four 400 m² plots on 14 % slope. Hedgerow treatment replicated twice. Control and mulch treatment not replicated.	Kiepe (1995)
Alley cropping and crop yield responses in the Philippines.	Four treatments and three replications. Plot size 6 m × 12 m on 17 % slope.	Comia et al. (1994)
Assessment of tillage erosion on steep slopes in northern Thailand.	Soil movement by manual tillage was measured in farmers' fields on five slopes (32 % to 82 %) in plots each 4 m × 4.5 - 8 m. No information on number of replications.	Turkelboom *et al.* (1997)
Role of human activities in land degradation in Rwanda.	11 plots, 20 m × 5 m on 60 % slope, 11 treatments (*i.e.* no replications).	Lewis and Nyamulinda (1996)
Influence of slash and burn practices on soil and water loss in Spain.	Three plots each 4 × 20 m established on 30 % slope. Control treatment not replicated, burn treatment replicated twice.	Soto *et al.* (1995)
Effect of vetiver grass live barriers in India.	Research plots 1.5 m × 8 m. No information on number of replications.	Kon and Lim (1991)
Effect of vetiver grass live barriers in India.	Research plots 20 m × 100 m. No information on number of replications.	Bhardwaj (1996)
Effect of vetiver grass live barriers in India.	Randomised block design, three replications. Plots 25 m × 3 m.	Subudhi *et al.* (1998)
Effect of vetiver grass live barriers in Tanzania.	Four treatments in a randomised block design with three replications. Plots each 4 m × 20 m.	Nehmdahl (1999)

(Contd.)

(*Contd.*)

Type of experiment	Trial design	Source
Effect of contour hedgerows on soil physical, chemical and hydrological properties in Nigeria.	Six treatments with no replications on a 7% slope. Each plot measured 10 × 70 m.	Lal (1989b)
Effectiveness of mesh-bags for calculating soil loss in Peru.	Four plots each 5 × 10 m on a 2% slope.	Hsieh, Y-P (1992)
Soil erosion and crop productivity research in South America.	Plots each about 50 m². Three replications of each treatment.	Tengberg *et al.* (1998)
Contour hedgerows and soil conservation in Peru.	Two treatments replicated three times on a 15% to 20% slope. Plots measured 30 × 5 m.	Alegre and Rao, (1996)
Effect of soil conservation technologies on soil loss, runoff and crop yield in Ethiopia.	Plots 6 m × 30 m. Four–six treatments at seven sites in Ethiopia.	Herweg and Ludi, (1999)

Box 3.5   Steps in construction of research plots

- Hillsides were manually cleared of secondary vegetation.
- Contours were marked out on the cleared hillsides with an A frame
- Research plots were mapped out on the hillsides (see Plate 9).
- The upper and lateral sides of the plot were demarcated by 30 cm wide metal sheeting embedded in the ground (approximately 18 cm above and 12 cm below ground level) to prevent surface leakage underneath and overtopping.
- Pieces of metal sheets overlapped by approximately 10 cm and were joined by rivets and then covered with an oil-based sealant. This was reapplied when necessary.
- Buffer strips were left at the sides and top of the plots to reduce border effects and to facilitate access to the plots.
- A 0.5 m deep drainage channel was constructed above each of the plots to divert surface water flowing down from higher ground.
- Plot boundaries and drainage channels were inspected on a regular basis and repaired when necessary. Vegetation in the buffer strips was cut frequently to reduce the risks of any interference with the experiment.

splits were planted at a distance of 3-5 cm. Vetiver grass barriers were pruned three times per growing season and following local practice, cuttings were laid above the barrier.

Table 3.5   Summary of main trial design at Santa Rosa Experimental Site

Treatment	Number of replications on shallow slope (35-45 percent)	Number of replications on shallow slope (65-75 percent)	Total number of replications
Control	6	6	12
Live barriers	6	6	12

3.4.6   Addressing Farmers Concerns: Measuring Maize Yields

Farmers are much more interested in agricultural production *per se* than the amount of soil eroded (see section 2.5.3). The experiment at Santa Rosa was designed to address the impact of vetiver grass barriers on maize production as well as soil loss. Following local practices, the research plots were planted with maize twice a year. Maize was sown in June (the *primera*) and September (the *postrera*) each year and harvested in September and January respectively (the exception was 1998 when Hurricane Mitch destroyed the *postrera*). Farmers planted maize at a regular spacing with 32 rows per plot, so that in those plots with live barriers no land was lost to agricultural production. All maize used at the site was the local variety and was sown at a characteristic density.

For assessment, each research plot was divided into four sections (each 6 × 5 m) with eight rows of maize per section. Maize yields were calculated for each crop row by weighing the maize cobs with a hand balance and multiplying this by the average dry weight of maize per cob. The lower boundary of the research plot was demarcated by a runoff collecting trough (see below). The upper boundary was demarcated by metal sheeting. To avoid edge effects, maize data from the lowest and uppermost four rows in the plot were not included in the analysis.

3.4.7   Use of Barrels and Catch-pits in Estimating Soil and Water Loss at Santa Rosa

The conventional research method to collect soil and water is to channel runoff into a collecting recipient such as a calibrated barrel or tank (e.g. Kiepe, 1996; Lal, 1989; Rüttimann et al. 1995) (see Plates 7 and 8). When large amounts of runoff are expected, some device is used to separate a known fraction that is then stored (Hudson, 1995:169). After each runoff event the amount of soil and water in the recipient(s) is then calculated. This can be done by taking an aliquot sample or installing a device such as a permeable cloth that separates the sludge from the supernatant. Barrels and tanks have been extensively used in soil erosion experiments worldwide (see Box 3.6). A simplified collecting barrel known as a Gerlach trough has also been widely used.

Plate 7   Runoff plots have been used worldwide in experiments to estimate soil loss. China. (Hellin, J.)

Plate 8 Runoff plots in China (Hellin, J.)   **Plate 9** Establishment of runoff plots in Honduras. (Hellin, J.)

Box 3.6 A selection of soil erosion experiments using collecting barrels

- Hudson (1995:173-175) illustrates the use of collecting barrels in Indonesia, Nigeria, Thailand and Taiwan.
- Kiepe (1996) used barrels to collect runoff as part of research into the effect of barriers of *Cassea siamea* on hillsides in Kenya.
- In Nicaragua, runoff from research plots designed to document the effect of live barriers of *Gliricidia sepium* was collected in barrels (Mendoza, 1996).
- Arévalo-Méndez (1994) used collecting barrels in Güinope, Honduras as part of research on the effect of cover crops on soil and water loss.
- Banda et al. (1994) used concrete pits as opposed to barrels to collect runoff as part of soil erosion research in Malawi.
- Collecting tanks were used by Rüttimann et al. (1995) in Switzerland.
- In Nigeria, Lal (1989) used a sample storage tank to collect runoff.

The use of barrels can, however, be time-consuming and expensive and catch-pits have proved to be an alternative and much simpler method for measuring soil loss (but not water loss) under different agronomic treatments

(Hudson, 1993 and 1995:162). Runoff is channelled into the catch-pits where it can be stored for several weeks or months. Plastic-lined catch-pits have been the most widely used (see Box 3.7). Small holes are punctured in the plastic so that the water slowly dissipates and the sediment load is retained until end of season measurements can be made. A visual display of the amount of soil lost enables farmers and researchers to detect changes in erosion resulting from a change in land use or management. Plastic-lined catch-pits cited in the literature vary from 0.40 × 0.40 × 15 m (Howeler, 1987) to 5 × 8 × 0.75 m (Sombatpanit et al. 1992). There are several advantages and disadvantages with the use of plastic-lined catch-pits (see Box 3.8).

Box 3.7   Previous use of plastic-lined catch-pits in experiments worldwide

- In Thailand, Sombatpanit et al. (1992) used plastic-lined catch-pits immediately below two 5 × 71 m plots on a 40 percent slope.
- In Colombia, Howeler (1987) used plastic-lined catch-pits in a series of on-farm trials to compare soil loss under different *Manihot esculenta* (cassava) cropping systems. Plots sizes ranged from 20 × 10 m, to 15 × 10 m, to 10 × 10 m and slopes varied from 15 to 45 percent.
- Chan et al. (1994) used plastic-lined catch-pits in Malaysia to compare soil losses from various tillage practices and cassava-crop combinations. The plots were established on a uniform slope of 6-11 percent and each plot measured 10 × 10 m.
- A number of national research institutes in Thailand, Indonesia, Vietnam and China have used plastic-lined catch-pits to compare soil loss from research plots with different cassava cropping systems (Howeler, 1994; Howeler et al. 1996; Howeler, 1995).

Box 3.8   Advantages and disadvantages of plastic-lined catch-pits

- Soil accumulation is a convincing visual demonstration of soil loss from plots under different agronomic treatments (Howeler et al. 1996; Hudson, 1993:21). A visual display of erosion is a component of efforts to develop simple quantitative tools that farmers can use to monitor changes under different land management regimes (Doran and Parkin, 1994; Pieri et al. 1995:25).
- Catch-pits are easy to establish and operate and labour costs are relatively low.
- The amount of soil lost is only known when the catch-pits are cleaned out. No data are available on soil loss from individual storm events. Also there is no record of runoff per event or in total.
- There may be spillage from the catch-pits if dissipation is slow. The amount of soil lost, however, will be minimal because a high

proportion of the sediment lost from a research plots is in the form of aggregates. These settle to the bottom of a catch-pit rapidly (Sombatpanit et al. 1992).

- When calculating the dry weight of accumulated soil, the sampling procedure must be such that the differences in soil moisture in the accumulated sediment are accurately represented. This may involve taking numerous soil samples from the accumulated soil in each catch-pit.
- Solar radiation may cause the plastic sheets to disintegrate (Chan *et al.*, 1994).

It might have been better to have installed collecting barrels in each of the 24 plots at Santa Rosa but, of course, this would have entailed a huge investment. It was, therefore, decided to use a combination of collecting barrels and catch-pits. Runoff was collected in barrels in eight plots, and in the remaining 16 plots, catch-pits were used. Barrels were placed at the bottom of the plots below ground surface. A collector trough channelled the runoff from the plots into a calibrated barrel with an overflow pipe that delivered 1/8 of the volume to a second barrel (see Figure 3.1).

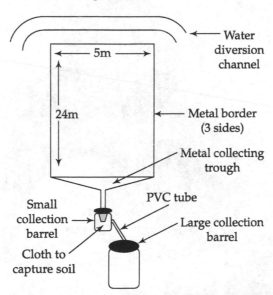

Fig. 3.1   Design of research plot and collecting barrels at Santa Rosa, Honduras

Sediment concentrations from aliquot samples are normally used to compute soil loss (Mutchler et al. 1994; Dangler and El-Swaify, 1976; Lal, 1989c). The use of aliquot samples can, however, be inaccurate because large particles of soil settle quickly and are hard to keep in suspension while the sample is taken (Stocking, 1995; Herweg and Ostrowski, 1997:11). Soil loss is therefore underestimated. Lang (1992) and Zöbisch et al. (1996) have tested in laboratory conditions the accuracy of aliquot samples. The former concludes that the method should not be used to estimate the sediment concentration of runoff water and suggests that researchers choose a method that separates sludge from the supernatant. This advice was interpreted for the calculation of soil and water loss at Santa Rosa. Details of the method are provided in Box 3.9 (also see Plates 10 and 11).

Plate 10   Preparation of platforms for location of barrels to capture runoff from experimental plots in Honduras. (Hellin, J.)

Plate 11   Use of barrels to estimate soil and water loss in Honduras. (Hellin, J.)

Box 3.9   Capturing runoff in barrels and estimating soil and water losses at Santa Rosa, Honduras

- At the bottom of each plot there was a triangular metal collecting trough. This was 5 m wide at the bottom of the plot and narrowed where the water passed down a PVC pipe into the barrels.
- The metal trough was covered with plastic sheeting (which rested on a wooden lattice) to prevent rain entering. The trough was made of three pieces of zinc lamina that were joined with rivets and subsequently welded along all the joints.
- At the junction of the trough and end of the plot, a 20 cm deep narrow trench was dug. A lip was moulded to the trough at approximately 90 degrees to the plane surface and was buried in the trench. Soil level at the bottom of each plot was flush with the surface of the trough.
- A 2 cm metal mesh was placed between the plot and trough (buried alongside and to the same depth as the lip of the trough). This was designed to prevent vegetation entering the trough and blocking the PVC pipe in large runoff events.
- Runoff from the collecting box passed into a 200-litre barrel. There were eight overflow holes, one of which lead into an 800 litre barrel in the case of exceptional runoff events.
- Based on Williams and Buckhouse (1991) a 35-litre container was placed within each of the 200 litre barrels. These containers captured the runoff from small rainfall events and could be more easily emptied and recorded. In larger events they simply overflowed into the surrounding barrel.

- Some soil did accumulate behind this mesh during heavy rains but after each runoff event, a sharp stick was run along the bottom of the mesh from inside the plot in order to "push" this eroded soil into the trough. This soil was brushed down to the PVC pipe. Water from the barrel was used to clean the trough and wash away the accumulated soil into the collecting cloth in the barrel.
- The majority of the sediment load was collected in a permeable cloth placed inside the 200 litre barrel The cloth was removed and weighed after each runoff event. A sample of the sediment in the cloth (approximately 80 g) was weighed, dried at 105° C for 24 hours, and weighed again. The total dry weight of the sediment load was then calculated.
- The total volume of water and sediment that passed through the cloth was measured by emptying the contents of the barrels either by removing the 35-litre container, or from a tap at the base of the 200- and 800-litre barrels.
- In the case of 50 percent of runoff events, a 1.0 litre aliquot sample was taken. This was filtered through Whatman Filter papers number 1 (see Tadesse and Morgan, 1996; Robinson et al. 1996) in order to calculate the amount of sediment per litre of water. The total amount of sediment in suspension was subsequently calculated. The average sediment load per litre of runoff was used to calculate the sediment load in runoff for those events when an aliquot sample was not taken.

Two factors were considered when designing the size and capacity of the barrels and catch-pits: their ability to handle the maximum possible rate of flow and also to store the maximum probable quantity of runoff. At Santa Rosa, the size of the catch-pits was based roughly on calculations of rainfall amount, intensity, potential runoff and infiltration and evaporation from the catch-pits. Each catch-pit measured approximately 1.0 - 2.5 m in depth × 1.5 × 1.5 m.

One thing that was not planned for in the design of the catch-pits was the amount of rain that would fall during Hurricane Mitch. The hurricane struck Honduras in October 1998 (see section 5.6) and the catch-pits were destroyed before the accumulated sediment could be weighed, so the soil loss data for 1998 were lost. Construction of the catch-pits and the methods used to estimate soil loss are detailed in Box 3.10 (also see Plates CP 4 and 12).

Box 3.10   Steps in the construction of plastic-lined catch-pits at Santa Rosa, Honduras

- Each catch-pit was covered with plastic sheeting. The sheets were overlapped by 0.5 m and secured with duct tape. The plastic sheets were replaced at the beginning of each rainy season.

Plate 12 Use of catch-pits to estimate soil loss in Honduras. (Hellin, J.)

- Nails with aluminium washers were hammered in along the length of the tape to prevent it peeling away.
- At the junction of the bottom of the plot and the plastic sheeting a small trench was dug and the plastic was buried 10-cm deep.
- Around the catch-pit a wall (over 10 cm high) was constructed from soil and stones. This was covered with plastic to prevent water and soil other than that from the plot entering the pits.
- Drainage channels were excavated around the catch-pits to reduce the risk of water entering.
- Holes at a density of 50 per m^2 were made in the plastic sheeting in the bottom of the pit with a nail to allow the water to drain away slowly.

In 1996, due to high rainfall, the catch-pits contained considerable amounts of water for almost the entire season. The total amount of accumulated sediment in the catch-pits was therefore calculated at the end of the rainy season. In 1997 the effect of *El Niño* severely reduced rainfall and as a result it was possible to weigh the sediment in August and in December. The accumulated soil in the catch-pits was removed and weighed. In 1996 because of the large volume of soil that had accumulated, five 1000 g samples were collected and weighed. The samples were dried at 105° C for 24 hours, and then weighed again. The total dry weight of the sediment load was then calculated. The same procedure was followed in 1997 but because there was less accumulated soil, fewer samples per catch-pit were collected.

3.4.8 How Good are the Soil Loss Data?

Soil loss and runoff are not directly measured values, they are always computed by measuring other variables like water depth or sediment concentration. Computed values for soil loss and runoff cannot, therefore, be validated and considerable uncertainty is involved in this procedure (Herweg and Ostrowski, 1997:1). The procedures to calculate soil and water loss are based on a number of assumptions. These include that the water collected is equal to the effective runoff from the plot; that the sediment collected is equal to the soil loss from the plot; and that external or internal sinks and sources of water and material are negligible (Herweg and Ostrowski, 1997:11).

Measurements always involve error and, therefore, equipment must be properly designed, installed and operated (Hudson, 1993:26; Roels, 1985). Errors were minimised at Santa Rosa by careful construction and mainte-nance of the runoff plots, barrels and catch-pits. During the experiment, the author had some concern about the accuracy of plastic-lined catch-pits i.e. what proportion of sediment that enters the catch-pit is captured. There is a dearth of literature on this subject. So, two small experiments were carried out that confirmed the catch-pits as a low cost and effective method to assess soil loss (see Box 3.11).

Box 3.11   Experiment at Santa Rosa to assess the accuracy of catch-pits

In 1997 two experiments were carried out at the trial site to test the accuracy of the plastic-lined catch-pits. In the first experiment five holes each 2.5 m deep × 1 m × 1 m were lined with plastic. A known volume of water (750 litres) and a known wet weight of soil (45 kg wet weight with three soil samples taken to calculate moisture content) were mixed and poured, over a period of two days, into the catch pits to simulate runoff during/following a heavy storm. The experiment took place during the dry season. The soil that had been retained in the bottom of the catch-pits was collected, weighed and samples were collected so as to calculate the dry weight of the accumulated soil.

The experiment was repeated during a three-month period in the rainy season. A known amount of soil was mixed with water and poured into each of the five catch-pits on 12 occasions. In total 36.1 kg (dry weight) of soil was poured into each of the catch-pits during this second experiment. A trench surrounding each catch-pit prevented the entry of any runoff. At the end of the rainy season, the soil that been retained was weighed and samples were taken in order to calculate dry weights. The percentage of sediment retained in each of the five catch-pits in both experiments varied from 95 percent to 99 percent.

3.5   MEASURING RAINFALL

Chapter 2 outlined how farmers often cite irregular rainfall as a greater threat to agricultural production than soil erosion *per se*. In any study designed to look at the impact of a soil conservation technology, it is critical to measure rainfall. Throughout the lifetime of the experiment at Santa Rosa, rainfall amounts and intensities were measured using a tipping-bucket rain-gauge (see Boxes 3.12 and 3.13) (also see Plate 13).

Box 3.12   Measuring rainfall amounts at Santa Rosa, Honduras

A tipping-bucket rain-gauge (ELE DRG-52) was used to measure rainfall amounts and intensities at Santa Rosa. The rain gauge was placed on a concrete platform on the crest of a hill at an elevation of 100 m.a.s.l and amongst the research plots. The rain-gauge had a catchment area of 324.3 cm^2, its bucket had a rolling tungsten carbide pivot system and a magnetic reed switch to count the pulses. The bucket tipped at 0.2 mm intervals and each tip caused the momentary closure of a reed switch. The duration of each closure was 50-100 mS. Accuracy is +/- 1 percent. Data were recorded in event mode on a data logger with time and date being stored for future retrieval and subsequent analysis of rainfall intensity as well as totals over a period.

Box 3.13   Calculating rainfall intensity indices at Santa Rosa, Honduras

Recorded data were downloaded directly to a personal computer every 2-3 weeks during the rainy seasons. The rainfall erosion indices EI_{30} and KE>25 were calculated for almost all rainfall events over the three year period using the procedures outlined by Hudson (1995:81). EI_{30} is the product of the kinetic energy of the storm and the 30-minute intensity. The latter is the greatest average intensity experienced in any 30-minute period during a storm. KE>25 is an orosivity index that consists of the total kinetic energy of all the rain falling at more than 25 mm hr^{-1}. This intensity is the practical threshold separating erosive from non-erosive rain (Hudson, 1995:79).

In the case of KE>25 and for all individual storms, the amount of rain that fell in 10-minute periods was generated. The total amount of rain that fell at intensities greater than 25 mm hr^{-1} for each 10-minute period was then calculated and placed into three categories of rainfall intensity (25-50 mm hr^{-1}, 50-75 mm hr^{-1} and > 75 mm hr^{-1}). The amount of rain in each category was multiplied by the energy for that category expressed in joules m^2 mm (figures supplied by Hudson, 1995:81). These figures were then added together to give the total energy of the storm in joules m^2.

In the case of EI_{30} and for individual storms (with the exception of those few events when the rain gauge failed to record, see below), the amount of rain that fell at 30-minute intervals was generated. The most intense 30-minute interval was selected and the intensity (I_{30}) was expressed in mm h^{-1}. Rainfall during the storm was placed in the four intensity categories. These were the same used to calculate KE>25 with the addition of a 0-25 mm hr^{-1} category. The total energy of the storm expressed in joules m^2 was calculated (see above). This figure was then multiplied by I_{30} and, following Hudson (1995:81), divided by 1000 to give convenient erosivity units.

On a few occasions two or more erosive events occurred in the period between measurements of soil and water loss. When this occurred, erosivity data from the events were added together. A rain-gauge that is read manually was also installed next to the tipping bucket rain-gauge in the event of the latter failing to record data. The former was read at 0600 if it had rained in the previous 24-hour period. Data on rainfall totals from the manual rain-gauge were used when the tipping-bucket rain gauge failed to record during a few rainfall events in June 1996, October 1997 and May 1998.

Plate 13   Tipping-bucket rain-gauge to measure rainfall amounts and intensities in Honduras. (Hellin, J.)

Rainfall totals in 1996, 1997 and 1998 were 3037, 1614 and 3175 mm respectively. Annual rainfall in the area varies greatly from year to year and there was considerable variability in rainfall with little rain in 1997 due to the effect of *El Niño* and excessive rainfall at the end of 1998 due to Hurricane Mitch. Before Hurricane Mitch struck at the end of October 1998, rainfall in

1998 had been similar to 1996, which in turn had been a particularly wet year. Rainfall characteristics are summarised in Table 3.6 along with total rainfall for the three years prior to the start of the experiment

Table 3.6   Rainfall characteristics at Santa Rosa (data do not include rainfall during and after Hurricane Mitch that struck Honduras at the end of October. Rainfall characteristics during the Hurricane are reported by Hellin et al. 1999b). From Hellin and Haigh (2002).

	1993	1994	1995	1996	1997	1998
Total rainfall (mm)	1379	1426	2527	3037	1614	2218
Rainfall intensity (joules/m²)*	n/a	n/a	n/a	31,252	16,076	30,875

*Hudson (1995:79) defines an intensity of 25 mm hr^{-1} as a practical threshold separating erosive from non-erosive rain. In Table 3.6, rainfall intensity is based on erosive rainfall events where rainfall intensities equalled or were greater than 25 mm hr^{-1} for any 10-minute interval during the storm. Calculations are based on Hudson (1995:81). The few rainfall events for which erosivity data are missing (due to an error in the automated rain gauge) have been omitted.

Monthly rainfall totals for the period are also displayed in Figure 3.2. The effect of Hurricane Mitch, when almost 25 percent of annual rainfall (747 mm) fell in a 48-hour period, is clear as is the impact of *El Niño* in terms of reduced rainfall in May, July and August in 1997. The monthly totals give more indication of the challenges facing smallholder farmers in terms of too much and too little rainfall at critical periods of the growing season.

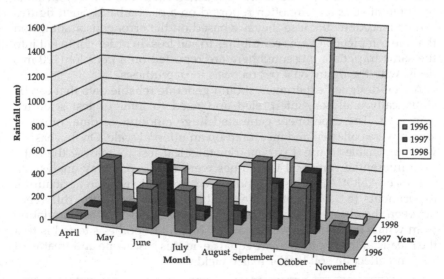

Fig. 3.2   Monthly rainfall (mm) 1996-1998 at Santa Rosa

3.6   IMPACTS OF CONVENTIONAL AGRICULTURE

3.6.1   Presentation of Data

Many researchers and development practitioners believe that the application of soil (and water) conservation technologies is critical to the mitigation of the soil and land degradation that threatens hillside agriculture. The experiment at Santa Rosa was designed to look at the impact of vetiver grass barriers, employed as a typical soil conservation technology, on soil loss and maize productivity (see Plates CP 5 and CP 6). The purpose of the experiment was to explore why farmers might be reluctant to adopt soil conservation technologies.

Firstly, however, it is necessary to examine soil and water loss and maize production under conventional agriculture. This section asks two fundamental questions, both raised earlier in the book:

- Is soil loss under conventional agriculture on steeplands severe enough that farmers should be expected to recognise this as a serious problem?
- Is the soil erosion that occurs a genuine threat to resource-poor farmers' livelihoods?

Some technical notes of the statistical tests used in this are outlined in Box 3.14.

Box 3.14   Technical notes on the presentation and analysis of data from the field experiments at Santa Rosa

Careful reading of results published in the soil conservation literature shows that not all the published results are as reliable as they suggest. Much of the data reported in the published literature come from plots of different sizes and are often reported on a per ha basis. Such figures can be inaccurate because they are based on the erroneous assumption that the complete plot area contributes to soil loss. In order not to fall into the same trap, data presented here are reported on a per plot (120 m^2) basis, with the figure on a per ha basis in parenthesis.

A well-designed experiment should generate reliable data that can be statistically analysed, interpreted and used to infer or test scientific principles. The experiment generated huge amounts of data, much of which were collected and measured on an interval scale. On this basis it was reasonable to use a parametric statistical test to analyse the data. Two inferential statistical techniques were used - one way analysis of variance (ANOVA) and t-tests, both are commonly used in agricultural experiments (c.f. Mead et al. 1996). These test whether the difference between the observed sample means is likely to exist in the population from which the samples were drawn. The null hypothesis tested is that the means in the population are equal i.e. in this case slope and treatment have no effect on soil loss or maize yields.

3.6.2 Soil and Water Loss Under Conventional Agriculture

Inequalities in land distribution have forced millions of resource-poor farmers, especially in the tropics, to cultivate steeplands. In the experiment at Santa Rosa, local farmers managed and followed the same agricultural practices used on their own plots. In this context, soil loss data from the control plots at Santa Rosa can be considered as representative of those occurring under normal husbandry in farming communities throughout the area.

Figure 3.3 shows annual soil loss (kg/plot) over a three-year period 1996-1998. The results are based on data from 12 control plots where soil was collected in barrels and catch-pits, although in 1998, the data are only from the four plots with barrels because the catch-pits were destroyed during Hurricane Mitch. Figure 3.3 demonstrates:

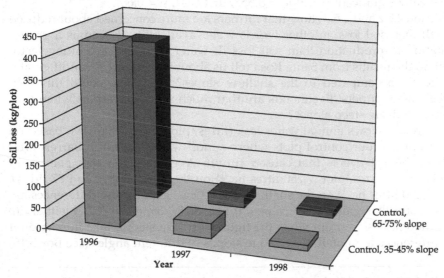

Fig. 3.3 Annual soil loss (kg/plot) from twelve control plots 1996-1998 (1998 data pre Hurricane Mitch)

- There was very little difference in annual soil loss between the steep (65-75 percent) and shallow (35-45 percent) slopes, indeed soil loss was marginally greater from the shallow slopes. Over the three-year period the average annual soil loss was 203 kg/plot (16.9 t ha^{-1}) from the shallow slope and 184 kg/plot (15.3 t ha^{-1}) from the steep slopes. Analysis shows that these differences are not significant (F value = 0.044, df 1/26, p = 0.836).
- Soil loss was substantially less in 1997 and 1998 (until Hurricane Mitch) than in 1996. This was largely due to less rainfall and fewer intense storms in 1997 (see Table 3.6) and more ground cover in 1998 (Hellin, 1999b).

The data from the barrels allow us to look at soil loss during the three-year period, 1996-1998, on a per runoff event basis rather than merely on a yearly basis. The data from the rain-gauge show that between 1996 and 1998 (before Hurricane Mitch) there were 352 rainfall events. Runoff occurred during 191 of these events. There was no rainfall amount and intensity threshold that separated runoff and non-runoff events. This was because of the effect of extraneous factors such as antecedent soil moisture and variations in soil cover afforded by the maize plants in different stages in the growing cycle.

Average soil loss per runoff event (for the four barril plots only) was 1.9 kg/plot (0.16 t ha^{-1}) on the steep slopes and 1.3 kg/plot (0.11 t ha^{-1}) on the shallow slopes. In this case, soil loss per runoff event was greater on the steep compared to the shallow slopes but once again these differences are not statistically significant (F value = 3.553, df 1/734, p = 0.60).

Section 2.5.2 and 2.5.3 show that farmers are more concerned about reduced rainfall than soil loss, in other words water availability is seen as a greater limitation to production than soil loss. If water is indeed a limiting factor what do the results from Santa Rosa tell us about water loss? Is runoff greater on the steep compared to the shallow slopes? And if so could this help explain why farmers' livelihoods are that much more precarious when they start to cultivate steep slopes?

Figure 3.4 shows annual water loss (litres/plot) from 1996-1998 based on data from the four control plots where runoff was collected in barrels after each of the 191 events that caused runoff. Average water loss per runoff event was 93 litres/plot (7,750 litres ha^{-1}) on the steep slopes and 87 litres/plot (7,250 litres ha^{-1}) on the shallow plots. Once again, these differences are not statistically significant (F value = 0.141, df 1/735, p = 0.707). Neither are the differences significant when the figures are adjusted to take into account the differences in rainfall delivered to slopes of different angles (see Box 3.15)

Box 3.15   Rainfall depth and a slope correction factor

At Santa Rosa, research plots were located within two different slope categories (35-45 percent and 65-75 percent). Based on the assumption that the rain falls vertically, the amount of water delivered to each plot during a rainfall event will vary depending on slope. Rainfall depth was adjusted for the slope of each research plot by multiplying rainfall by a slope correction factor (Smith, 1997:15; Bryan, 1979). This factor was based on the following trigonometric relationship: Correction factor = (L cos ^{0}) / 1 where L is the slope length of the research plot and $^\circ$ is the slope angle of the research plot in degrees.

Water loss for each runoff event was subsequently expressed as a percentage of the rainfall depth. Results of the ANOVA of the runoff

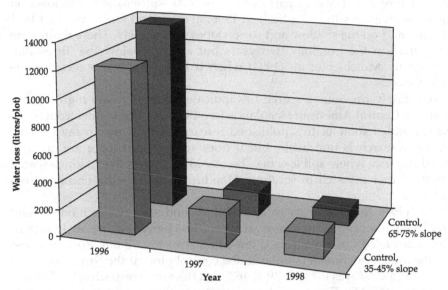

Fig. 3.4 Annual water loss (litres/plot) from four control plots where runoff was collected in barrels 1996-1998 (1998 data pre Hurricane Mitch)

percentage (as opposed to water loss) per runoff event show that the average runoff percentage per event was 2.60 on the steep slopes and 2.02 on the shallow slopes. The differences, however, are not significant (F value = 2.523, d/f 1/735, p = 0.113).

The literature suggests that soil and water loss should be greater on steep slopes (see section 3.2.3). This is largely because the erosive forces - splash, scour and transport - all have a greater effect on steep slopes. In addition, many tropical soils are characterised by weak structure and, under the impact of high-intensity rains, are prone to slaking, crusting, compaction and a rapid loss of infiltration capacity. The absolute, although not significant, differences in soil and water loss reported above suggest that some deposition of eroded soil and infiltration of overland flow may be taking place within the research plots. In this case, soil and water loss as determined by barrel and catch-pit measurements would not reflect the total amount of soil eroded and volume of overland flow.

There is, of course, a danger though that we can get too caught up in the analysis of these data and lose sight of the objectives of the research. The experiment at Santa Rosa was designed to help understand the rationale behind farmers' non-adoption of soil conservation technologies. So, what do the results from Santa Rosa tell about whether farmers really ought to be able to recognise that soil erosion is occurring?

Mutchler et al. (1994) assert that annual soil loss rates of 20 to 40 t ha^{-1} are difficult to notice even by trained observers. Figure 3.3 shows that in 1997 and 1998 (pre Hurricane Mitch) average annual soil loss at Santa Rosa on both slopes was less than 3 t ha^{-1}. In 1996, average soil loss was 36.2 t ha^{-1} and 33.0 t ha^{-1} on the shallow and steep slopes respectively. These figures are greater than in the previous two years but are still below the 'threshold' identified by Mutchler et al. (1994) when it might become a recognisable phenomenon.

Interestingly previous research has indicated that there are high rates of soil loss in Central American steeplands (see Box 2.9). The results from Santa Rose contradict some of this published research. This is not to say that the previous research is unreliable, but it does suggest that there are areas in Central America where soil loss may be considerably less than some research suggests. Other research in southern Honduras supports the findings from Santa Rosa.

In the 1990s, the University of Texas A&M conducted research on soil and water losses from three catchment areas on 55-63 percent slopes at an experimental site called Los Espabeles, a few kilometres from Santa Rosa. Soil loss from the mulch catchment (similar to the control plots at the Santa Rosa site) from 1993 to 1997 was 0.7, 0.4, 39.1, 46.2 and 0.3 t ha^{-1} respectively (Thurow and Smith, 1998:15). These results are similar to those at Santa Rosa in 1996 and 1997 when soil and water losses were being recorded at both sites simultaneously.

In the context of the results from Santa Rosa and Los Espabeles, it becomes much easier to understand why Honduran farmers seldom cite soil erosion as a problem: they are very likely to be unaware that it is happening. In fact, for two of the three years and Santa Rosa and three of the five years at Los Espabeles, soil loss rates were likely to be below any soil loss tolerance values that might be expected for this environment.

The difficulty of recognising erosion rates of less than 40 t ha^{-1} yr^{-1} may well explain the fact that farmers in southern Honduras often refer to rocks 'growing out of the hillside'. Farmers cannot see erosion occurring so the exposure of rocks can only 'logically' be explained by the fact that they are growing. Of course, erosion is not the only reason rocks 'grow' out of the soil, most cultivation – hoes and tillage work – that loosens the soil also helps bring stones to the surface.

If resource-poor farmers cannot reasonably be expected to recognise that erosion is occurring, is it any wonder that they do not identify soil erosion control as a priority and, subsequently, see recommended soil conservation technologies as a waste of time? The counter argument is of course that even though farmers do not identify soil erosion as a problem, they ought to establish soil conservation technologies because although soil erosion may be largely 'invisible, it remains a long-term threat to their already precarious livelihoods. Farmers' livelihoods in Honduras are intricately linked to maize production and this is where we next turn.

3.6.3 Is Soil Erosion a Threat to Farmers' Livelihoods?

The predicament faced by farmers who are forced to cultivate increasingly steep slopes is clearly illustrated by an analysis of maize yields from Santa Rosa (see Plate 14). Figure 3.5 shows that maize yields are greater on 35-45 percent than 65-75 percent slopes for each of the five harvests from 1996-1998. Over a three-year period the average yield per harvest was almost 1,895 kg ha^{-1} on the shallow slopes and just over 1,109 kg ha^{-1} on the steep slopes. These differences are highly significant (F value = 25.644, df = 58, p< 0.000) as are the differences for each of the years 1996, 1997 & 1998 (Hellin, 1999b).

Plate 14   Maize yields from the runoff plots demonstrated the predicament faced by farmers who are forced to cultivate increasingly steep slopes. Honduras. (Hellin, J.)

What are the implications of these results for smallholder farmers in Honduras? Is their quest for food security threatened by the fact that they being forced to cultivate even steep slopes? In much of Honduras the average size of a farm (made up of a number of plots) is less than 2.0 ha (Hellin, 1999b). Kass (1999) reports that on average a Central American farming household (5-6 members) needs 1,800 kg of maize per annum for subsistence needs. Based on a yield per harvest of 1000 kg ha^{-1} and two harvests per

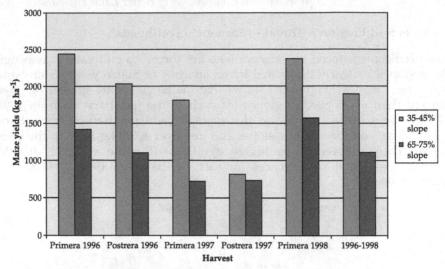

Fig. 3.5   Average maize yields (kg ha⁻¹) on control plots on steep and shallow slopes from 1996 to 1998. Primera and Postrera are the first and second maize crops respectively.

annum, the results from Santa Rosa suggest that for many smallholder farmers (with about 2.0 ha of land), maize production on steep slopes should, in theory, be above subsistence levels.

On the other hand, Bunch (1999b) believes that for many farmers farming steep slopes, maize production is actually below subsistence levels. Bunch also cautions that some official data on agricultural yields can be very inaccurate. He reports that the average maize harvest in Honduras is approximately 840 kg ha⁻¹, and that resource-poor farmers at the bottom end of the scale actually produce only between 400 and 650 kg ha⁻¹. When post-harvest losses from insect attacks are taken into account, it is clear that maize yields are not sufficient to meet household needs.

It is also possible that maize yields at Santa Rosa are a little higher than elsewhere in the region because of greater annual rainfall due to the local topography and the fact that prior to the establishment of the field trials, the land had not been cultivated for several years and been colonised in this period by herbaceous and woody plant species. It is also possible that the quality of the land at Santa Rosa is better than that being cultivated in neighbouring communities. The most important thing, however, is not the absolute figures for maize yields at Santa Rosa, but rather the relative values between steep and shallow slopes. The data clearly indicate that productivity on steep slopes (the land increasingly being farmed by smallholder farmers) is problematic, and that over a three-year period, yields are almost 50 percent less than those from shallower slopes. So, at the very least, the farmer's margins of error are being reduced along with the capacity to generate a surplus in good years.

The results from Santa Rosa suggest that significantly reduced maize yields on the steep slope *may* be due to the marginally greater soil losses compared to the 35-45 percent slopes. Although, the critical differences in the soil quality may be due to long-term slope processes more than the changes that have followed the relatively brief period of recent cultivation. Nevertheless, even the soil loss differences between steep and shallow slopes are not significant, the data presented above suggest that during the lifetime of the experiment the average soil loss per runoff event was 1.9 kg/plot (0.16 t ha^{-1}) on the steep slopes and 1.3 kg/plot (0.11 t ha^{-1}) on the shallow slopes. Of course, it is possible that the most erodible material from the steep slopes may already have been lost. This could, perhaps, account for the similarities in soil loss data from the steep and shallow slopes: in the case of the former there is basically less material to erode.

To test this hypothesis, a series of soil depth measurements were made towards the end of the experiment. Figure 3.6 shows reduced soil depth on the steep compared to the shallow slopes and also that, despite great variability, soil depth increases downslope. Intuitively, these differences could be attributed to the cumulative effect of increased soil loss over many years. The A horizon provides the growing medium for subsequent crops: the greater the depth of the A horizon, the more productive the slope. In this context, reduced maize production on the steeper slopes could indeed be attributed to the consequences of soil erosion in the past.

The results from Santa Rosa suggest that farmers should be concerned about cultivating ever-steeper slopes because of reduced maize production associated with steepland cultivation. The data seemingly strengthen the argument in favour of promoting soil conservation technologies. On shallow slopes, these technologies have been shown to reduce soil loss and periodically contribute to increased productivity. The critical question is whether they are effective on the steep slopes increasingly being brought into production and whether it is in the farmer's interest to invest resources in their establishment and maintenance?

3.7   EFFECTIVENESS OF SOIL CONSERVATION TECHNOLOGIES

3.7.1   Vetiver Grass Live Barriers, Soil Erosion and Water Loss

Farmers' interest in adopting and adapting soil conservation technologies will be enhanced if recommended technologies can be shown to reduce soil and water losses *and* contribute to increased productivity (Pellek, 1992; Bunch, 1982:99). Figure 3.7 shows that annual soil loss (kg/plot) for 1996-1998, based on data from 12 live barrier plots where soil was collected in barrels and catch-pits, are similar for steep and shallow slopes. Over the three-year period the average annual soil loss was 186 kg/plot (15.5 t ha^{-1}) from the shallow

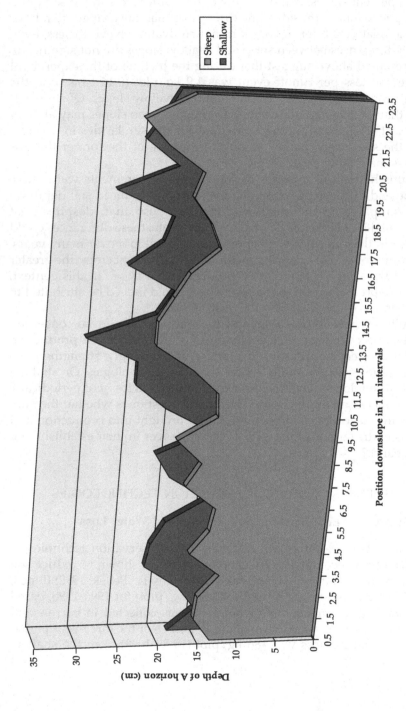

Fig. 3.6 Depth of A horizon (cm) on steep and shallow slopes for the control plots only

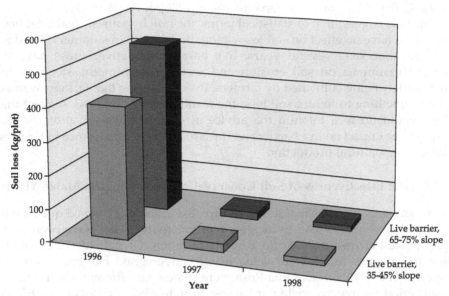

Fig. 3.7 Annual soil loss (kg/plot) from twelve live barrier plots 1996-1998 (1998 data pre Hurricane Mitch)

slope and 237 kg/plot (19.7 t ha^{-1}) from the steep slopes. Analysis shows that these differences are not significant, (F value = 0.222, df = 26, p = 0.642).

It is obviously worthwhile comparing these results to soil loss from the control plots (Figure 3.3 and section 3.6.2). The figures are broadly similar although the average soil loss from the steep slope plots supposedly protected with live barriers is higher than the corresponding control plots (19.7 t ha^{-1} and 15.3 t ha^{-1} respectively)! These differences are again not significant.

Meanwhile, data from 191 rainfall events that caused runoff over the three-year period demonstrate that average soil loss per runoff event on the steep slopes planted with live barriers was 1.9 kg/plot (0.16 t ha^{-1}). This is the same as that recorded on the corresponding control plots. Soil loss on live barrier plots on the shallow slope was 1.6 kg/ha (0.13 t ha^{-1}) as compared to 1.3 kg/ha (0.11 t ha^{-1}) in the control plots. Once again, on one of the slopes, this time the shallow slope, soil loss from the protected plots is higher, although not significantly so, compared to the unprotected control plots. Soil conservation technologies are supposed to reduce soil loss and these results are, therefore, rather unexpected.

Similarly the water loss for each runoff event expressed as percentage of the rainfall depth (see Box 3.15) show that on the live barrier plots the average runoff percentage per event was 3.01 on the steep slopes and 2.50 on the shallow slopes. The differences between the slopes are not significant (F value = 1.565, d/f 1/749, p = 0.211) but the figures are higher (although not significantly) than the corresponding figures for the unprotected control plots (2.60 and 2.02 on the steep and shallow slopes respectively).

The data suggest that in statistical terms, the null hypothesis that treatment and slope have no effect on soil loss cannot be rejected. In summary, the data indicate that, over several years, live barriers of vetiver grass have no significant impact on soil erosion and water loss from the steep slopes increasingly being cultivated by farmers. Even assuming that farmers wanted to do something to reduce soil loss, the results from Santa Rosa suggest that they would do well to shun the advice of some soil conservationists! But what of the impact on live barriers on the issue of most importance to farmers, namely agricultural production?

3.7.2   The Effectiveness of Soil Conservation Technologies: Maize Yields

Comparatively less research has been directed at the more vexed question of whether live barriers lead to increased productivity. There have been some exaggerated claims: in a World Bank publication, Greenfield (1989) claims that the average increase in yields using vetiver grass has been over 100 percent. The results from Santa Rosa were somewhat different. Over a three-year period the average yield per harvest from live barrier plots was 1,876 kg ha^{-1} on the shallow slopes and just under 1,080 kg ha^{-1} on the steep slopes. These differences between slopes are highly significant (F value = 28.810, df = 58, p< 0.000). The results also happen to be almost identical to maize yields on the control plots (see Figure 3.5).

In fact over a three-year period and across both slopes, the average yield per harvest was 1500 kg ha^{-1} on the control plots and 1477 kg ha^{-1} on the live barrier plots (Hellin, 1999b). There was only one harvest out of five when maize yields on live barrier plots were significantly greater than yields from the control plots. In the *postrera* of 1997 maize yields on live barrier and control plots, across both slopes, were 956 kg ha^{-1} and 775 kg ha^{-1} respectively. These differences were significant (p = 0.004). These results are discussed in more detail below.

For the other four harvests, the differences between live barrier and control plots are minuscule. In 1996, the difference in the average maize yield per harvest between control and live barrier plots was only 17 kg ha^{-1} (1746 kg ha^{-1} and 1729 kg ha^{-1} respectively) while in 1997 the difference is only 11 kg ha^{-1} (1089 kg ha^{-1} and 1078 kg ha^{-1}). In all cases, the marginally higher yields came from the control plots! Given this, the case in favour of taking land out of production and investing resources in establishing soil conservation technologies looks very weak indeed.

3.7.3   The Dangers of Scouring and Deposition

There is another concern with the use of live barriers. Live barriers reduce slope length and angle. This in turn reduces the capacity of runoff to move soil particles down the slope. Intuitively, there should be less soil movement between live barriers and mobilised soils should be trapped up-slope of the barriers. Soil augur measurements towards the end of the experiment show, as expected, deposition of soil above the barriers. However, they also demonstrate that scouring occurred below the barrier, see Figure 3.8 and Plate 15. The deposition and scouring effect was not evident in the control plots (c.f. Figure 3.6 above).

Plate 15   Cross-section of a three-year old vetiver grass live barrier showing soil deposition above and scouring below the barrier. Honduras. (Hellin, J.)

A detailed analysis of per-row maize yields and soil depth measurements at Santa Rosa shed more light on the relationship between live barriers and maize productivity. The analysis also demonstrates the importance of the dangers of scouring and the importance of soil cover (see Chapter 4).

Soil erosion would be expected to be less in the upper alleyway zone. Scouring, however, does occur in the zone immediately below the artificial barriers because the barrier ensures that there is little enrichment of soil from above (Lal, 1982; Garrity, 1996; McDonald et al. 2002). The result is soil accumulation above the barrier and soil loss below the barrier. Hence, while there is no overall difference in soil loss between control and live barrier plots, there is a difference in the patterns of soil erosion and deposition within plots. This loss of soil depth immediately below each live barrier is a concern because of the impact that this has on maize yields, particularly in years with reduced rainfall.

In 1996 and 1998, there was little pattern in per-row maize yields and no significant difference in the yields recorded on corresponding rows on control and live barrier plots (Hellin and Haigh, 2002b). However, in 1997, and particularly during the *postrera*, maize yields in live barrier plots varied systematically from a high immediately above the barrier to a low immediately

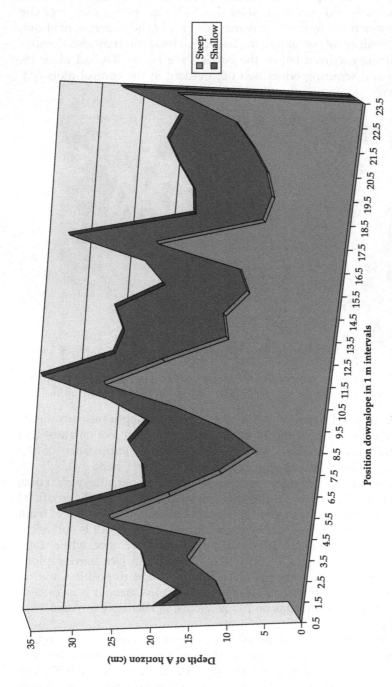

Fig. 3.8 Depth of A horizon (cm) on steep and shallow slopes. Live barrier plots only. Barriers established at 6, 12 and 18 cm.

below. There was no similar pattern on parallel control plots. Table 3.7 shows per-row maize yields during the *postera* in 1997. These data come from all the plots but exclude data from the uppermost and lowermost four crop rows because there were no live barriers at either extreme of the plots.

Table 3.7   Maize yields (per 5-m crop-row) from *Postrera* harvest 1997. Compares yield from live barrier plots, row 1 (immediately below) through row 8 (immediately above a barrier) with yields on parallel conventionally farmed plots with no live barriers. From Hellin and Haigh, 2002b

Both slopes, *postrera* 1997 only	Live Barrier Yield (grams/row)	Control Yield (grams/row)	*t*-value	Significance p =
Row 1	262	321	1.587	0.177
Row 2	379	283	-2.581	0.012
Row 3	379	330	-1.505	0.137
Row 4	356	309	-1.387	0.170
Row 5	362	261	-2.989	0.004
Row 6	377	269	-3.076	0.003
Row 7	418	290	-3.270	0.002
Row 8	480	286	-4.338	0.000

Maize yields in the four rows above the barrier (rows 5, 6, 7 and 8) were significantly greater ($p \leq 0.01$) than those from corresponding rows in the control plots. Yields in the four rows below the barrier (rows 1, 2, 3, and 4) were similar to those in the control plots. As a result, in the *postrera* in 1997, maize yields from the live barrier plots were 23 percent greater than those from control plots (34 percent and 12 percent higher on shallow and steep slopes respectively). As reported above, in the other harvests, there were no significant differences in maize yields between live barrier and control plots.

At Santa Rosa most erosion took place in 1996 and yet maize yields only became skewed in live barrier plots in the *postrera* in 1997, when there was much reduced rainfall. Maize yield data suggest that, in 1997, the increased *postrera* maize yield from the crop rows above the live barriers were linked to soil accumulation. The special feature of 1997 was that it was an exceptionally dry year, caused by *El Niño*. Rainfall in 1997 was approximately 60 percent of normal annual rainfall.

In 1996 and 1998, when rainfall was not a limiting factor, there were no significant differences in per-row and, hence, total maize yields between live barrier and control plots. In 1997, productivity gains in live barrier were more pronounced in the *postrera* than in the *primera* harvest because the former is sown just a month before the end of the rains. Hence, moisture availability was a limiting factor for a large proportion of the growing season. If soil fertility were the limiting factor, maize yields in crop rows immediately above a barrier should have been higher compared to those from corresponding positions in the control plots across all harvests. In four out of five harvests these differences were not apparent. This suggests that it was soil moisture, held in soil accumulated above each live barrier that was responsible for the increase in productivity (Hellin and Haigh, 2000b). Shaxson and Barber (2003) detail the importance of soil moisture to crop production as opposed to soil fertility *per se*.

In 1997, skewed maize yields in the *postrera* come from reduced production in the upper crop rows where scouring had occurred and increased production in the lower crop rows (just above the barrier) where soil was deposited. Similar patterns have been reported in Ethiopia (Herweg and Ludi, 1999) and the Philippines (Garrity, 1996). Lewis (1992) observes a similar situation in Rwanda where in response to exposure of poor sub-soils (and subsequent reductions in yield) along the back portions of terrace benches, farmers remove a portion of the terrace berm. The soil from the berm is then spread over the lower bench, especially along the back portions.

The key issue is the determination of those circumstances where yields below a barrier are so reduced that they cancel out yield increases elsewhere on the plot, perhaps leading to an overall reduction in yields (cf. Garrity et al. 1997). This was not apparent at Santa Rosa, but in regions where soils are shallower and live barriers in operation for longer periods, the potential seemed greater. In these cases the scouring effect may become so pronounced that yields in the upper alleys drop to levels that make farmer-adoption of the technology even less probable.

There is also the danger that any potential benefits from greater soil moisture in accumulated soil above a barrier are not fulfilled because of competition for water between the live barrier and agricultural crops. In this case, there can actually be a reduction in yield in the first one of two rows above a live barrier. This has been reported by some researchers in the case of vetiver grass (Smyle and Magarth, 1993; Nehmdhal, 1999; Dalton and Truong, 1999) and also napier grass (*P. purpureum*) (Fujisaka, 1997; Howeler, 1987; Hellin and Larrea, 1998).

The results from Santa Rosa suggest that one of the reasons for the ineffectiveness of live barriers on steep slopes to control soil and water loss *and* increase production is the scouring effect below live barriers. The results also illustrate one aspect of the difficulty of convincing farmers that it is in their best interest to establish and maintain soil conservation works. To be attractive to farmers, soil conservation technologies have to offer sufficient and obvious benefits (especially increased productivity) to compensate the farmers for increased labour input. These results demonstrate that these benefits may not be apparent routinely when technologies, such as live barriers, are used in isolation. For example, when commenting on the live barrier technology, farmers in Güinope voiced concerns about reduced yields below the barriers (Hellin, 1999b).

The results from Santa Rosa suggest that the problems caused by soil redistribution in the barrier-protected plots might be offset by the use of productivity-enhancing and soil generating strategies such as cover crops and green manure crops. These are detailed in Chapter 4. The results also raise more critical issues such as the validity of the relationship between soil erosion and productivity, a relationship that has been used to justify soil erosion control initiatives worldwide. This likewise is discussed in Chapter 4.

3.7.4 Where Do We Go from Here?

Soil conservation technologies are promoted as an effective way to control soil loss. Results from the trial site show that vetiver grass live barriers may not be effective in controlling soil and water loss on the slope angles being cultivated by resource-poor farmers in Honduras and the rest of Central America. More importantly, they do not seem to be effective in contributing to increased productivity. This could largely explain why smallholder farmers worldwide have been reluctant to adopt the technologies, despite their being enthusiastically promoted by people whose livelihoods do not depend on whether the same technologies work or not.

The results from Santa Rosa differ from some of the research reported in Table 3.2, which demonstrates that in some cases yields increase following the adoption of soil conservation technologies such as live barriers. A critical question is why vetiver live barriers-one of the most heavily promoted soil conservation technologies worldwide—were not effective at Santa Rosa? The question has to be asked because the results from Santa Rosa have profound implications for soil conservation initiatives and subsistence farming worldwide. If research, that mimics local agricultural practices and that is conducted under controlled conditions with several replications, cannot show that a technology works in controlling soil and water loss and increasing agricultural yields, on what grounds should it be promoted?

3.8 SOIL CONSERVATION AND SITE VARIABILITY

One explanation for the unexpected results from Santa Rosa is that there were insufficient replications to account for the variation in soil characteristics found through out the site. Soils in the tropics are highly variable and there is often variability within a distance of a few metres (Cassel and Lal, 1992; Siderius, 1987). This variation can have a huge impact on soil erosion because the amount of soil erosion at any location is determined by the interaction of climate, topography, vegetation and soil (Bryan and Luk, 1981).

Results from individual plots at Santa Rosa confirm that there was considerable variation in soil and water loss due to site heterogeneity. There was, though, much less variation in maize yields between plots. In some cases, individual plots lost significantly more (or less) soil and water than adjacent plots under the same treatment. For example, adjacent control plots on one of the 35-45 percent slopes showed over a three-fold difference in soil loss over the three-year period (486 kg plot^{-1} and 139 kg plot^{-1}). This raises several fundamental questions:

1. Was the site heterogeneity at Santa Rosa more extreme than has been the case on other research sites?
2. Did site heterogeneity, and the subsequent variation in soil and water loss from individual plots at Santa Rosa, mask the effectiveness of the live barriers in controlling this loss?
3. In order to account for this site heterogeneity, would a better designed experiment i.e. one with many more replications, have demonstrated that live barriers are indeed effective at controlling soil loss at the slope angles increasingly being cultivated by smallholder farmers?

It is surprisingly difficult to answer the first question because, in much previous research, when replications are used, the usual practice is to report the arithmetic mean of the measurements. It is very rare for reports to include any information on the variance of the measurements. Hence, published results may well not reveal the degree of variation that exists among the research results (Hudson, 1993:31). By virtue of this, it is questionable whether some of the soil conservation technologies being promoted really are as effective at controlling soil erosion as is often portrayed in the literature.

Some new caution might be recommended in accepting results from research that indicates that a particular technology is effective at controlling soil loss. The reality might be that its supposed effectiveness is based on mean values that say little about on-site variation, or a spurious non-replicated experimental design that just happens to assign the treatment in question to a plot that inherently suffered from less erosion than an adjacent plot.

In some of the experimental designs reported in the scientific literature, and summarised in Table 3.4, researchers did not use any replications. There is indeed a body of evidence that suggests that the variation in soil and water loss recorded at Santa Rosa is not as uncommon as some of the literature suggests (see Box 3.16). If a non-replicated trial had been established at Santa Rosa, it is perfectly possible that the live barrier treatment could have been located where soil loss is inherently small and conversely the control plot where it is inherently large. In this case, this project would have demonstrated happily that live barriers significantly reduce soil loss and that all is well in soil (and water) conservation.

Box 3.16   Research demonstrating that variation is all too common

Bryan and Luk (1981) carried out laboratory experiments on soil erosion under simulated rainfall on three Canadian soils. In the experiment great care was taken to hold all factors contributing to soil loss and runoff as constant as possible. There was much variation in soil and water loss and the authors concluded: *"experimental conditions employed represent the most constant practicably achievable in laboratory soil erodibility studies. The variability is thus virtually unavoidable and unlikely to be corrected for and hence is described as **inherent** soil loss variability. In field conditions where most factors cannot be so closely controlled, somewhat higher variability and prediction error may be expected"* (Bryan and Luk, 1981:260).

Meanwhile, field research by Rüttiman et al. (1995) and Wendt et al. (1986) also revealed considerable variation in soil and water losses on replicated research plots in Switzerland and the United States respectively. Rüttiman et al. (1995) concluded that the results represent the arbitrary natural magnitude of soil erosion variability. The authors attributed the variability to differences in soil texture; soil type; slope angle and shape; occurrence of subsurface flow; surface roughness; soil compaction; and degree of soil cover. Nearing (2000) points out that there is a dearth of knowledge about natural variability between plots that have the same treatments.

In response to the second question posed above, whether site variation at Santa Rosa masked the effectiveness of the live barriers, attempts were made in the statistical analysis to account for on-site variation. See Box 3.17.

Box 3.17   Accounting for on-site variation at Santa Rosa

Covariates were used in the statistical analysis. Covariates are quantitative predictor variables that, in this case, reflect site heterogeneity. In the statistical analysis, treatment effects are adjusted to allow for the effect of the covariates and, by so doing, so the precision of the analysis can be increased (Sinclair, 1989).

The following covariates were used singularly and in combination: bulk density; infiltration rates; and runoff percentage (based on average water loss expressed as the percentage rain depth for all runoff events over a six-week period at the beginning of the experiment before the live barriers were established). The use of these covariates did not improve the significance levels (Hellin, 1999b).

Variation in erosion rates can also be affected by particle size distribution. Generally, sand and silt tend to increase erodibility, while clay decreases it (Hudson, 1995:94; Lal, 1990b:61). Data on particle size distribution at Santa Rosa were used in a Spearman's correlation. This is a non-parametric test and is a measure of association that can be used to examine relationships between variables. Spearman's rank-order correlation coefficient measures the strength and direction of the association.

> None of the relationships between soil and water losses at Santa Rosa and a number of variables based on soil, clay and sand variables are significant and the null hypothesis, that there is no association between two variables, cannot be rejected (Hellin, 1999b).

Critics of the research at Santa Rosa may argue that the effectiveness of live barriers in controlling soil and water loss on steep slopes was not borne out by the results because insufficient numbers of replications were used to fulfil the primary objective of controlling and quantifying variation. The critical point, though, is that if it takes so much effort to demonstrate the differences scientifically, how can farmers be expected to see these differences for themselves. Furthermore, it is far from clear that more replications would have had and any significant impact on the results. A salutary lesson for those who argue that the inherent variability of a site can be accounted for by increasing the number of replications is that in the United States, Wendt et al. (1986) carried out soil erosion research on 40 replicated runoff plots. Despite this, coefficients of variation in measured soil loss on individual storms ranged from 18 to 83 percent.

3.9   TIME FOR A RETHINK

From a research perspective and given the variation of soil and water loss at Santa Rosa and other research showing on-site variation (see Box 3.16), there are grounds for arguing that measurements of soil erosion under field conditions should be kept as simple as possible. An argument in favour of simplicity and a search for relative, as opposed to absolute differences, also seems sensible when the potential inaccuracies of measuring soil and water loss are taken into account (Herweg and Ostrowski, 1997: 23).

Field techniques should be modified so that more use is made of relatively simple and cheap catch-pits as opposed to more intricate collecting devices such as barrels. There are other collecting methods such as Gerlach troughs and mesh bags (Hsieh 1992) that could be used. Relative values of soil and water loss may be more useful than absolute values. Stocking (1993b) writes that absolute accuracy is a mirage and that with regards to estimating soil loss.

"Quick and dirty may, in fact, be a lot cleaner than heavyweight science might like us to believe ....especially as the natural environment is so variable and the general validity of a few data points must be questioned" (Stocking, 1993b:21).

The argument for using simple methods for estimating soil loss is also strengthened if a better land husbandry, rather than soil conservation, approach is adopted. Norman and Douglas (1994:71) argue that, from a better land husbandry perspective, it may not be necessary to have precise figures on soil loss before designing programmes that focus on improving soil quality.

Finally, and far more importantly when we consider the farmer - the ultimate beneficiary of most soil conservation initiatives, the most important issue is whether the technology works in the short-term and whether it is reliable, rather than why it does not work. The issues of experimental design, replications, on-site variation and averages make little sense to farmers:

> *"The farmer's concern is not with expected effects on yield for the parent population of study; rather he wants to know what effect using the studied treatment will have on his field … so telling the farmer about average yields or expected outcomes is of little use to him in helping him decide what treatment to apply to his fields"* (Suppe, 1988:10).

On this basis, it is perfectly understandable why farmers are more concerned about rainfall. At Santa Rosa and due to the effect of *El Niño*, total rainfall in 1997 was almost 50 percent lower than in 1996. As a result, maize yields on the 65-75 percent slopes averaged 722 kg ha^{-1} and those on the shallow slopes, 1375 kg ha^{-1}. These represents a reduction of 45 percent and 37 percent respectively compared to yields in 1996. Farmers seldom deal with averages and are most concerned with the next harvest. In this context, farmers understandably associate reduced yields with reduced rainfall, even though the problem may be more to do with insufficiency of moisture in the rooting zone. Soil loss is clearly not an immediate and real threat to their livelihood security. Farmers cannot influence the amount of rain that falls, but as we will see in Chapter 4, they can improve the quality of their land so that their land absorbs and stores more water.

Commentators may also argue that the live barriers' ability to control soil and water loss and contribute to improved soil productivity will become clearer over time as more soil accumulates above the barrier. However, time is precisely what many resource-poor farmers do not have. As Bunch (1982:102), Hudson (1993b) and other researchers have pointed out, one of the criteria for an appropriate technology is that substantial results are evident after one or two years.

Farmers who choose not to adopt technologies such as vetiver grass live barriers, may well be making a very sensible decision. In fact, their sensible decisions may well have an important message for soil conservation outsiders, namely that their focus, some would say obsession, with soil loss is profoundly misplaced. The results from Santa Rosa support the argument that we need to rethink our entire approach to reducing soil and land degradation. The conventional soil conservation approach is not working at any level. There is an urgent need for a change in direction, one that contributes to the livelihood security of the millions of smallholder farmers who are being forced to cultivate marginal steeplands. There is an alternative, even better, approach to improving land management and this is explored in Chapter 4.

4

Better Land Husbandry:
A New Paradigm

"Land husbandry is less of a closely defined discipline than it is a philosophy whose practical expression–by both farmers and advisers–is both science and art" (Chinene et al. 1996:40).

4.1 SOIL CONSERVATION VERSUS BETTER LAND HUSBANDRY

Anyone who has seen pictures of denuded and barren hillsides could be forgiven for believing that the 'need to stop erosion' is the obvious solution to the problem of land degradation and that if farmers are reluctant to adopt recommended soil conservation technologies, they need to be persuaded otherwise. An analysis of the conventional soil conservation approach, however, suggests that the concept of soil conservation *per se*, with its focus on the control of soil erosion may be both technically flawed and inappropriate to social and economic conditions in rural communities.

The conventional soil conservation approach links loss of productivity to soil erosion, seeks to control soil loss through the transfer of technologies and tends to eschew active farmer participation let alone farmer direction. It often has a negative image for farmers due to its implied need to control and restrict present land use in order to preserve soil for the future. The conventional approach tends to focus on capturing eroded soil. However, erosion is a consequence rather than a cause of soil degradation and is a clear indication that soil quality has already deteriorated.

Soil conservation measures designed to control soil loss are unlikely to lead to sustained and improved productivity because, by focusing on particle soil loss and nutrient depletion, they fail to improve the quality of soil that determines future productivity. The better land husbandry approach recognises the importance of soil quality. It offers a practical holistic framework within which land management issues can be analysed and practical approaches to maintaining and improving soil quality formulated and implemented (Shaxson et al. 1989; Norman and Douglas, 1994; Hudson and Cheatle, 1993; Moldenhauer and Hudson, 1988).

With its focus on soil quality, better land husbandry strives to integrate farmers' concerns about productivity with soil conservationists' concerns about minimising runoff and avoiding erosion. Well-selected improvements in management can help farmers fulfil their aims of more reliable yields and greater production per unit area, while simultaneously increasing the conservation-effectiveness of their existing farming or land-use systems.

From the technical perspective, what is involved is the active management of rainwater, vegetation, slopes and soils via the use practices that, together or individually, are productivity-enhancing and conservation-effective. This may entail the use of a combination of agronomic, biological and mechanical practices (Herweg and Ludi, 1999). Cross-slope technologies such as live barriers may prove to be appropriate in some circumstances, especially if species are used that make a direct contribution to the farming system (Hellin and Larrea, 1998) but in other circumstances they will not be appropriate.

The focus of improving land husbandry, though, is directed at achieving a self-sustainable system by seeking to improve soil quality via soil protection, incorporation of organic matter, and the use of soil organisms. The aim is to achieve and maintain optimum soil conditions – in physical, chemical, biological and hydrologic terms – for root growth and for the acceptance, transmission and retention of water. The focus is 'conservation' in terms of prolonging the life of resources.

The objective of any management inputs, within a better land husbandry framework, is to enhance the recuperative capabilities of soil with regards to the inter-related factors of physical architecture, biological balances and nutrient content. The narrow concept of 'soil' as an inert substrate is eschewed in favour of seeing soil for what it is – a living and self-renewing resource. The philosophy is very different to that behind conventional soil conservation practices. Soil conservation structures (even biological ones such as live barriers) are only sustainable in the sense that they can be repaired and managed. They are not self-sustainable because sustained investment in management of erosion control structures may be an expense that future land-users will not meet (Haigh, 2000b).

4.2 NEW PERSPECTIVES ON SOIL EROSION AND LAND DEGRADATION

4.2.1 The Raison D'être

For many years, soil conservation programmes have been based on the assumption that runoff is the main cause of erosion, and that runoff and erosion are inevitable consequences of farming and the principle causes of land degradation. Smallholder farmers have been seen as major contributors to land degradation by virtue of their cultivating marginal areas such as steeplands.

Although the quantity of soil eroded per unit of time and per unit of area may not necessarily be greater in the tropics than in temperate regions, development professionals have been very concerned about soil erosion in steeplands. This is largely because the consequences of erosion, in terms of a reduction in productivity and the impact that this has on farming communities, are generally more profound in the former than in the latter. Low productivity of the exposed subsoil and low inputs used in farming are the reasons that erosion is considered more severe in soils in the tropics than in temperate-zone areas (Lal, 1990).

Development professionals have perceived the challenge largely in terms of a need to control runoff on agricultural lands in order to prevent loss of soil through accelerated erosion (Douglas, 1993). These same professionals have identified a host of soil conservation technologies, including cross-slope barriers, as the most effective way to meet this challenge. As we saw in Chapter 2, the problem is that farmers have not adopted these technologies as readily as expected. Furthermore, as Chapter 3 showed, there can be very sound reasons for this non-adoption, mainly that technologies do not lead to the productivity gains that are so important to resource-poor farmers' livelihoods.

The development community continues to argue about soil erosion and what are the most effective technologies to arrest the movement of soil down a slope. Whilst these debates are interesting and intellectually challenging, others argue that many of them are superfluous to the challenge of reducing soil and land degradation (Shaxson et al. 1989). Rather, they argue that the entire soil conservation approach is misguided because it is tackling the symptoms rather than the causes of land degradation.

4.2.2   Soil Erosion: Confusion Over Causes and Symptoms

As we saw in Chapter 3 (section 3.2.2), soil erosion represents the balance between the power of the rain to cause erosion (erosivity) and the ability of the soil to resist the rain (erodibility). The extent and severity of erosion depends on a complex interaction between soil, climate, topography, land use and the farming system (Lal, 1990b:494). Figure 4.1 illustrates that the more water the soil absorbs through infiltration, the less runoff and erosion. Soils vary widely in their ability to retain water and the fate of precipitation once it arrives at the soil surface is intimately linked to soil structure (Cassel and Lal, 1992). If a soil is crusted or compacted, it is less able to transmit water through the soil surface and there is reduced infiltration and increased runoff, as shown in Figure 4.1, Figure 4.2 and Figure 4.3.

The kinetic energy available in falling rain is substantially higher than that of runoff (c.f. Section 3.2.2). The impact effect of raindrops is, therefore, the first and most important stage in the erosion process. A specific form of compaction occurs when the surface aggregates disintegrate under the impact of raindrops. Pore spaces are filled with fine particles. The crust that forms from this aggregate breakdown impedes water infiltration, leading to greater runoff and erosion (Hallsworth, 1987:11; Lal, 1990b:65).

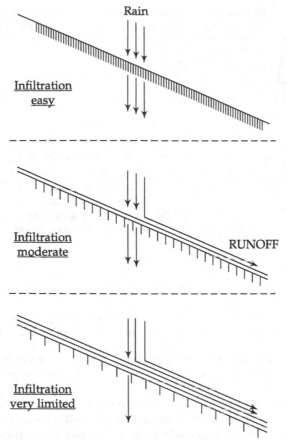

Fig. 4.1 Condition of the soil surface determines the division of rainwater between infiltration and runoff (from Shaxson, 1999:22)

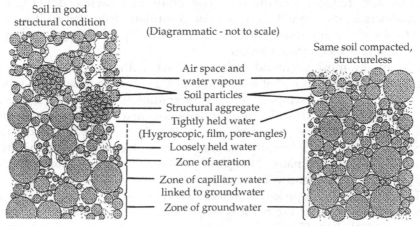

Fig. 4.2 Loss of soil voids affects soil infiltration capacity and its quality as a rooting environment (from Shaxson, 1999:18)

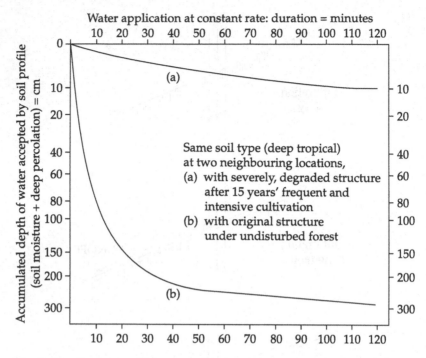

Fig. 4.3   Infiltration rates of the same soil type: a) with a severely degraded structure, and b) in an un-degraded state under forest (from Shaxson, 1999:21)

The onset of soil erosion is, therefore, a consequence of reduced soil cover that allows high-energy rainfall to impact the soil surface directly. Reduced porosity in the surface layers subsequently causes more run-off (Shaxson et al. 1997). Where surface runoff is a problem, it indicates that the soil is unable to absorb water and that much of the rainfall is 'ineffective' in terms of plant growth and regular streamflow. The challenge facing development practitioners is, therefore, to improve the condition of the soil so it is better able to absorb, retain, release and transmit water. By so doing, there will be less runoff and less surface erosion.

One of the most profound technical flaws with the conventional soil conservation approach is that efforts to capture soil once it has been eroded are dealing with the symptom of land degradation rather than the cause. Accelerated runoff and erosion are consequences of a declining soil structure and the loss of effective pore space. Cross-slope barriers, such as those used at Santa Rosa and in literally thousands of soil conservation initiatives worldwide, divide the natural length of a hillside slope into shorter sections and limit the volume and velocity of runoff.

This reduction in slope length reduces the chances of runoff gathering into constricted flow lines and so reduces the probability of rills and gullies forming (Shaxson et al. 1989:45). Cross-slope barriers can, therefore, reduce runoff and erosion and, by trapping sediment, reduce the effective slope of the cultivated land between the strips (Shaxson et al. 1989:57). The same barriers, however, act only against runoff and do little to protect the inter-row areas from rainfall impact. It is what happens in this inter-row area that is so critical to farmers' livelihoods because it determines future agricultural yields.

4.2.3 Questioning the Soil Erosion and Productivity Relationship

Chapter 2 showed that while development practitioners often focus on the need to combat soil erosion, farmers seldom see the need to reduce soil loss. Development practitioners are partly motivated by an underlying assumption that there is a relationship between soil loss and productivity. Specifically, they assume that soil erosion will result in reduced soil fertility and, hence, a reduction in productivity. See Box 4.1.

Box 4.1   Assumed or 'established' relationship between soil loss and productivity

- Hillel (1991:161) refers to a yield reduction of about 40 percent when 15 cm of top soil is removed while Ellis-Jones and Sims (1995) assume a loss of productivity of 15 percent for each 5.1 cm of soil lost.
- Wiggins (1981) in El Salvador uses an estimate of 2 percent loss of productivity per cm of soil based on regressions from observed yield differences in maize fields in the United States.
- In Kenya, Pagiola (1994) relates soil loss to yield decline using an experimentally derived relationship between topsoil loss and yield.
- Lutz, et al. (1994) contend that without soil conservation measures, maize yields would decline by 20-25 percent in sub-humid hillsides of Honduras while coffee yields would decline by 10 percent in the highlands of Costa Rica.
- Pimentel *et al.* (1995) state that soils that suffer severe erosion may produce 15-30 percent lower maize yields than soils that have not been eroded.
- Tengberg et al. (1998) report on research from South America that indicates that there is a logarithmic form of relationship between yield and soil loss i.e. that for some tropical soils there is an initial large decline in yields with the first five cm loss of topsoil and that further erosion has only a modest impact.

The assumed relationship between soil loss and productivity has arisen because declines in soil productivity are often equated with losses of soil particles through erosion and plant nutrients with them. As a result, many soil conservation programmes, based on the principle that runoff and erosion are the principle causes of land degradation, seek to control soil and water loss.

The problem is that whilst the quantities of soil lost or captured by soil conservation technologies may excite researchers and the readers of the journals in which they publish, farmers have a much more pressing concern. They need to feed their families. This is the crux of the problem–the strong focus in soil conservation programmes on the quantity of soil lost has meant that less attention has been directed to what is really important for smallholder farmers.

Admittedly, there is evidence, presented in Table 3.2, that in some circumstances, live barriers are effective at reducing soil loss. Furthermore some of the research data shows that these technologies also lead to an increase in productivity. There is, however, much evidence to the contrary (see Box 4.2). The conclusion of one soil conservation expert should make us pause for thought: *"fifty years of soil conservation research have still not managed to establish clear quantitative relationships between soil losses and crop yields"* (Shaxson, 1999:2).

Box 4.2   Research evidence demonstrating no clear relationship between soil loss and productivity

In Ethiopia although cross-slope oil conservation technologies significantly reduced soil loss compared to control plots over a 3- to 5-year period, there was no increase in productivity (Herweg and Ludi, 1999). Similarly, Alegre and Rao (1996) report that over a 5-year period, contour hedgerows in Peru significantly reduced soil and water loss but there was no crop yield increase.

In the United States, Stone et al. (1985) demonstrated the difficulty of establishing a link between erosion and productivity. Water-holding capacity was slightly higher on more eroded Piedmont soils in North Carolina, and maize yields were higher on the moderately eroded sites in comparison to the slightly and severely eroded areas. This was particularly true in a dry year.

The existence of research that fails to demonstrate a clear quantitative relationship between soil losses and crop yields should come as no surprise. It would be strange if such a relationship could be established. This is because the better the quality of soil, in terms of its physical, biological and chemical status, the more productive it is, irrespective of how much soil has been eroded provided only that it retains a sufficient root zone. In summary, the relationship between soil loss and productivity has absolutely no scientific grounding precisely because the loss of soil productivity is much more important than loss of soil itself (Norman and Douglas, 1994:7).

The complexities in the relationship between soils and plants and their management under a variable climate also make it difficult to monitor the impact of erosion on productivity (Pierce and Lal, 1994; Daniels et al. 1985). Crop yields under field conditions can rarely be related to any individual factor. Erosion rates are poor proxies for impact in terms of changes in yield (Stocking, 1995b). Even under controlled conditions, it is difficult to identify quantitatively the soil properties that regulate productivity for a given soil volume. In addition, the effects of erosion are cumulative and are often not observed until long after accelerated soil erosion begins. Furthermore, eroded fertile top soil can be deposited in other parts of a watershed leading either to an increase in productivity (Enters, 1998:45) or in some cases burying (and destroying) growing crops.

Shaxson (1997) points out that in some circumstances the subsoil exposed by surface soil loss may be of equal or better quality than the topsoil. In these cases, soil erosion may be expected to lead to sustained or increased productivity respectively rather than the loss in productivity that we have been led to believe ought to be the result of soil erosion (see Figure 4.4). Sustained or increased productivity caused by soil erosion is not a hypothetical scenario: Lutz et al. (1994) cite the example of Tierra Blanca in Costa Rica where extremely high erosion rates have very little effect on productivity. This is because these soils are deep and fertile and, hence, the soil's productivity is largely unaffected by soil loss.

4.2.4 Clarifying Causes and Consequences

Indeed we have got in a real muddle about the causes and consequences of soil and land degradation. Furthermore, the obsession with soil loss and the need to reduce soil erosion has distracted attention from the importance of soil productivity, a productivity that can be evaluated in terms of the environment it provides for root growth and development (Pierce and Lal, 1994).

Traditional cross-slope soil conservation technologies such as rock walls and live barriers may lead to an improvement in yield in narrow strips where eroded soil has been captured. This improvement, however, is seldom sufficient to compensate farmers for the land taken out of productivity by the technology and the establishment and maintenance costs (Douglas, 1993). Furthermore, the cross-slope technologies do not address the critical issue of minimising rainfall erosivity and maintaining soil architecture in the inter-row areas (Shaxson, 1993). Climatic droughts may be unavoidable in smallholder agriculture but to worsen the situation by poor land husbandry that reduces infiltration (and so reduces potential soil moisture and possible groundwater) is inexcusable.

Agricultural yields are more related to the quality of the soil remaining than the quantity of soil eroded. If there is no real link between soil loss and productivity, then it is clearly time for a major rethink of the entire conventional soil conservation approach. If erosion is a consequence of land degradation then a more logical approach would be to direct efforts to tackling the cause of this degradation. If farmers are primarily concerned with stable and

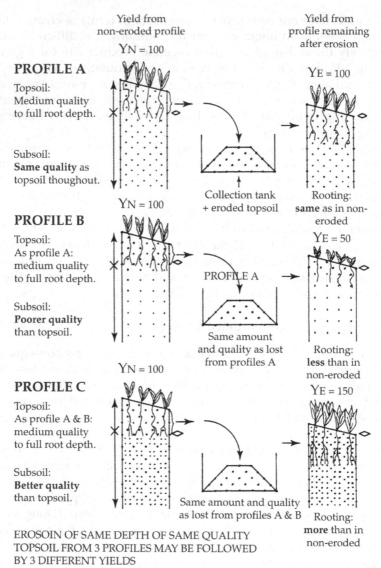

Yield from
non-eroded profile
YN = 100

Yield from
profile remaining
after erosion

PROFILE A

Topsoil:
Medium quality
to full root depth.

Subsoil:
Same quality as
topsoil thoughout.

YN = 100

Collection tank
+ eroded topsoil

YE = 100

Rooting:
same as in non-
eroded

PROFILE B

Topsoil:
As profile A:
medium quality
to full root depth.

Subsoil:
Poorer quality
than topsoil.

YN = 100

PROFILE A

Same amount
and quality as lost
from profiles A

YE = 50

Rooting:
less than in
non-eroded

PROFILE C

Topsoil:
As profile A & B:
medium quality
to full root depth.

Subsoil:
Better quality
than topsoil.

YN = 100

Same amount and quality
as lost from profiles A & B

YE = 150

Rooting:
more than in
non-eroded

EROSOIN OF SAME DEPTH OF SAME QUALITY
TOPSOIL FROM 3 PROFILES MAY BE FOLLOWED
BY 3 DIFFERENT YIELDS

Fig. 4.4   Crop yield from land after erosion is related to the quality of soil remaining
rather than the quantity of soil eroded (from Shaxson and Barber, 2003:34)

economic production then new initiatives should ensure that farmers benefit
from adopted technologies and modifications to their farming practices. Hence,
we need to focus less on capturing soil once it has been eroded and more on
maintaining or enhancing the quality of soil in the entire inter-barrier area.

Cross-slope soil conservation technologies may have a role to play in
preventing off-site sedimentation and the long-term loss of rooting depth, but
they are better deployed as part of a package of measures that augment soil

depth and enhance the biological, chemical and physical health of the soil. Since the live barrier *per se* does little to improve agricultural productivity, its opportunity cost to the farmer is high. It would be better if the barriers themselves contributed to farm income, for example, live barriers of species such as vetiver grass could be replaced by sugar cane and fruit trees (Hellin and Larrea, 1997).

There is plenty of evidence that suggests alternative ways to deal with soil and land degradation. The better land husbandry approach is one such alternative. With its focus on soil quality and active farmer participation, it has much potential to contribute to a more productive agriculture and sustainable rural livelihoods. The focus of this approach is to maintain and improve soil quality and it is to soil quality that we now turn.

4.3 IMPORTANCE OF SOIL QUALITY

4.3.1 Soil as a Living and Self-renewing Resource

Natural soils are complex, dynamic, evolving, biologically controlled open systems rather than an inanimate collection of minerals (Haigh, 2000). Soils are composed of different sized inorganic mineral particles (sand, silt and clay), reactive and stable forms of organic matter, water, gases, and a variety of living organisms (earthworms, insects, bacteria, fungi etc.) that rely on organic matter produced by green plants for their sustenance (Hallsworth, 1987:ix).

Soils form slowly through the interaction of climate, topography, living organisms and mineral parent material over time (Doran et al. 1996). This, however, is the geological rate and with the addition of organic matter and a nurturing approach, the process is quicker. Viewed from this perspective it is easier to appreciate the critical importance of soil quality, otherwise known as soil health (see Box 4.3).

Box 4.3   Soil quality

Soil quality is the capacity of a soil to function within ecosystem boundaries to sustain biological productivity, maintain environmental quality, and promote plant and animal health (Doran and Parkin, 1994:7). Soil quality is defined by soil function and represents a composite of its physical, chemical, biological and hydric properties (Doran et al. 1996). These properties provide a medium for plant growth and biological activity; regulate and partition water flow and storage in the environment; and serve as an environmental buffer in the formation and destruction of environmentally hazardous components. These soil functions form the basis for evaluating soil quality (Warkentin, 1995).

A healthy soil is a productive soil, one that provides a suitable environment for the development and growth of roots (Pierce and Lal,

---

1994; Gruhn et al. 2000:3). Soil serves as a medium for plant growth by providing physical support, water, nutrients and oxygen for roots. The suitability of soil for sustaining plant growth and biological activity is a function of physical properties (structure, porosity and water-holding capacity) and chemical properties (ability to supply nutrients and pH of soil) many of which are a function of soil organic matter content. Microbial decomposition and chemical reactions are, in turn, responsible for recycling organic matter into carbon dioxide and water, in addition to degrading chemical pollutants (Doran and Parkin, 1994).

What also needs to be remembered is that soil systems are also self-regulating, they utilise their own materials, water and energy from the sun and constitute a fertile substrate for the initiation and maintenance of life (Hillel, 1991:24). Soil is a renewable resource and the sustainability of its productivity depends, in technical terms, on its power of auto-recuperation after damage, the potential for which resides primarily in its organic constituents and biological processes (Shaxson, 1999:104).

We are faced with increasing soil and land degradation and the need to feed increasing number of people from a smaller per capita land resource. As the conventional soil conservation approach proves to be less popular with farmers than anticipated, it is no surprise that increasing numbers of development professionals have now identified "soil quality management as a central natural resource issue for sustainable agriculture in the developing world" (Scherr, 1999:48). The growing concern with maintaining and improving soil quality reflects the increasing realisation that much growth in food production will have to come from productivity gains (see Chapter 1).

A soil management system is sustainable only when it maintains or improves soil quality (Larson and Pierce, 1994). The desirable outcome of any agricultural system is, therefore, to utilise the soil's capacity to act as a self-sustaining system. Successful shifting cultivation systems are an example of this desirable outcome. The key feature of successful shifting cultivation systems is the length of the regenerative 'bush-fallow' period. If this period is sufficiently long before the cycle is started again, vegetation can be restored along with abundant biological activity, soil physical and chemical fertility. While shifting cultivation systems may in fact be breaking down because the bush-fallow period is being shortened, there is still much to learn from these one-time examples of sustainable agriculture systems.

From a practical point of view, there are four main ways of increasing the effectiveness of recuperative periods (Shaxson, 1999:48). The key to all is an increase in organic matter and biological activity.

- Increase the duration of the recuperative periods.
- Reduce the duration of the crops or grazing periods that damage the soil.

- Reduce severity of the damage, for example, by reducing the frequency and type of tillage.
- Manage the vegetation and crops in order to produce moist, well-aerated and nutrient-rich soils that will promote the biological transformation of organic materials into soil organic matter for the improvement of soil structure.

As pressures on the land increase and farmers are forced by necessity to reduce or abandon fallow periods, the first option becomes less practical. Farmers may also face difficulties in implementing the second and third options. It is the fourth option that offers the best possibility of making agriculture sustainable.

The key to sustainable agriculture is to acknowledge the importance of maintaining and improving soil quality and to appreciate that soil is a living self-renewing resource. Interactions among the components of soil, rather than the components themselves, have the greatest impact on the functioning of soil systems. There is, therefore, also a need to appreciate that soil fertility, as productive potential, lies in the combination of physical, biological, chemical and hydrological components of the soil and in their spatial arrangement, not just in the contents of nutrients and water alone (Shaxson et al. 1997).

4.3.2   Soil Biology

Soil biology is often the neglected component of soil quality, despite its critical importance to maintaining and enhancing that quality. A healthy soil is a biologically active soil with a large and diverse microbial biomass pool ranging from micro to macro-organisms (Eash et al. 1994; Lavelle et al. 1992). Soil organisms range in size from a few micrometers (protozoa) to several centimetres in diameter (large snails) or over a metre in length (large earthworms). The soil microbial community can contain as many as 10,000 different species in a single gram of soil (Turco et al. 1994) and each hectare may contain up to 1 million earthworms, 7 million arthropods, and larger organisms such as snails, slugs and mammals (Haigh, 2000).

Soil organisms are also the key architects in nutrient turnover and organic matter transformation (see Box 4.4). Hence, the abundance and diversity of organisms in the soil affect its physical structure, nutrient availability and moisture status (Haigh, 2000; Sherwood and Uphoff, 2000). As such, soil organisms have been referred to as *"subsurface workers, who perform many soil-improving activities without cost to the farmers"* (Shaxson and Barber, 2003:41). They deserve more attention than they currently receive.

Box 4.4   Soil organisms and release of nutrients

The microbial biomass is largely responsible for mineralisation and turnover of organic substrates that release many of the nutrients that are subsequently used by plants. Essential parts of the global calcium, nitrogen, phosphorous and sulphur cycles, along with water cycles, are carried out in soil largely through microbial and faunal interactions with soil physical and chemical properties (Doran and Parkin, 1994). Soil enzymes also play an important role in soil microbial activity, by acting as catalysts that speed up chemical reactions. The activity of 50 to 60 enzymes has been identified in soils (Dick, 1994).

Macrofauna such as earthworms, ants and termites are responsible for a range of soil functions beyond that of the soil microbial community. They function as secondary consumers, feeding largely on bacteria and fungi and hence contribute to the turnover of microbial biomass and their associated nutrients (Doran et al. 1996). In addition, many chemical properties of soils required for plant growth are affected strongly by biotic processes (van Breeman, 1993; Logan, 1990; Hurni et al. 1996:26). Dominguez et al. (2004) argue that earthworms can increase the leaching of water and nitrogen to greater soil depths.

Well established vegetative cover and the activities of these soil organisms, both plants and animals, enhance a soil's capacity to establish and reconstitute a spatial architecture that provides a suitable environment for the growth and development of roots (Shaxson, 1993b; Lal, 1990b:84) (see Box 4.5). For example, free-living nematodes play an important role in the biotic regulation of soil nutrient cycling processes and, through their feeding and excretory activities, they increase the turnover and mineralisation of nutrients that have been immobilised by the soil micro-flora. The enhanced mineralisation brought about by nematode feeding can increase the availability of nutrients for plant growth (Bohlen and Edwards, 1994).

Box 4.5   Soil macrofauna and soil structure

Soil macrofauna can substantially modify soil structure through the formation of macropores and aggregates (Hauser, 1993; Linden et al. 1994; Lavelle et al. 1992). For example, earthworms affect the soil through their burrowing, fecal excretion (casts), feeding and digestion. Their feeding habits help homogenise topsoil and in the case of surface feeders, incorporate large amounts of surface litter into deeper soil levels. Earthworm's digestive process also releases nutrients and fragments of plant residues, leaving behind fertile casts and mucus burrow linings (Doran et al. 1996).

Macropores created by earthworms are usually many times larger in diameter than the remaining soil pore network and soil that has been digested by worms is highly charged with soil-stabilising organic compounds and is much more water-stable than undigested soil (Haigh,

2000). Higher stability of soil aggregates, derived from earthworm excrements and increased infiltration capacity by worm channels, tend to decrease erosion. Additionally, earthworm burrow walls are nutrient-rich and may promote root growth and provide a mechanism for enhanced nutrient uptake by crops (Linden et al. 1994). It is, therefore, no coincidence that farmers worldwide have credited earthworms with indicating good soil quality.

Although, as we will see below, residue cover on the soil surface reduces the energy of raindrops, it also provides a substrate for soil fungi (Eash et al. 1994). Fungal hyphae can physically stabilise soil particles into larger aggregates (van Breeman, 1993). There is also evidence that organic inputs lead to an increase in populations of fungiverous and bactiverous nematode groups (Bohlen and Edwards, 1994).

Despite the importance of soil organisms, farmers' agricultural practices often reduce the diversity of soil organisms. Biotic activity is readily altered by crop residue management, tillage practices and agricultural chemicals (Lal, 1990b:84). Erosion results in loss of organic carbon and, therefore, a loss of substrates for soil organisms. Meanwhile, pesticides, including fumigants, fungicides, herbicides and insecticides, are used in agriculture because they are toxic to some organisms. Many metabolic processes are common to all cellular organisms and it is therefore not surprising that pesticides often display toxic effects on non-target organisms including micro-organisms (Sims, 1990). However, farmers' practices can also increase the diversity of soil organisms.

4.3.3 Organic Matter

Plant growth is the result of a complex process whereby the plant synthesises solar energy, carbon dioxide, water and nutrients from the soil. Between 21 and 24 elements are necessary for plant growth. The primary ones are nitrogen, phosphorous and potassium and when insufficient, crop growth is limited. Agriculture is a soil-based industry that extracts nutrients from the soil. As a result effective and efficient approaches to slowing that removal and returning nutrients to the soil are required in order to maintain and increase crop productivity on a sustainable basis (Gruhn et al. 2000:1).

In subsistence cropping systems of the tropics and sub-tropics, organic matter is the major source of nutrients and an important reservoir of nutrients that would otherwise leach from the soil. Doran et al. (1996) argue that soil organic matter replenishment is the cornerstone to regenerating soil quality. Anderson and Ingram (1993:62) report that organic matter is not a well-defined entity. It consists of a wide range of compounds from cellular fractions of higher plant, microbial and animal origin, through low to medium molecular weight organic substances with a known structure, to high molecular weight humus compounds whose structure is ill defined.

As suggested in section 4.3.2, the decomposition of dead organic matter by soil organisms releases many of the nutrients that are subsequently used by plants. Soil nutrient availability changes over time and the complex biological and chemical interactions involved in the continuous recycling of nutrients into and out of the soil are not fully understood. In general, though, the content of organic matter depends on the relative rates of supply of litter production and of decomposition and these processes are biologically-mediated (van Breeman, 1993).

For a cultivated soil, the amount of organic matter in any horizon is determined by the gain from humification of crop residue and loss through microbial oxidation (mineralisation) and erosion (Lu and Stocking, 1998:18). In the humid tropics, organic matter is oxidised much faster than in temperate zones. This is due to high temperatures at the soil surface in places where there is little vegetation cover. It has been calculated that the rate of decomposition can increase 2 to 3 times with every 10^0 C increase in mean annual temperature (Lal, 1990b:91). The result is that the organic matter content of soils (upon which several desirable characteristics of soils depend) is reduced (Chinene et al. 1996:14).

Despite not being well defined, soil organic matter plays other critical roles in plant production. It is also a substrate for soil organisms (a soil with high organic matter content is favoured by earthworms); an ion-exchange material; and a factor in soil aggregation and root development (Sanchez, 1987). Organic matter is critical to another component of soil quality, that of soil architecture: organic matter stabilises soil structure and decreases a soil's susceptibility to crust formation and surface sealing. All other factors being equal, soils with high organic matter are less susceptible to erosion than soils with low organic matter content.

When soil organic matter falls below about 2 percent there is a marked decrease in soil structure stability and soils that are of low erodibility under natural conditions many then become unstable and erodible if poorly managed (Cassel and Lal, 1992:68). Higher contents of organic matter and fine materials also generally increase the water- and nutrient-holding capacity of soil (FAO, 1995:5). Soil organic matter is positively correlated with, and makes the greatest contribution to changes in available water capacity (Lu and Stocking, 1998:20).

Organic matter also contributes to cation exchange capacity (CEC), which enables the soil to buffer nutrient concentrations in solution (Doran et al. 1996; van Breeman, 1993; Shaxson, 1993). Loss of organic matter can, therefore, reduce the soil's ability to retain base cations and to buffer acidity. Excessive soil acidity is detrimental to plant growth and to microbiological processes in the soil (Logan, 1990). There is also evidence that the ability of agricultural crops to resist or tolerate pests and diseases is linked to optimal physical, chemical and mainly biological soil properties (Altieri, 2002).

It is clear that increasing organic matter content leads to an improvement in soil quality. In general, though, farmers' agricultural practices, such as excessive tillage and burning, lead to a reduction rather than an increase in organic matter content. The loss of soil organic matter from the long-term cultivation of tropical soils will, therefore, degrade soil conditions for crop production (Logan, 1990). The great advantage is that farmers worldwide recognise the importance of organic matter and this is a very strong foundation on which better land management initiatives can be built. See Box 4.6.

Box 4.6   Farmers' appreciation of the importance of organic matter

Research that the author carried out in Honduras demonstrated that farmers recognise the importance of organic matter and organic fertiliser in terms of giving the land 'strength', improving soil structure and conserving soil moisture. Farmers recognise that it is preferable to leave the land in fallow in order to increase the 'strength' of the land through the accumulation of organic matter but land shortages have forced farmers to reduce fallow periods.

Honduran farmers also cited land with a high organic matter content as one of the key indicators of good quality land and they recognised that the destruction of organic matter was one of the biggest disadvantages of burning plots prior to sowing. Lawrence (1997) also notes that farmers in the lowlands of Bolivia, who know very little about soil conservation and land management, were knowledgeable about the role and value of organic matter.

Building on farmers' appreciation of the importance of organic matter is an effective way to improve soil quality. There are also indigenous practices to build on. Some farmers in Honduras have also traditionally dug a small trench at the end of the harvest, filled it with old maize stalks and then covered it with soil. The maize stalks decompose and provide a good growing medium for the subsequent crop. The principle behind this practice is similar to the planting holes described in semi-arid regions of West Africa (Critchley et al. 1994).

Despite this appreciation of organic matter, farmers' agricultural practices can easily lead to a reduction in organic matter. Tillage accelerates organic matter decomposition due to the physical disruption of aggregates, increased aeration and warming of the soil that encourages microbial activity (Doran et al. 1996). Also when farmers burn their fields prior to planting, organic matter is destroyed and volatile constituents like carbon and nitrogen literally go up in smoke. Calcium, potassium and phosphorous are also lost in the burning process (Critchley and Bruijnzeel, 1996).

However, there is an arsenal of agricultural practices that can maintain and increase organic matter content. These include the use green manure crops, cover crops and the abandonment of the practice of burning. An advantage of not burning is that nutrients are slowly released as the organic material gradually decomposes.

4.3.4   Soil Architecture

Soil organisms and organic matter content are key determinants of soil quality but, for a greater understanding of the linkages between soil quality and sustainable agriculture, we need to look more closely at the environment that soil provides for root growth and development. This leads to an examination of soil architecture, for it is soil architecture – the structure that preserves the open spaces in the soil – that provides the habitat for soil life. The issue is very closely linked to water availability (see also section 4.3.5).

Although the productive potential of soil is determined by the quality of soil remaining, actual yields are determined by the complex interaction of a number of factors including soil quality, crop and land management system, and climate (Bunch, 1982:2; Enters, 1998:8; Clark, 1996:14). An undue focus on the quantity of soil lost has obscured the important contribution to declining productivity arising from the collapse or destruction of soil-structural units and subsequent shortage of soil moisture (Shaxson et al. 1989:22).

All soil biological and chemical activities are dependent on an adequate level of soil water (Anderson and Ingram, 1993:3) and scarcity of water is likely to become the most important factor in limiting agricultural production (Leslie, 2000). Consequently, increased global agricultural outputs will increasingly have to come from rain fed rather than irrigated land because the resource-poor can not afford expensive but efficient irrigation technologies such as drip irrigation. Hence, we need to value rainwater more as a free (albeit limited) resource and focus on better efficiency in its use for plant production.

A soil aggregate is the basic component of soil structure and the building blocks of soil architecture. Soil aggregates are formed by physical, chemical and biological processes that encourage the association of clay particles into domains, domains and silt particles into micro-aggregates, and micro-aggregates and sand particles into aggregates (Lal, 1990b:62; Eash et al. 1994). The extent of aggregation within a soil is the controlling factor for soil porosity (Karlen and Stott, 1994). Pierce and Lal (1994) define soil structure as having three components. The first is structural form governed by the geometry of the soil pore space. The second is aggregate stability and the third is structural resilience, which is the ability of a structure to re-form once it has been degraded. The amount of organic matter in the soil and the activity of soil organisms largely determine aggregate formation, aggregate stability and structural resilience.

The ability of a soil to store and transmit water, and hence affect root development, is a major factor regulating water availability to plants (Doran et al. 1996). However, it is not changes in soil structure i.e. the arrangement of the soil's solid particles *per se* that adversely affect water movement and retention, rather it is changes in the size and continuity of the soil's pores, the soil architecture. As in a building, all the interesting events—gas exchange, root exploration, water movement and retention—occur in the spaces or voids of the architecture and not within the physical properties themselves (Shaxson, 1999:19). See Box 4.7.

Box 4.7   Soil architecture and the importance of pores and spaces

In order to provide good anchoring, water-holding capacity, water transmission and aeration, the rooting system must involve a heterogeneous pore system (van Breeman, 1993; Karlen and Stott, 1994; Cassel and Lal, 1992). The heterogeneous pore system partly results from the growth and decay of plant roots, the activity of soil flora and fauna, and clay shrinkage (van Breeman, 1993; Haigh, 2000). A well-aggregated soil maintains a structure that allows a variety of pore spaces and soil pores are maintained by the presence of clays and organic compounds. Pores and spaces in the soil architecture are important for:

- Infiltration, absorption, transmission, retention and supply of soil water;
- Providing an optimum balance between air and moisture in the rooting zone of the soil;
- Gas-exchange in relation to root growth and functioning of roots (Lavelle et al. 1992; Lal, 1991; Shaxson, 1993; FAO, 1995:8).

The amount of unoccupied space provided by pores and voids in a soil normally ranges from about 25-60 percent (Haigh, 2000). Coarse pores, those greater than 75 microns in diameter, control infiltration rates, whilst fine pores between 0.1 and 15 microns control the capacity of a soil to store plant-available water (Karlen and Stott, 1994). Pores of the rooting system must be at least 0.1 mm in diameter to allow initial penetration of root tips (van Breeman, 1993).

If the majority of pore-spaces is very small, the water may be held so tightly that plants are unable to extract much of it. If, however, the soil has a predominance of large pore-spaces, much of the rainfall may enter the soil and pass through the soil profile so quickly that little of the water is made available to the plant. In this case the root-zone may dry out and the subsequent water stress in the plants will have an adverse impact on plant growth. A wide range of pore sizes in a stable soil architecture is, hence, desirable for enabling water retention and transmission.

The term 'soil architecture' reflects this change in emphasis from the solid constituents of soil to the spaces (the distribution of the spaces and pores between soil particles) rather than the term 'soil structure'. Soil architecture has a direct impact on root growth and is, therefore, a critical factor in determining soil productivity. Soils in good physical condition are loose, moist and well aerated with well-connected macropores that allow roots to grow unimpeded (Cassel and Lal, 1992).

Soil architecture has a major impact on plant production because plant roots are sensitive to excessive water in the soil. Also, most terrestrial plants are not able to transfer oxygen internally from their aboveground parts to their roots at a rate sufficient to supply the demands of root tissues. The rate of diffusion of oxygen through water-filled soil pore space is about one ten-thousandth the rate through air-filled pores, resulting in much slower rate of oxygen supply in wet soils (Sims, 1990). Adequate respiration of roots requires that the soil itself be aerated (Hillel, 1991:43).

Pore spaces can be reduced, however, following mechanical soil compaction, tillage and trampling (Shaxson, 1988) and also from the impact of raindrops. Soil compaction causes an increase in particle-to-particle contact within the soil, leading to a reduction in the number of macropores an increase the number of small pores. This increases the proportion of water in the soil held at high tensions and hence unavailable for absorption by plant roots. Compacted soils also have fewer water-stable soil aggregates, lower infiltration rates, and reduced water-holding capacities (Haigh, 2000), raising the degree of soil saturation and decreasing soil aeration. Compacted soils also exhibit greater resistance to root penetration. Compaction also affects chemistry and nutrient availability by encouraging anaerobic rather than aerobic soil conditions and unfavourable microbiological processes such as de-nitrification rather than nitrification (Haigh, 2000).

Soil architectural stability, especially pore stability in the top few millimetres of the surface, is of critical importance because it determines how rainwater is partitioned between infiltration and runoff (Shaxson, 1999:18) (see Figure 4.1 and Figure 4.5). As discussed earlier in this Chapter, there is greater runoff if the top few millimetres of the soil are degraded by high impact rain drops. Further degradation of the soil's surface layers is often caused by compaction from animal hooves and also the use of agricultural machinery, which can also lead to an impermeable sub-surface plough-pan, disc-pan or hoe-pan or a more general sub-surface compaction.

Surface crusts are characterised by a high density of fine pores and water transmission properties that are 2 to 3 orders of magnitude lower than the underlying soil. These not only affect root growth and development but they also interfere with seedling emergence, leading to poor crop stands (Karlen and Stott, 1994), especially in soils with weak structure (Cassel and Lal, 1992).

4.3.5   Importance of Soil Moisture to Agricultural Production

Declining soil structure and the loss of effective pore-spaces within the soil architecture may have a more detrimental long-term effect on plant growth

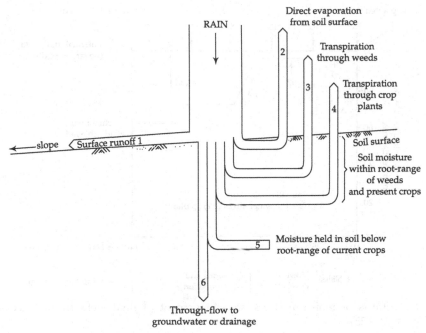

RAIN

Direct evaporation
from soil surface

Transpiration
through weeds

Transpiration
through crop
plants

slope ◄ Surface runoff 1

Soil surface

Soil moisture
within root-range
of weeds
and present crops

Moisture held in soil below
root-range of current crops

Through-flow to
groundwater or drainage

Fig. 4.5   Possible destinations of rainwater (from Food and Agriculture Organization
of the United Nations, 1995:4)

than erosional loss of soil particles through its direct and indirect effects on
root growth and functioning (Shaxson, 1993b). The reduction in plant-available
water capacity is often the most important factor affecting soil productivity
in many soils (Scherr, 1999.49). Often, the issue is related less to an absolute
insufficiency of rainfall and more to a deterioration in soil quality that leads
to an insufficiency of moisture in the rooting zone.

The importance of soil moisture is demonstrated by the fact that successive
periods of water stress during the life-cycle of plants result in a cumulative
lessening of the potential final yield (see Figure 4.6). Shaxson et al. (1989:39)
argue that where yield benefits are claimed from soil conservation technologies,
such as those detailed in Chapter 2, they are usually more closely associated
with water conservation and improvement in soil moisture availability than
with any savings in soil and plant nutrients.

The link between soil quality, soil degradation and changes in productivity
are key components of the better land husbandry approach. There is ample
research evidence to support this link. For example, Mokma and Sietz (1992),
conducted a study in the United States to compare soil properties of areas
exhibiting slight, moderate and severe erosion and to determine the effect of
past erosion on maize yields. In 1988, the yields of all three erosion phases
were less than yields for the previous three years because of reduced rainfall
and soil moisture in the early growing season.

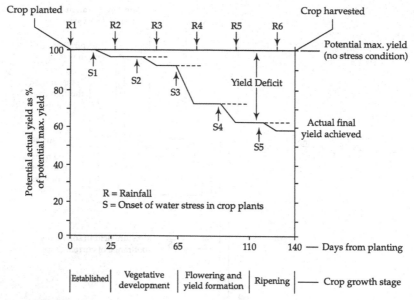

Fig. 4.6   Effects of moisture stress on achievement of final yield (from Shaxson, 1999:23)

A similar situation occurred at Santa Rosa in Honduras where the author conducted research. In 1997, maize yields at the trial site (and throughout Honduras) were significantly reduced (see Table 4.1). Although there had been much soil erosion in 1996 at Santa Rosa, low maize yields in 1997 were due to reduced rainfall caused by *El Niño* and not to the quantity of soil lost the previous year. This assertion is reinforced by the fact that in 1998, above average rainfall meant that maize yields in the first harvest, the *primera,* on both slopes were significantly greater than the *primera* in 1997 and similar to the *primera* in 1996.

Table 4.1 clearly shows that in drought conditions, soil moisture rather than soil fertility proves the limiting factor to productivity. In these conditions, the first symptoms of soil degradation, in terms of reduced productivity, may result from plants being deprived of rainwater. Loss of internal voids and surface crusting of the soil can, therefore, be more important than a reduction in plant nutrients as a result of soil loss (Shaxson, 1993; FAO, 1995:5). Water is more likely to become a limiting factor if the soil's water storage capacity is reduced due to poor management resulting in a deterioration in soil architecture.

For many resource-poor farmers worldwide, the main concern is productivity harvest-by-harvest. Farmers do not deal with averages. The data from Santa Rosa in Table 4.1 show that the average yield per harvest on steep slopes over a three-year period was 1100 kg ha^{-1}. This is likely to be of less

Table 4.1 Summary of analyses (*t*-tests) of maize yields (kg ha^{-1})* and slope, 1996-1998 from Santa Rosa, Honduras. Results are for both treatments and the *primera* and *postrera* except 1998 (*primera only*) From Hellin, 1999b

Year	Slope (percent)	Mean yield (kg ha^{-1})	*t*-value	Significance
1996–1998	35-45	1883	13.331	≤ 0.000
	65-75	1093		
1996	35-45	2160	10.216	≤ 0.000
	65-75	1315		
1997	35-45	1375	9.198	≤ 0.000
	65-75	722		
1998	35-45	2348	8.507	≤ 0.000
	65-75	1392		

concern and interest to a farmer than the fact that in 1997 maize yields per harvest on the steep slopes averaged 722 kg ha^{-1} (see Table 4.1). This represents a 33 percent reduction in yield. The data cannot be used to demonstrate a clear link between soil loss and productivity. It is, therefore, understandable why farmers identify drought and erratic rain as one of their priority problems rather more than soil erosion. The results confirm the need for a rethink about land management strategies.

4.3.6 Soil Quality and Farmers: Linking Production and Conservation

A focus on soil quality, therefore, requires a holistic approach to soil management and agriculture, one in which a soil's inherent productive and recuperative capacities are enhanced (Sherwood and Uphoff, 2000). As farmers decide what happens on their land, strategies to improve soil quality need to be developed in partnership with them. The great advantage is that many concepts of soil quality accord with farmers' strong affinities with the land and unconscious resonance with the vitality of the Earth (Shaxson, 1994).

Hence, a deterioration of soil architecture may adversely affect productivity more than soil loss *per se*. Of course, if a soil is more productive it is because it also provides a suitable environment for root growth and development. Soil factors that favour root growth, such as increased organic matter content and improved soil architecture, also favour better water relations in the soil and the conservation of soil and water *in situ* (Shaxson, 1993).

---

*Unlike the case with soil loss and runoff data maize yields can be reported on a per ha basis because the yield data are based on a uniform density of maize throughout the plot. The possibility of an edge effect as investigated in the *postrera* of 1996. Maize was harvested from an inner 3 × 22 m core and also from the outermost 1 m of each plot. Analysis (*t*-test) showed no significant difference between the yields.

An improvement in the soil conditions for rooting along with improved soil structure and soil water infiltration capacity, may achieve both increased soil conservation and production (Shaxson, 1988). This provides a route to reconciling farmers' concerns about productivity and increased plant production with conservationists' concerns about minimising runoff and avoiding erosion, (Shaxson, 1997; Norman and Douglas, 1994:44) (see Figure 4.7). Conventional soil conservation programmes typically identify conservation as a precursor to increasing yields. The better land husbandry argument is that conservation can be achieved as a consequence of soil improvement and increased production per unit area (Shaxson, 1997). As Hudson (1988) writes, improved production should lead to better erosion control instead of the other way around.

Land husbandry is not a synonym for soil conservation. The latter often has a negative image for farmers due to its implied need to control and restrict present land use in order to preserve soil for the future. The concept of good land husbandry moves on from the restrictive approach of erosion control to a more positive and forward-looking style of soil- and land-improvement, where conservation is achieved as an inevitable by-product (Shaxson, 1997; Douglas, 1993; Hudson, 1988). The next section takes us from the theory of maintaining and improving soil quality to the practice.

4.4   MAINTAINING AND IMPROVING SOIL QUALITY

4.4.1   Think Like a Root

Shaxson (2004) urges better land husbandry practitioners to *think like a root*. From the viewpoint of a root, the quantity and quality of the root-environment that remains behind is of far greater significance for its future growth and development than the quality and quantity of that which has been eroded away. Progress may be made, first by identifying in what conditions of the soil plant roots function optimally, and then taking measures to improve the state of the roots' habitat.

Unlike the conventional soil conservation approach, better land husbandry is preventative rather than corrective. From a philosophical perspective, the key concern is to sustain the dynamic linkages among the living and non-living components of land-use systems, within a unifying thought-context of the Earth as a self-sustaining organism (Shaxson, 2004). Farmers are receptive to this approach, not only because it leads to improvements in productivity but also because the focus on soil quality and soil moisture availability, rather than soil loss, is one that they understand and appreciate.

The recovery of degraded soil is due not only to plant colonisation but also to favourable soil conditions and the enhanced fertility that soil organisms promote (Turco et al. 1994). The regenerative and buffering capacity of soil

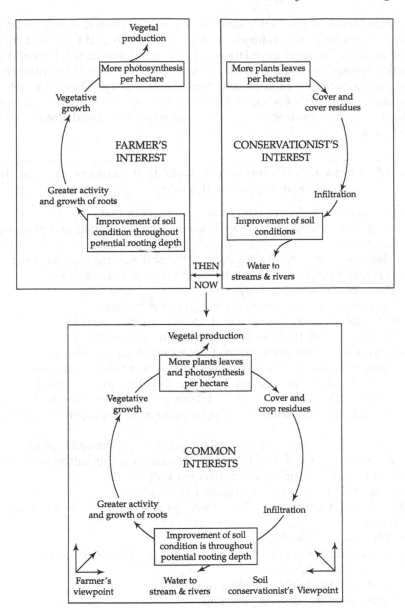

Fig. 4.7 Farmers and conservationists have similar concerns but different viewpoints. A focus on soil quality can bring these concerns and viewpoints together (from Shaxson, 1997).

depends primarily on biological processes (Sherwood and Uphoff, 2000). In almost all agricultural situations, biological processes over time are the only means by which the soil can be re-structured so as to provide a suitable environment for plant growth (Shaxson, 1993).

A central tenet of the better land husbandry approach is improved soil quality via a biologically healthy soil. A prerequisite is that there is sufficient organic matter to support populations of soil organisms (FAO, 1995:5). From a practical perspective, the first of many important steps is to enhance the organic matter content of soils and biological activity. These are detailed below and a summary of some of the most appropriate practices appears in Box 4.8. Ultimately, the objective is to enhance the soil's productive capacity. See Figure 4.8.

Box 4.8   Summary of simple practices that farmers can use to maintain and improve soil quality

Farmers should be encouraged to adopt practices that:
- Produce biomass (utilise crops, varieties, crop rotations and planting densities)
- Maximise the percentage cover of the soil surface (do not remove residues from fields, do not burn and do not graze animals)
- Promote biological activity
- Fulfil much of the potential for nutrient recycling.

Many farmers live in less-favoured areas that have poor infrastructure and market access. In these cases, it is often uneconomic for farmers to use high levels of external inputs. Low-external-input technologies are, however, often labour-intensive and this can be an important constraint on their uptake. The challenge is to develop low-external-input technologies that boost labour and land productivity (Barghouti and Hazell, 2000). In this context the simplest and most appropriate practices are:
- Use of cover crops and leave plant residues on the soil surface
- Zero tillage or reduced tillage, especially in combination with soil cover (such as a green manure/cover crop)
- Crop rotation and use of organic fertilisers

Individually and used together, these practices have several effects including
- Better soil architecture
- Better infiltration of rain water and, hence, reduced runoff
- Reduced weed problems
- Increased availability of plant nutrients
- Fewer pest and disease problems
- Avoidance of excessive surface temperatures
- Reduction in loss of soil moisture from evaporation
- Increase in the amount of organic matter and biological activity in the soil.

4.4.2   Soil Cover and Soil Quality

As discussed in Chapter 2, farmers seldom see soil erosion as a problem and

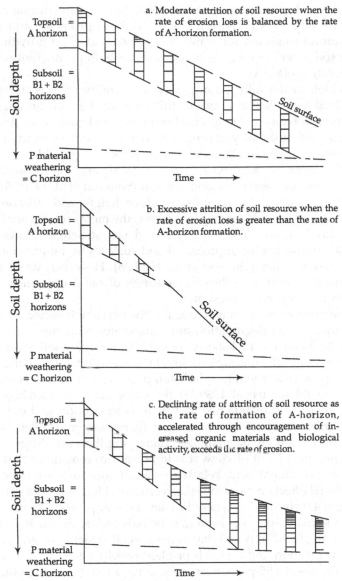

Fig. 4.8 Significance of A-horizon management in maintaining soil productive capacity (from Shaxson et al. 1989:27)

are, therefore, unlikely to adopt a technology or practice unless it addresses their main concern, namely increased and regular production. Providing better soil cover can achieve just this. Soil cover not only protects the soil surface from the impact of raindrops, it can also reduce soil-moisture losses, ameliorate soil temperature, suppress weeds, and increase biological activity. By conserving soil moisture for use by plant roots and providing suitable conditions for soil organisms, it improves soil quality. Agricultural production subsequently improves because a soil cover both promotes and maintains soil in optimum condition for water acceptance and plant growth. As such, increasing soil cover is a farming practice that is both productivity-enhancing and conservation-effective.

A soil's biological health and organic matter content affects changes in soil structure (and hence soil quality). In this context, if a soil can degrade over time, it can with biological help also be regenerated over time. An objective of the better land husbandry approach is to minimise the severity of damage and assist soil recuperation as soon and as quickly as possible.

Soil detachment by raindrop splash and transportation by surface runoff are the two main processes of water erosion (Shaxson et al. 1989:35). In terms of the factors involved in soil erosion, slope length and steepness can be modified, for example by terracing. However, the most critical variables under control of the land user are soil cover and soil structure (Shaxson, 1988). Sealing and compaction of unprotected soil surfaces by raindrop impact can occur in a few minutes (Shaxson et al. 1989:38). However, when the soil is covered and protected from the erosive forces of raindrops, this sealing and compaction is mitigated or prevented.

Soil surface cover, be it living or dead, is the best single factor for protecting the soil surface from degradation and, therefore, reducing erosion (Foster et al. 1982); Kellman and Tackaberry, 1997:264). Low-lying soil cover dissipates the erosive energy of raindrops by breaking them up into smaller droplets whose energy is insufficient to splash soil particles or to cause compaction at the soil surface (Shaxson et al. 1989:35). By protecting a soil from high-intensity raindrops, porosity is maintained and there is less runoff and erosion. Even where water accumulates on leaves and forms larger droplets, these also have less energy and are normally less damaging than raindrops.

From a practical point of view, it is important to remember that the effect of cover is not linear and relatively small amounts of cover have a disproportional effect on reducing splash erosion (Hudson, 1992b:94; Stocking, 1994). Figure 4.9 shows that where low-level cover protects about 40 percent of the soil's surface, splash erosion may be reduced by as much as 90 percent (Shaxson et al. 1989:37). A soil that is deep, well structured, and covered by protective vegetation and a mulch of plant residues (soil in a natural state) will normally absorb 95 percent or more of the rainfall. On the other hand, a soil that is denuded of vegetative cover and deprived of a surface mulch may absorb less than 80 percent of the rain, and in extreme cases less than 50 percent (Hillel, 1991:97).

For example, in the highlands of Papua New Guinea soil erosion is not a problem nor is it perceived to be a problem despite the fact that the climate is wet all year and farmers cultivate steep slopes, semi-continuously, as part of a system of shifting cultivation (Sillitoe, 1993). The reasons are linked to the characteristics of the soil, rainfall intensities, and local farming practices that afford good cover and soil protection.

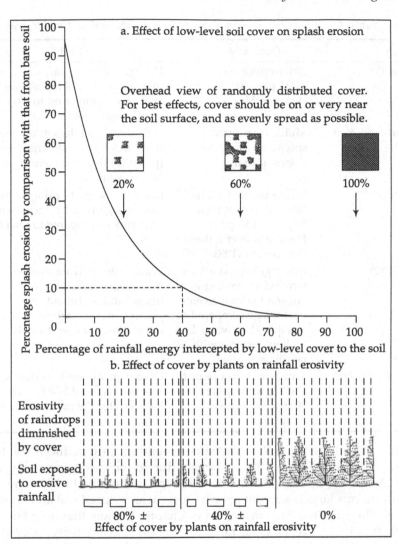

Fig. 4.9 Impact of low-level cover in reducing splash erosion and rainfall erosivity (from Shaxson et al. 1989:36)

According to Lal (1982), a mulch of crop residue of 4 to 6 t ha^{-1} can effectively control soil erosion on steep slopes of up to 15 percent and for open-row crops such as maize. While on slopes up to about 20 percent, a 40 percent, ground cover has been found to reduce soil losses to about 10 percent of the losses that occur from bare soil. Table 4.2 confirms the dramatic impact on soil and water loss when there is soil cover.

Table 4.2　Effectiveness of soil cover in reducing soil and water loss

Source	Treatment	Effect
Maass (1992)	Litter-mulch on 31-51 percent slopes in Mexico.	During the first year, soil loss from maize treatment without mulch was 99.8 t ha^{-1}, compared to 5.0 t ha^{-1} in mulch treatment.
Omoro and Nair (1993)	Mulch from three tree species on 15 percent slopes in Kenya.	Soil and water loss from the plots with mulch were significantly lower (p = 0.10) than the control plot over two seasons.
Smith (1997)	Mulch used on 0.2 ha field catchment on a 60 percent slope in Honduras over a three-year period (1993-1995).	Burning the plots prior to planting maize, significantly increased (p < 0.10) runoff compared to mulch only.
Lal (1982)	Maize grown on a 5 ha terraced watershed in Nigeria under conventional tillage compared to an equivalent watershed without terraces with a no-tillage system.	Water runoff from conventionally managed catchment was 60 to 70 times, and sediment load 5 times greater, than catchment with soil cover. Data for one storm in 1979.
Soto et al. (1995)	Two plots burnt prior to sowing with rye on 30 percent slope in Spain.	Over a nine-month period soil loss was 46 t ha^{-1} and 57 t ha^{-1} compared 1.4 t ha^{-1} for the undisturbed control plot.

Similarly, at the Santa Rosa test site in Honduras, the reduced soil loss in 1998, reported in Chapter 3, was largely due to increased soil cover in that year. As shown in Table 3.6, reduced rainfall caused by *El Niño* and fewer intense storms largely account for why there was reduced soil and water loss in 1997. Precipitation data, presented in Chapter 5, show that until Hurricane Mitch struck in October 1998, rainfall amounts and intensity in 1998 were similar to 1996. Soil loss, however, was much reduced in 1998 compared to 1996. Farmers who assisted in the management of the trial site attributed reduced soil loss in 1998 to the fact that there was more weed cover in 1998 than in the previous two years. This was due to the fact that light rain at the beginning of the rainy season in 1998 meant that weed seeds were not washed away, they germinated on the slopes and the farmers did not clear them. Greater soil cover meant that more of the soil surface was protected the soil from the impact of high-energy raindrops.

4.4.3　Soil Cover: Complementing Farmers' Needs and Resources

One aim of the better land husbandry approach is to introduce land management practices that seek to recreate the vegetative ground cover found in forests rather than leaving fields bare (see Plates 16 and CP 7). In this case, a protective cover of litter and vegetation over the soil surface contributes to a high infiltration capacity of the soil, stable porous soil architectural conditions and high biological activity. The more rainfall that is absorbed, the more is available for plant growth and regular stream-flow.

Plate 16   Fields without soil cover are more prone to erosion. Ecuador. (Hellin, J.)

In general, more attention needs to be paid to reducing the effect of erosive rain by improving soil cover and maximising infiltration by improving soil structure in areas where productive plants are growing (Douglas, 1993; Shaxson, 1992; Lal, 1988). The advantage of this approach is that cover can be provided in a variety of ways. It is not a question of introducing a new technology and subsequently stimulating farmer adoption, but rather one of developing what farmers are already doing. Hallsworth (1987:87) confirms that traditional conservation practices, many used for hundreds of years, often use soil cover to reduce erosion rates.

Increasingly, during land preparation farmers slash the cover and green manure crops and then sow their plant crops directly through the mulch. These slash and mulch systems, which are found worldwide, and are good examples of sustainable agriculture. They could make significant contributions to increased food production in developing countries without having to incur in excessive quantities of external inputs. The Quezungual System in Honduras and use of zero tillage in South America are examples of where slash and mulch systems have paid dividends for the participating farmers. See Chapter 5.

Slash and mulch methods also discourage farmers from burning their fields prior to sowing. Fire is one of the oldest tools humans have used in managing agriculture (Thurston, 1992:70). Slash and burn systems using fire have been in existence since the Neolithic era and are still used extensively in the tropics. However, fire destroys built-up reserves of soil organic matter and kills soil organisms, so degrading the soil.

Doran et al. (1996) argue that soil organic matter replenishment is the cornerstone to regenerating soil quality. The promotion of cover crops is a first step towards improving soil quality. The ideal situation is to encourage farmers to stop burning or to burn less frequently so that there are more plant residues and so that cover crops can be grown. Even though farmers often burn their fields to control pests and diseases and reduce labour costs in preparing fields for sowing, farmers do show a readiness to desist from the practice if and when they see advantages in so doing. One such advantage is increased productivity from improved soil quality.

As farmers' understanding of soil as a living resource develops, their appreciation of the importance of organic matter and their understanding of the need to enhance soil quality grows. There are gaps in their knowledge that have to addressed, especially soil biology (with the exception of earthworms, few farmers appreciate the importance of soil biology and this is one area where researchers and extension agents could contribute to farmer knowledge). Indeed, the concept of soil quality does have the potential to bring together the conservation objectives of outsiders and the production objectives of farmers. The Quezungual system described in Chapter 5 clearly illustrates what can be achieved, in terms of improved soil cover, reduced soil loss and improved productivity, when farmers stop burning.

Meanwhile, in Guaymango, El Salvador farmers who were persuaded to stop burning and to leave plant residues on the soil surface found that they were able to grow maize and sorghum on 45 percent slopes with few erosion problems and, more importantly, yields also increased by 20-43 percent in less than three years.

Farmers' preference for following recommendations to stop burning, as opposed to adopting soil conservation technologies, is illustrated in Figure 4.10. The figure, based on the results from a soil conservation project in Honduras, clearly shows an increase in the area of cropping land where farmers have abandoned the practice burning. Figure 4.10 contrasts with Figure 2.2 that show the same farming communities rapidly abandoning the soil conservation practices such as live barriers once the project stopped offering direct incentives. In summary, management for soil conservation is as much a question of encouraging vegetative growth, in some cases by not burning cultivated areas, as it is constructing physical soil conservation structures (Stocking, 1994) (see Box 4.9).

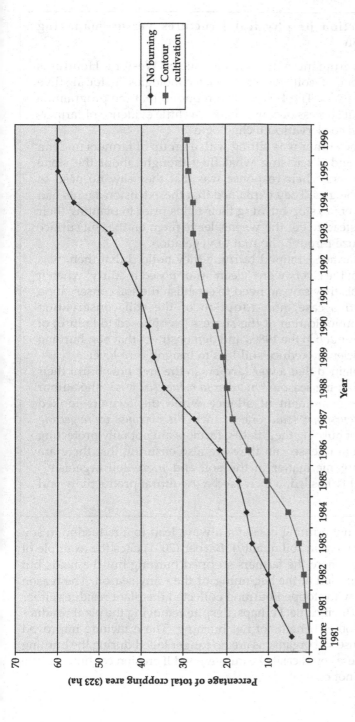

Fig. 4.10 Adoption and abandonment of agronomic measures in La Paz, Honduras. Project duration 1980-1991. Direct incentives provided 1984-1991. Number of farms = 147 (from Hellin and Schrader, 2003).

Box 4.9   Construction of physical structures versus managing vegetation

In the 1980s, government extension agents in Western Honduras promoted a number of soil conservation technologies including live barriers and stone walls. The FAO (1994) reported that the programme had not been a great success because there was little evidence of farmer-adoption of the soil conservation technologies.

One afternoon the author was sitting with a group of farmers under a large mango tree and discussing what they thought about the stone walls and live barriers. Their response was that they saw no need to establish the technologies. They explained that the extension agents had also encouraged them to stop burning their fields prior to planting their maize crop and instead to cut the weeds, leave them on the soil surface and to sow the maize through the mat of vegetation.

As soon as the farmers stopped burning they noticed that there was hardly any erosion: the rivers were 'clean' as opposed to 'dirty' when it rained. On this basis they saw no need to establish the soil conservation technologies. In this case, non-adoption of the soil conservation technologies was an indication of the success (as opposed to failure) of the extension programme in the 1980s and demonstrates that non-burning alone may be sufficient to reduce soil loss to insignificant levels.

The farmers explained that fewer farmers in the area now burnt their fields before the planting season. *"In order to reduce soil loss?"* the author asked. There was a moment of silence while the farmers looked incredulous. *"Of course not"* said one of them *"it's because we're getting better yields"*. By not burning their fields, farmers are not only protecting the soil surface from intense rain they are also ensuring that there are higher levels of organic matter in the soil and more soil organisms. Improved soil quality has led to increased agricultural productivity and production.

Of course, merely not burning does not always lead to a reduction in soil loss (and an improvement in soil quality). Barber (2001) cites the example of an area in Nicaragua where the farmers stopped burning but the fields but still left the fields denuded at the beginning of the rainy season. The reason was that the farmers who stopped burning collected the plant residues either to sell or feed them to their cattle. Perhaps, despite removing the plant residues there would still be other benefits of not burning. These include improved biological heath because soil organisms are no longer killed during the burning process. However, these could easily be outweighed if erosion continues while weeds and pests are not controlled.

4.4.4 Cover Crops and Green Manure Crops

The most effective way to provide ground cover is via the use of green manure and cover crops. These plants, mostly leguminous, can be inter-cropped with agricultural crops such as maize and beans, or grown during the dry season. In recent years, much attention has been directed at the use of green manure and cover crops and also the application of organic manure (see Box 4.10). Cover crops and green manure crops can include legumes that provide nitrogen to plants via nitrogen fixation. They are also of great benefit in weed control since the space, light, moisture and nutrients they need for their development reduces the growth of weeds (Erenstein, 2003). In zero tillage systems, the mulch that results from pruning, chemical or manual control of the cover crops, can significantly reduce the weeds.

Box 4.10   Use of cover crops and green manure crops (from Bunch, 2003 and 1997)

There has often been confusion between the terms cover crop and green manure. Most of the time they are used inter-changeably in the literature. Traditionally, green manure crops provide material that is incorporated into the soils in its green or mature stage with the objective of enriching the soil. In recent times, the term green manure crop has also referred to plants that are left on the ground as green or dry material, again with the objective of fertilising the soil. Cover crops are planted to cover the soil regardless if they are incorporated or not in the future. However, although they are used mainly to cover and protect the soil surface, they can also be incorporated as green manure crops.

Over the last decades, numerous farmers worldwide have used different species of leguminous green manure and cover crops in their farming systems. The species are frequently food crops themselves and include cowpeas or rice bean (both *Vignas*), lablab beans (*Dolichos lablab*), scarlet runner bean (*Phaseolus coccineus*), or fava beans (*Vicia faba*). Farmers also use species that are important as an animal feed, such as sweet clover (*Melilotus albus*) and velvet bean (*Mucuna pruriens*). Cover crops are also a way of reducing natural fallow periods to one or two years, as with *Tephrosia candida*.

Perhaps the second most common use of green manures and cover crops is in weed control. In South-east Asia, a perennial species of the velvet bean is use to improve fallow and to control weeds. More modern practices include using jack bean (*Canavalia ensiformis*) tropical kudzu (*Pueraria phaseoloides*) and perennial peanuts (*Arachis pintoi*) under a variety of plantation crops, including coffee, citrus, and African oil palm. The velvet bean is also used to control imperata grass (*Imperata* spp.) and this practice is spreading rapidly throughout Benin, Togo, and

Columbia. Velvet bean and jack bean are used to control paja blanca (*Saccharum* spp.) in Panama and to combat nutgrass in several other countries

Another potential benefit that will probably acquire more significance as experience increases, is the use of green manures as animal feed. Most green manures and cover crop species, with the major exception of *Melilotus albus* cannot be grazed well, but many can be used for cutting and carrying even after months of drought, the most notable examples of this type being *Lathyrus nigrivalvis* and lablab bean (*Dolichos lablab*). Seeds also provide fodder, one good example being the seeds of the velvet bean which in Campeche, Mexico are cooked for half-hour, mixed with an equivalent amount of maize and then ground into pig feed. The University of Yucatan calculated that this velvet bean feed cost less than commercial feeds per unit of weight gained.

Well over 125,000 farmers are using green manure and cover crops in Santa Catarina, Brazil. Green manure and cover crops are equally popular in neighbouring Parana and Rio Grande do Sul. In Central America and Mexico, an estimated 200,000 farmers are using 20 traditional systems involving some 14 different species of green manure and cover crops, including *Mucuna* (Eilittä et al. 2003), and organisations from Central Mexico to Nicaragua are promoting their use in at least 25 additional systems. Across the ocean in West Africa more than 50,000 farmers have adopted *Mucuna* spp. or *Dolichos lablab* as green manure crops in the last eight years. Cover crops are also used in East Africa (Gicheru et al. 2004).

The reason for using green manure and cover crops, in addition to their other purposes, is to produce *in situ* large quantities of organic matter for soil improvement, which is then usually applied to the soil surface. In this way, harvests can be raised from 20 percent to 50 percent a year, while the land is being used, and in some case fallow periods can be avoided totally. In the Andes, for example, using sheep manure has, in some cases, led to an increase in potato yields in the Andes from about 2 to 5 t ha^{-1}. In other cases when farmers grow lupine as a green manure and then mix it with the soil in order to increase the amount of organic matter and nitrogen, potato yields can rise to 8 t ha^{-1} (Ruddell and Beingolea, 1999).

4.4.5   Managing Agricultural Crops to Provide Cover

Whilst the preference would be for farmers to stop burning and to use cover crops and practice conservation tillage (reduced tillage, minimum tillage and no-till), there may be rational reasons why they do not. Much can still be achieved. Ground cover with a high degree of contact with the soil surface such as plant residues, cover crops and mulching material – known as 'surface contact cover' – can protect the soil surface from the direct impact of raindrops and reduces the velocity of overland flow.

The canopy provided by crops can also protect the soil. The level of management exercised by the farmer affects the yield of a crop and simultaneously affects the amount of cover provided by crop leaves. Improvements in crop husbandry practices such as early planting and changes in crop density can reduce splash erosion and improve water infiltration by providing more soil cover (Astatke et al. 2003; Sillitoe, 1993; Gardner and Mawdesley, 1997; Douglas, 1993) (see Figure 4.11).

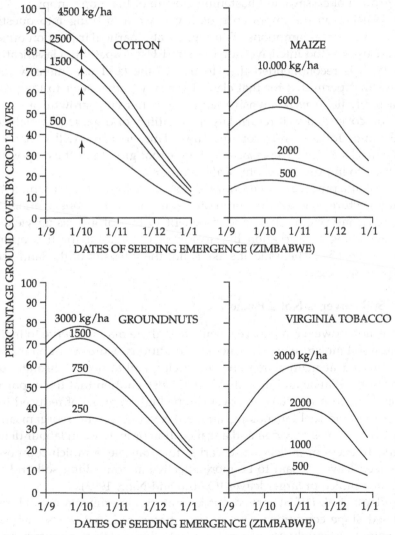

Fig. 4.11 Effect of crop management on mean seasonal percentage ground cover (from Shaxson, 1999:64)

Figure 4.9 indicates that small changes in density of cover cause big changes in erosion because erosion depends on the amount of soil exposed. If a crop is planted at a different density, after *x* weeks, the crop with density *a* may cover 60 percent of the ground, leaving 40 percent uncovered, while a crop with density *b* may cover 90 percent of the ground, leaving 10 percent uncovered. Hence, while the cover provided by density *b* is increased by 50 percent, the amount of exposed soil subject to erosion is reduced by 75 percent (Hudson, 1995:267). Meanwhile, Howeler (1987) reports that in Colombia, fertiliser application reduced soil losses due to erosion, by enhancing early plant growth of cassava and hastening closure of the crop canopy.

By focusing on the importance of cover, we also bring into question a number of other 'assumptions'. Sometimes, at a particular location outsiders see land use as 'too intensive' for the Land Use Capability classification of that site. It is recommended then to reduce the land use intensity until it matches that permitted for that class. Ironically, the answer to the problem may actually lie in *more* intensive land use. Better crop growth can provide more soil cover and will return larger quantities of organic materials to the soil through foilage and crop residues. This, in turn, will enhance soil architectural stability. The associated better root growth of the crop will also contribute to improved soil physical conditions.

Once again, practices that encourage greater yields can also promote better soil architecture that in turn, promotes soil and water conservation. Since improved conservation is an added benefit to that of increasing yields, the term 'achieving conservation by stealth' is often used for this approach (Shaxson, 1999:54). The issue at stake is not the intensity of the land use but the type of land use.

4.4.6   Soil Cover is Not a Panacea

Cover is not, however, a panacea. Soil cover alone may not serve to maintain optimum soil moisture or soil structural conditions. More attention may need to be directed at maintaining and improving soil structure throughout the rooting zone. Blaikie and Brookfield (1987) also caution that the opportunity cost of using crop residues for cover can be high in terms of reduced fodder for livestock and fuel for cooking, and that cover crops and reorganisation of crop planting times may demand a reallocation or increase in labour demand. The effectiveness of mulches also varies, for example, a mulch composed of small-sized leaves is said to be more effective at controlling soil and water loss than a mulch of larger leaves (Omoro and Nair, 1993).

De Ploey et al. (1976) carried out laboratory experiments on soil loss on simulated slope conditions. The authors discovered that coarse debris and grass covers, as roughness factors, induce hydraulic discontinuities, activate local turbulent flow and so cause slope erosion on slopes greater than 15 percent. These results have been confirmed by other laboratory research (Ligdi and Morgan, 1995). There is, therefore, a danger that soil cover may, in some cases, exacerbate soil loss rather than reduce erosion.

Mulches can also have other negative effects including disease enhancement by preserving excessive soil-moisture levels during rainy periods; preventing seedling emergence when the mulch is too thick; and, in some cases, allowing allelopathic problems (Sanchez, 1987). Plant and litter biomass on steep slopes can also trigger mass movements, as has been reported in northern Honduras where *Mucuna* spp. are used as a cover crop (Buckles et al. 1998:66). This was especially the case during Hurricane Mitch in 1998 (see Chapter 5). Humphries et al. (2000) have also reported that the productivity of the *Mucuna*/maize system in northern Honduras has been further undermined by the invasion of a persistent weed (*Rottboellia cochinchinensis*) that reduces the productivity of the system and increases labour costs.

In Brazil, no-tillage, which is associated with the use of cover crops, has led in a few areas to soil compaction, restricted root growth and reduced infiltration (Busscher et al. 1996). Although, as we will see in Chapter 5, no-tillage has in the majority of case been a great success, which accounts for the quite remarkable spread of this practice throughout parts of Brazil, Argentina and Paraguay.

4.5   LESSONS FOR OUTSIDERS

4.5.1   Working with Farmers for Better Land Husbandry

The concept of soil quality is the starting point for better land husbandry. The concept of 'husbandry' is completely meaningless without the direct involvement and active participation of farmers. Hence, better land husbandry has agro-ecological, social and economic principles. The approach is holistic, its sees farmers as part of the solution, seeks to work with them to improve soil quality and, by focusing on the integration of soil quality-enhancing activities with production farming system activities, it aims to improve productivity and achieve conservation simultaneously.

The emphasis is on biological rather than mechanical strategies and it is as much about people as it is about land. Better land husbandry is not only about soil quality and productivity, it is also about increasing the capacity of rural people to be self-reliant and resilient in the face of change, and about building strong rural organisations and economies (Pretty and Shaxson, 1998; Haigh, 2004).

The approach signifies a move away from looking at erosion in terms of 'what' is happening to questioning 'why' it is happening (Norman and Douglas, 1994:9). The loss of spaces within and between the structural units of soils and its particles is the cause of increased erosion and runoff (Shaxson, 1993). Soil erosion is a symptom of soil degradation and symptoms have causes (Norman and Douglas, 1994:45). Examining the 'why' component encourages an analysis of the agro-ecological, social and economic reasons why farmers' land management practices are leading to a deterioration of soil architecture.

Examining the 'why' component also facilitates the identification of appropriate strategies to combat soil degradation. These strategies ought to be developed in partnership with farmers as farmers are highly unlikely to change what they are doing purely for 'soil conservation'. If a farmer-first approach is adopted, it becomes clearer that the challenge is to get soil conservation integrated with plant production rather than vice versa (Shaxson, 1988).

As we saw in Chapter 2, there are host of reasons why farmers may allow their land to be degraded. For example, farmers burn their fields even though it damages the soil's organic structure and encourages erosion. They do so because it is a labour-saving way to prepare their fields for planting; because they cannot afford to buy fertiliser to ensure the higher yields that burning can give in the first one or two years; and because it helps control pests and diseases.

Better land husbandry is largely concerned with the factors that affect farming communities' interaction with the land (Norman and Douglas, 1994:133). Several problems, though, may have to be tackled before farmers are prepared to start on-farm conservation actions. Generally, these problems are the result of the agro-ecological, social and economic constraints that prevent farmers from attaining their production goals and targets. In better land husbandry, the 'land' component explains the technical aspects of improving productivity and sustainability but the 'husbandry' component, the people part, concerns the activities of smallholder farmers without whose participation there would be no husbandry.

4.5.2   Selecting Technologies and Practices

There are many actions that farmers can take to improve soil quality. To a development practitioner, it may not always be clear which practices to promote for the best. From a purely conservation-effective perspective, Table 4.3 shows a number of farming practices that, if well managed, can contribute substantially to maintaining and improving soil quality.

Table 4.3 Conservation effects of some good farming practices. Conservation effect, ✓ = yes, x = no, ( ) = possibly. From Chinene et al. 1996: 117

Recommended practice	Rain/ Wind erosivity	Soil erodibility	Infiltra- tion capacity	Soil permea- bility	Runoff control	Wind control
Timely planting for early cover and subsequent management to maintain it	✓	x	x	x	x	x
Intercropping	✓	x	x	x	x	x
Mulching	✓	(✓)	✓	(✓)	x	x
Crop rotation with grasses, biennials and perennials	✓	✓	(✓)	✓	(✓)	✓
Enriched fallows	(✓)	✓	✓	✓	✓	(✓)
Contoured strip farming	x	x	✓	(✓)	✓	x
Composting and manuring	x	✓	✓	✓	(✓)	x
Hedgerow intercropping	✓	(✓)	(✓)	✓	✓	✓
Reduced- or zero-tillage	x	(✓)	(✓)	(✓)	✓	✓
Shelter belts	✓	x	x	x	x	✓

Adopting a wider perspective, Bunch (1982:98-116) has identified a number of agro-ecological, social and economic criteria for choosing an appropriate technology or practice (see Box 4.11). These criteria provide a framework within which the appropriateness of a better land husbandry strategy can be appraised. It is clear that the better land husbandry approach with its focus on soil quality and emphasis on practices that are both productivity-enhancing and conservation-effective, meets all of these criteria. In summary the better land husbandry approach is very likely to be acceptable to farmers.

Box 4.11   Criteria for choosing appropriate technologies for resource-poor farmers (adapted from Bunch, 1982:98-116)

1. **Is the technology/practice recognised by the poorest farmers as being successful?**
   - Does the technology meet a felt need? Farmers are only likely to adopt a practice if it is seen to address a priority problem.

- Is the technology financially advantageous? Farmers are more interested in an innovation if it leads to a substantial increase in food supply or income.
- Does the technology bring recognisable success after one or two years? If not, it is unlikely to be popular with farmers.
- Does the technology fit local farming patterns?

2. **Does the technology/practice deal with those factors that most limit production?**
   - An innovation must deal with the limiting factor in the local farming system if it is to increase the system's productivity. Is the main limiting factor to productivity soil erosion or are there other more important factors, for example soil moisture?

3. **Will the technology/practice benefit the poor?**
   - Is the technology relatively free of risk? Resource-poor farmers cannot afford to take the same risk that more prosperous farmers can.
   - Is the technology simple to understand? Smallholder farmers have much empirical knowledge of soils and climate etc. A technology that makes only a few changes in traditional farming systems will be able to utilise local knowledge. In addition, farmers' self-confidence will grow when working with crops and techniques they know.

4. **Is the technology/practice aimed at adequate markets?**
   - Prices at harvest time are the only ones relevant to resource-poor farmers unless they can store the harvest. Is there a market for any surplus production?

5. **Can the technology/practice be communicated efficiently?**
   - Technologies that fail to arouse people's enthusiasm will spread only as far as the paid extension agents personally take them. If farmers are enthusiastic about a technology, its use will spread.

4.5.3   Hoodwinking Farmers?

The better land husbandry approach has not, however, gone unchallenged. Stocking (1993) poses a moral conundrum:

"Is 'Land Husbandry' past its sell-by date? The move to land husbandry was a reaction by professionals to the rejection of soil and water conservation programmes by local people. Shaxson calls it "conservation-by-stealth" meaning that we can slip our objectives of conservation by camouflaging them under rural development. Is that not sneaky? Neocolonialist? Paternalist? Bribery? Shouldn't honesty prevail?" (Stocking, 1993:25).

Better land husbandry cannot, and is not designed to tackle all facets of development. The examples in the following chapter, and a host of better land husbandry initiatives documented by Pretty and Hine (2001), however, have demonstrated that an approach that unashamedly links conservation and development can achieve much by increasing peoples' access to basic goods and services through increasing incomes and facilitating wider, more equitable social participation by vulnerable, marginal groups.

Furthermore, Bunch (1982:224) stresses that farmers see life as an indivisible whole. They intuitively know that 'development' means more than just raising productivity. With its focus on enhancing farmers' capacity for learning and action, better land husbandry can contribute to increasing natural and social capital. Pretty (1998b:84) refers to this as the 'sustainability dividend'. Few would argue that farmers should be forced to adopt certain agricultural practices or that they be prevented from seeking off-farm income-generating activities. However, any participatory approach that leads to a more sustainable agriculture, far from being 'sneaky', is overtly contributing to development and offering farmers viable alternatives.

4.6   BETTER LAND HUSBANDRY: WHY HAS IT TAKEN SO LONG?

4.6.1   Rediscovering the Principles of Agriculture?

One of the surprising aspects of better land husbandry is that it has taken so long to emerge as a coherent and holistic approach to better land management. Hudson (1995:205) writes that the components of better land husbandry have been around for a long time: soil chemists know about the importance of plant nutrients; soil physicists know about compaction, aeration and porosity; the conservation engineer knows about increasing infiltration and reducing runoff, and farmers know about soil quality.

There is a sound argument for claiming that the better land husbandry approach is little more than a rediscovery. In the late nineteenth century, Tolstoy (1887), for example, used the character of Levin in *Anna Karenin* to discuss the reluctance of Russian peasants to adopt improved farming practices. He stressed the need to recognise Russian farmers' instincts and to organise the Russian system of agriculture accordingly: *"Levin had that winter begun writing a book on agriculture, the idea of which was that the temperament of the agricultural labourer was to be treated as a definite factor, like climate and soil, and that therefore the conclusions of agronomic science should be deduced not from data supplied by soil and climate only, but from data of soil, climate, and the immutable character of the labourer"* (Tolstoy, 1887:363).

There are several reasons why a holistic approach to agriculture came to be lost and why it has taken so long for the better land husbandry approach to be articulated. These range from the direction of modern agriculture, with its focus on manipulating rather than nurturing the soil, to the rise of 'specialisation' in agricultural education. Better land husbandry requires an inter-disciplinary approach, which does not fit well with current agricultural research paradigms. Furthermore, until recently, there has been a tendency to deify the knowledge of the university-trained specialist and dismiss the practical understanding of the professional farmer as 'traditional' or 'primitive'. The missing ingredient in bringing the better land husbandry philosophy together has perhaps been the farmer-first component. It is not until the issues of soil degradation and soil erosion are looked at from the farmers' perspective i.e. with a focus on soil productivity and its sustainability that the social and technical pieces of the better land husbandry jigsaw begin to fit together.

4.6.2   The Legacy of Modern Agriculture

Archaeologists believe that humans began crop production approximately 10,000 years ago (Thurston, 1992:9). For the first 9,900 years, agriculture functioned almost entirely on the internal resources available to it and depended on natural processes and ecological relationships for its productivity (Doran et al. 1996; Hallsworth, 1987:54). These internal resources include the sun, air, rainfall, plants, animals, soil, and humans. About 100 years ago, agriculture started to move beyond its internal resources to production systems based on external inputs such as fertilisers, pesticides and antibiotics (Doran et al. 1996; Pretty, 1995:26). By relying on external inputs, modern agriculture has effectively bypassed soil biological processes (Anderson and Ingram, 1993:1). This is the reason for its modern tendency to treat the soil as a physical medium for anchoring plant roots rather than a living ecosystem and the reason why little attention has been directed at the need to replenish organic matter and maintain complex communities of soil organisms (Doran et al. 1996).

The modernisation of agriculture has led to increases in agricultural productivity. However, these have often been at the cost of increased soil organic matter loss, soil erosion and the contamination of runoff waters. Conventional soil conservation technologies are designed to deal with some of the legacies of modern agriculture and may originate from a similar mindset. However, this abandonment of ecological principles to produce human food raises difficult questions about the long-term sustainability of agriculture (Doran et al. 1996).

The alternative is to return to a greater reliance on the internal resources available to agriculture (Pretty, 1995:9; Bunch, 1982:4). One aspect of this is that, instead of injecting prescribed nutrients into the soil in order to extract desired output from it, the soil should be managed in a way that will sustain and multiply its inherent productive and recuperative capacities. Several key interacting factors that maintain or enhance soil quality such as soil cover; soil architecture; diverse biology; soil organic matter; varying use and recuperative periods; and limited tillage have been known for a long time to be beneficial to agriculture. One of the consequences of the modernisation of agriculture has been that the value of these factors, when used in combination, has been neglected while soil management has been approached in a mechanical manner. The soil has been manipulated soil rather than nurtured.

It is an appreciation of soils as 'biotic constructs favouring net primary productivity' (van Breeman, 1993) that largely distinguishes the better land husbandry approach to soil management from the soil conservation one. Better land husbandry is, in many respects, a rediscovery of the type of agriculture practised for millennia. However, better land husbandry does not demand a purist approach whereby farmers desist from using all artificial fertilisers and pesticides. Instead, it entails a "third-way", where the reliance on these inputs is reduced. This approach is more likely to complement the needs and resources of millions of resource-poor farmers worldwide.

4.6.3 Threat to Established Research and Extension Agenda

Farmers' realities are diverse and complex. Hence, any comprehensive study of farmers' decisions *vis-à-vis* land management should be holistic and interdisciplinary. Farmers' realities are such that they need assistance from 'an ecology of disciplines' working together (Pretty and Shaxson, 1998). This represents a challenge to educated professionals who are often channelled by disciplinary training into narrow ruts while farmers manage and experience the whole of their farming system, (Chambers, 1993; Enters, 1994). Educated professionals also spend too much time in offices or research stations far from and isolated from farmers' realities (Bunch, 1982:4; Chambers, 1997:31).

It is interesting that although there has been growing interest in the concept of soil as a dynamic, living and natural body (Doran et al. 1994; Doran and Jones, 1996), the development of this interest has been held back by the absence, until recently, of indicators that can be used easily to assess soil quality and an absence also of standardised sampling methodologies (Doran and Parkin, 1994). This is in sharp contrast to the situation *vis-à-vis* air and water quality.

Without reliable indicators of the impacts of different land management systems on soil quality, the concept of soil as a living self-sustaining system was more easily neglected. Sherwood and Uphoff (2000) estimate that at least 90 percent of soil research conducted in the Twentieth Century was directed at chemical or physical aspects and less than 10 percent focused on soil biology (see Plate 17). This was due, in part, to the fact that it is easier to study soil chemistry and soil physics than soil biology. In addition, while the biological health of soil in general has been neglected, the importance of the microbial component of soil is often overlooked because it is largely invisible to the naked eye (Doran and Parkin, 1994; Grossman, 2003).

Inter-disciplinary approaches, such as land husbandry, have taken time to evolve because their implementation requires a new professionalism for agricultural scientists and extension agents, a professionalism with new

Plate 17   The use of soil quality field kits has enabled scientists and farmers to assess more easily the chemical, physical and biological health of soil. Honduras. (Hellin, J.)

concepts, values, methods and behaviour (Pretty and Chambers, 1994; Pretty, 1995:203). This new professionalism demands new skills in unfamiliar areas of the social sciences as much as those in the traditional skills of agriculture and soil science. Sillitoe (1998), with reference to studies of indigenous knowledge, has questioned the feasibility of interdisciplinary work partly in terms of handling the volume of information linked to a study of this nature. The examples of successful better land husbandry initiatives detailed by other researchers (e.g. Pretty and Hine, 2001; Pretty and Koohafkan, 2002) and outlined in the following chapter, however, demonstrate that with a judicious use of different methods, and allowing for time to build up relationships of trust with farmers, interdisciplinary work is both feasible and fruitful.

5

Better Land Husbandry in Practice

"Soil conservation has a somewhat negative image for farmers, due to its implied need to control and restrict present land use so as to preserve the soil for future. The concept of land husbandry should prove more acceptable to farmers as it has a more positive image of managing and improving the present use of the land for productive purposes on a sustainable basis – that is, husbanding the soil resource" (Douglas, 1993).

5.1 INTRODUCTION TO CASE STUDIES

This chapter uses case study material to illustrate a number of better land husbandry principles. These include: soil cover (examples from South and Central America); active farmer participation (example from Honduras); and soil moisture (examples from West Africa and Kenya). It ends with a cautionary note: that despite some claims to the contrary, better land husbandry does not guarantee protection against all of the soil loss and mass movements associated with extreme rainfall events (example of Hurricane Mitch in Honduras). This chapter is not designed to present detailed and numerous examples from around the world of better land husbandry in practice. For the reader who is interested in these examples, there are a number of accessible publications, for example Pretty and Hine (2001), Pretty and Koohafkan (2002) and Shaxson and Barber (2003).

5.2 ZERO TILLAGE IN SOUTH AMERICA

5.2.1 Land Degradation and Promotion of Green Manure Crops in Santa Catarina

Santa Catarina is located in southern Brazil and covers almost 10 million hectares. Soils in the State, derived from sedimentary rocks, are generally acidic with low base saturation, high saturation of aluminium and low phosphorous availability. Average annual rainfall ranges from 1,220 to 2,280 mm and there is no typical dry season (de Freitas, 2000). Summer is from late December to late March and winter is from late July – late September. About 2 million hectares in the State are cultivated and the most common crops are maize, beans, tobacco, cassava, potatoes and onions. Despite forming just over 1.0 percent of Brazil, Santa Catarina is the fifth largest food producer among the country's 26 states. In the 1970s, however, land degradation became a real problem. See Box 5.1.

Box 5.1   Land degradation in Santa Catarina, Brazil

In the 1970s farmers switched from animal traction to the use of heavy machinery such as tractors, disc ploughs and harrows. This resulted in declining soil quality and greater land degradation. Intensive use of heavy machinery led to the breakdown of soil aggregates and a subsequent reduction in soil pore spaces. The problem was exacerbated by the use of ploughs and harrows at the same depth, leading to compacted sub-surface layers.

Furthermore farmers used to leave the soil bare during the winter period between successive summer crops. Farmers either burnt the crop residues or incorporated them into soil with ploughs and harrows. Soils were, therefore, exposed to the erosive impact of raindrops and the sun's ultra-violet rays. The result was reduced infiltration and increased runoff and erosion. Plant root systems were unable to develop in the poorer quality soil and there was a reduction in oxygen availability and soil water movement. Crop productivity declined and production could only be maintained by an increased use of mineral fertilisers.

Faced with this problem of land degradation and from the late 1970s, the State government began to promote alternative land preparation and crop management systems. The campaigns focused on the use of green manure crops and minimum- or zero-tillage systems. The use of cover crops is facilitated by the fact that high rainfall is distributed fairly uniformly throughout the year. Few farmers knew about the use of green manure crops and where they were used, the practice was to incorporate them into the soil in order to improve soil fertility and consequently crop yields. The problem with this practice was that it often left the soil surface exposed to erosive rains. An alternative use of green manure crops, as we saw in Chapter 4, is to provide soil cover and to protect the soil surface from raindrop impact.

Green manure cover crops have proved to be very popular with local farmers and are now a fundamental component of minimum- and zero-tillage systems in Santa Catarina. In Brazil, zero tillage began with one farmer in 1972 who implemented the practice on 500 hectares of land (Shaxson and Barber, 2003). By 1980, some 200,000 hectares in Brazil were under zero tillage. By 1999, the figure had risen to 9 million hectares and, by 2001, to 13 million hectares with most of it in the southerly States of Paraná, Santa Catarina, Rio Grande do Sul and Mato Grosso do Sul. Much of the spread has been farmer-to-farmer particularly through the affiliated network of 'Friends of the Lands Clubs' and the State extension services.

Extension agents worked directly with farmers in the planning and implementation of the alternative land preparation and crop management systems and, hence, gained farmers' respect and confidence. Between 1994 and 1997, there was a 5.5 fold increase in the use of these conservationist tillage systems from 124,000 to 685,000 ha. In 1999, there were about

880,000 ha. The area under conservation management practices now totals about 60 percent of the State's cropped area.

The species commonly used as green manure crops include *Avena strigosa* (black oats) and *Vicia* spp. (Vetch) during autumn/winter (sown between March and July). *Cajanus cajan* (Pigeon pea), *Canavalia ensiformis* (Canavalia), *Crotalaria* spp. and *Stizolobium* spp. (Mucuna) are often used in the spring and summer (sown between September and December). Several of these are nitrogen-fixing leguminous species and the management of these cover crops is outlined in Box 5.2.

Box 5.2   Management of green manure cover crops in Brazil

- Planted green manure cover crops and weeds are controlled by a pre-sowing application of a non-pollutant desiccant herbicide. In some cases, this is a mechanical operation involving rollers with blades.
- Plant residues are distributed evenly and left on the soil surface, so protecting the soil from the direct impact of raindrops and the sun's rays.
- A specialised planter or drill cuts through the desiccated cover and residues that have accumulated on the soil surface, and sows seed (and fertiliser) directly into the soil. In some cases an animal-drawn tool is used to clear a narrow furrow in the cover into which the seed is sown. In both cases there is no need to use a plough.
- If the accumulated crop residues are not sufficient to smother weeds, subsequent weed control is carried out with some pre- but mostly post-emergence herbicides.
- Crop rotation is used in order to promote adequate biomass levels of permanent mulch cover, assist in control of weeds and pests and diseases, recycle nutrients and ameliorate soil physical conditions.

According to Shaxson and Barber (2003), the distinguishing principles of this type of 'conservation-farming' are:
- No mechanical damage of the root-zone
- Permanent soil cover provided by plant materials—residues of the previous crop, the growing cover crops and also the cover crops once desiccated
- Rotation of crops—both harvested and soil-improving green manure crops-including legumes for nitrogen-fixation and when appropriate, grasses for forage.

5.2.2   The Spread of Zero-tillage in Paraguay

The introduction of soybeans to the southern and eastern parts of Paraguay in the early 1970s, followed by wheat in the mid-1970s, was accompanied by the use of conventional mechanised soil preparation with disc ploughs and harrows. As in Santa Catarina, this lead to soil degradation. Following the example of the southern Brazilian states of Parana, Santa Catarina and Rio Grande, zero-tillage along with crop rotation practices were introduced to Paraguay in the late 1970s.

The area where zero-tillage was practised rose from 20,000 hectares in the early 1990s to 250,000 by the mid-990s. The Paraguayan government continued to promote the use of zero-tillage in the main grain producing departments of the country. By 1997, some 480,000 hectares was under zero-tillage. This figure represents approximately 50 percent of the total area cultivated mechanically in Paraguay.

5.2.3   Benefits of Minimum and Zero-tillage Systems in Brazil and Paraguay

The use of green manure cover crops have proved popular with local farmers in Brazil and Paraguay because of improvement of the soil resources, hence higher yields, along with lower costs and improved returns. Minimum- and zero-tillage systems have offered the possibility to diversify the farming system, reduce production costs and eliminate dependency on hired labour and machinery (see Box 5.3). There have, however, been some reports of greater pest and disease problems and, hence, a greater need for insecticides.

Box 5.3   Benefits of minimum- and zero-tillage systems to farmers and society in general (based on de Freitas, 2000; Sorrenson, 1997; Shaxson and Barber, 2003)

On-farm benefits
- Increased levels of soil organic matter in the upper layer of the soil along with increased biotic diversity (earthworms, fungi, bacteria)
- Better soil structure and stability of soil aggregates leading to higher infiltration rates
- Reduced soil erosion (by over 80 percent) and runoff (by over 50 percent)
- Increase in nutrient levels resulting in less fertiliser needed for same result
- Less variation in yields from year to year
- Better resilience of crops in drought conditions due to increased water-holding capacity of soil resulting in greater food security
- Economies of 10-20 percent in use of water in irrigation
- Higher sustainable yields - maize > 20 percent, beans > 37 percent, soya > 27 percent, and onions > 26 percent - resulting in less pressure to open new land
- Reduced costs (after only two years) for labour, fuel and machinery hours. For example, fossil fuel use reduced by 40-70 percent
- Heightened environmental awareness among farmers

Off-farm benefits
- Conservation of bio-diversity in soil, terrestrial and aquatic fauna and flora
- Reduced flooding risk due to greater rainfall infiltration
- Less sedimentation and infrastructure damage, e.g. silting of waterways and large dams
- Savings of up to 50 percent in costs of maintenance and erosion-avoidance on rural roads
- Zero-tillage has also had a major impact in reducing carbon emissions by comparison with conventional cultivation, by immobilising carbon in incremental soil organic matter and surface residues. This is estimated to be about 1 tonne of carbon fixed per hectare
- Virtual elimination of pollution of water by soil-applied chemicals and, hence, lower treatment costs for domestic water from surface water sources
- Elimination of cultivation-induced dust clouds in towns and cities

From the farmers' perspective one of the most important reasons for using a minimum- or zero-tillage system is that it leads to increased yields. Experiments conducted in Brazil over a six-year period in the late 1970s and early 1980s compared the yields from conventional tillage and zero-tillage. The results in Tables 5.1 and 5.2 demonstrate substantial increases in wheat and soya yields under zero-tillage systems. Meanwhile, Table 5.3 shows the increase in numbers of earthworms under zero-tillage and conventional cultivation systems in the State of Parana in Brazil.

Table 5.1 Yields of wheat, averaged across rotations, under two different soil-preparation methods, at Londrina, Brazil 1978-1983 (Derpsch et al. 1991, quoted in Shaxson and Barber 2003:91)

	WHEAT			
Year of harvest	Conventional cultivation with disk equipment		Zero Tillage	
	Tonnes/hectare	Relative	Tonnes/hectare	Relative
1978	1.36	100	1.81	133
1979	1.60	100	1.84	115
1980	2.25	100	1.97	87
1981	0.72	100	1.12	156
1982	0.39	100	0.86	220
1983	1.72	100	1.98	115
Mean yield	1.34	100	1.60	119

Table 5.2 Yields of soya, averaged across rotations, under two different soil-preparation methods, at Londrina, Brazil 1978-1983 (Derpsch et al. 1991 quoted in Shaxson and Barber 2003:91)

	SOYA			
Year of harvest	Conventional cultivation with disk equipment		Zero tillage	
	Tonnes/hectare	Relative	Tonnes/hectare	Relative
1979	1.43	100	1.99	139
1980	2.51	100	3.09	123
1981	2.03	100	2.86	141
1982	1.34	100	2.03	151
1983	1.45	100	1.90	131
1984	1.60	100	2.00	125
Mean yield	1.73	100	2.31	134

Table 5.3 Influence of different methods of soil preparation on population of earthworms, Parana, Brazil (FAO, 2001b, quoted in Shaxson and Barber 2003:04)

	Number of worms per sq. m to 30 cm depth	Number of worms per sq. m to 10 cm depth
Soil type	'Latossolo roxo'	'Terra roxa estruturada'
Zero Tillage	27.6	13.0
Conventional disc-plough & harrowing	3.2	5.8

In Paraguay, differences in crop yields and per crop fertiliser and herbicide usage (most significant items of farm costs) were studied on farms under both conventional cultivation and zero-tillage. Based on actual farmer experience, crop yields under conventional cultivation decline over a period of about 10 years by 5-15 percent, while over the same period and under zero-tillage, they increase by 5-20 percent. These trends in crop yields in Paraguay impact strongly on farm incomes (see Box 5.4). Furthermore, under zero-tillage there are also savings in herbicide and fertiliser inputs. These, in turn, impact on farm variable costs and profits.

Box 5.4   Zero-tillage, crop yields and farm income in Paraguay (based on Sorrenson et al. 2001)

In the case of one farmer in Paraguay, after six or seven years of zero-tillage, soybean yields averaged 2,780 kg/ha in an area that under conventional cultivation used to yield 800 kg/ha. Maize planted with the cover crop *Mucuna* spp. produced yields of 3,500 kg/ha. This is similar to levels achieved on virgin land following deforestation.

> After six-seven years of zero-tillage net farm income has almost doubled compared to conventional cultivation. Returns to labour are also higher under zero-tillage due to increased net farm incomes being achieved with lower labour inputs than conventional cultivation.

5.2.4 Obstacles to Adoption of Zero-tillage by Smallholder Farmers

Experience with green manure cover crops in Santa Catarina and other states in Brazil and Paraguay has shown that it is not enough solely to introduce new agricultural practices such as the use of green manure crops and minimum tillage. Despite the advantage of zero-tillage there are many obstacles to smallholder farmers adopting the technology. See Box 5.5. While zero-tillage has spread on tractor-mechanised farms, it has hardly reached small farms.

Box 5.5   Costs and inconveniences of zero-tillage (based on Sorrenson et al. 2001)

- Shortage of information on how to zero-tillage (hence government-funded extension services needed).
- A more complicated system of production that necessitates technical assistance.
- Access to long-term credit so farmers can get hold of manual and animal traction zero-tillage machines.
- Access to green manure crop seeds.
- Farm 'looks' untidy with many 'weeds'.

In order to meet some of these challenges and under the auspices of the World Bank, a Santa Catarina incentive fund, PROSOLO, was established to encourage farmers to work together and operate equipment and facilities collectively (World Bank, 2000). These incentives are justified where on-farm changes generate public goods such as decreased soil loss, control of pollution and other environmental benefits. The extension agents were, in turn, trained in more participatory approaches. One of the main constraints to rapid adoption of zero-tillage in Santa Catarina (and Paraguay) was the unavailability of the necessary machinery. One consequence of the project was that some 60 private companies started to manufacture minimum and zero tillage machinery for land preparation and sowing.

Low adoption rates of zero-tillage by smallholder farmers is a concern in countries such as Paraguay. Here, there are almost 250,000 small farms with fewer than 20 hectares. Whilst these farms occupy only 6 percent of the country's agricultural area, they generate 35 percent of the sector's output. The agricultural sector is the backbone of the Paraguayan economy. Paraguayan smallholder farmers often see the advantages of zero-tillage but can not afford the seeds of green manure crops or the purchase or rent of animal-traction zero-till machinery.

To encourage smallholder farmers, it may be necessary to incorporate loan repayment delay and loan write-off clauses (to cover possible crop failures due to circumstances beyond the control of the farmers, such as severe drought or pest attack), to eliminate the risk of financial ruin. It may also be necessary to improve farm administration, add value to harvested crops, identify alternative markets and diversify agricultural production. Farmers may also need to learn new crop management skills. As a result, participatory research and development, training and extension are needed to speed up the learning of these new skills.

In summary, adoption rates among smallholder farmers will require important policy and institutional reforms and considerable government financial support. Where Paraguayan farmers have adopted zero-tillage, they have received high-level technical assistance, initial stocks of green manure crop seed and free zero-till machinery and equipment (see Box 5.6). Furthermore, potential adopters have often had access to technical assistance and long-term credit at affordable rates in order to purchase a minimum of equipment and machinery.

Box 5.6   Reasons for popularity of zero-tillage among smallholder farmers in Edelira district, Paraguay (based on Sorrenson et al. 2001)

- Enthusiasm and dedication of a handful of extension agents.
- A small group of farmers interested in and convinced about the value of green manure crops and zero-tillage.
- Valuable technical and financial support provided by the MAG-GTZ Soil Conservation Project (under zero-tillage, net income in year 1 is likely to be negative because of high costs of fertiliser and seed).
- Market advice.
- Grain storage facilities (so farmers can store grain and sell it at more opportune periods).
- Formation of a zero-tillage small farmer association that was equipped (free of charge to the farmers) with a full complement of manual and animal-driven zero-tillage equipment.

5.3   SOIL COVER IN HONDURAS

5.3.1   Characteristics of the Quezungual System

Several of the indigenous agroforestry systems found in Central America have been documented   (Budowski, 1987; Kass et al. 1993; Current et al. 1995). A particularly interesting one is the Quezungual System that is used by smallholder farmers in western Honduras (Hellin et al. 1999) (see Box 5.7). The most distinct feature of the system is the existence of naturally regenerated and pollarded shrubs and trees, in association with more traditional agroforestry components such as high-value timber and fruit trees. Farmers' plots, therefore, have three levels: trees; pollarded trees and shrubs; and agricultural crops. A typical plot is approximately 600 m^2 with numerous pollarded trees and shrubs, and approximately 15 to 20 large trees: found on slopes from 5 percent to 50 percent, but more commonly on slopes of 10 percent to 25 percent.

Box 5.7   Department of Lempira, Western Honduras (from Hellin et al. 1999)

The Quezungual System is found in the southern part of the department of Lempira at altitudes between 200 and 900 metres above sea level. The region is mountainous and the vast majority of farmers cultivate steep slopes of 5 percent to 50 percent. Although the area has suffered from much land degradation, there is still a relative abundance of forest and trees in the landscape.

Annual precipitation varies from 1400 mm to 2200 mm per annum and the rainy season lasts from early May to the end of October. Many of the farmers are smallholders with land holdings of less than 2-3 ha. The traditional crops in the area include: *Zea mays* (maize), *Sorghum bicolor* (sorghum) and *Phaseolus vulgaris* (beans). Some farmers also grow *Oriza sativa* (rice), *Citrullus vulgaris* (watermelon), and *Saccharum officinarum* (sugarcane).

The system operates as follows: when a plot is cleared, farmers leave a few higher-value tree species and the remaining shrubs and trees are pollarded. Farmers manage the natural regeneration so as to achieve an ideal tree density for the type of crop being grown. The higher-value tree species most commonly retained are fruit trees such as *Byrsonima crassifolia* (nance) and *Psidium guajava* (guayabo), and timber species such as *Cordia alliodora* (laurel), *Diphysa robinioides* (guachipilin) and *Swietenia* spp. (caoba). Some of these, particularly the fruit trees, are also planted. Other tree and shrub species are pollarded in the dry season in order to reduce the risk of pests and diseases (Hellin et al. 1999; Welchez and Cherrett, 2002).

Three crops are traditionally found in plots with the Quezungual System: *Z. mays* (maize), *S. bicolor* (sorghum) and *P. vulgaris* (beans). Crops are rotated in order to reduce the incidences of pests and diseases. Practitioners of the system tend not to burn their fields prior to sowing. Instead, vegetation is cleared by hand with a *machete* although some farmers use a herbicide such as paraquat.

Seeds are either scattered (this is particularly common with *Phaseolus vulgaris*) or farmers make a small hole in the soil and mulch of cut vegetation, and sow a few seeds per hole. This is the most common way of sowing *Zea mays* and *Sorghum bicolor*. Farmers manage the trees and shrubs to ensure that there is optimum shade for the agricultural crops. Pollarded and pruned material is spread throughout the plot as a mulch, although woody material, which is suitable for firewood, is collected (see Plates 18 and 19).

Plate 18   Plate 19

Plates 18 and 19 The Quezungual agroforestry system in Honduras provides plenty of soil cover. This protects the soil against intensive rainfall and contributes to improved soil quality. (Hellin, J.).

5.3.2   Reasons for Adopting the Quezungual System

The majority of smallholder farmers adopt the Quezungual System in order to meet household subsistence needs for fruit, timber and firewood, as well as for basic grains. Farmers also take advantage of market opportunities once immediate household subsistence needs have been met. Fruit trees, particularly, are seen as a non labour-intensive cash crop. The advantages of the system most frequently mentioned by farmers are outlined in Box 5.8.

Box 5.8   Advantages of the Quezungual system (from Hellin et al. 1999)

- Agricultural production is greater due to increased levels of organic matter from the pollarded material. More organic matter means better soil structure and higher soil moisture levels.

- Mulch provided by the pollarded material helps protect the soil surface from intense rainstorms and there is less soil erosion.
- Farmers obtain firewood and fruits from the trees and shrubs. Occasionaly a timber species is also felled for house construction.
- Establishment and maintenance of the Quezungual system does not require much labour.
- Plots with the system can be cultivated for longer periods than is the normal practice before they have to be left in fallow.

Due to its many advantages, the system has spread from farmer to farmer, especially since the early 1980s when the Government of Honduras initiated a programme to encourage farmers not to burn their fields. However, the Quezungual system is actually uncommon in areas where land is still sufficient for farmers to practice a slash and burn agriculture. Those practising the system tend to be smallholder famers with fewer than 2.5 hectares of land living in areas where land scarcity has forced them to intensify agricultural production. Farmers may, therefore, be reacting to land scarcity in ways described by Boserup (1965).

The fundamental aspect of the system's popularitiy is that, although it contributes to soil conservation, practitioners do *not* see the system as a soil conservation practice *per se* – the retention of humidity and sustainability of the system are seen as far more imporant attrributes. This supports the argument that farmers will more readily adopt a practice when it addresses their main concern, namely increased production, and that the most effective way of doing this is to focus on improving soil quality.

In the case of the Quezungual system, the cover provided by the pollarded material and the cessation of the practice of burning means that levels of organic matter in the soil increase, there are greater numbers of soil organisms, and the mulch protects the soil from erosive rain. Consequently there is much less soil erosion and (more importantly for the farmer) increased and more stable and reliable yields.

5.4   FARMER PARTICIPATION IN HONDURAS

5.4.1   Land Degradation and Out-migration

Much has been written about the Integrated Development Programme in Güinope (1981-89), which was a collaborative effort between World Neighbors, the Honduran Ministry of Natural Resources and the Association for the Co-ordination of Development Resources (ACORDE) (e.g. Bunch, 1999; Pretty, 1995:220). Prior to the arrival of World Neighbors, the Güinope region suffered from land degradation and rural poverty was endemic (see Box 5.9).

Box 5.9   Güinope region in southern Honduras

The Güinope region in southern Honduras lies between 500 and 1800 metres above sea level and covers an area of 204 km². The population is approximately 5,500 of which 80 percent is engaged in agriculture. Annual rainfall is 1100-1300 mm and the main rainy season is from May to October. An impenetrable subsoil underlies the 15-50 cm deep top soil and farming takes place on slopes which are often 15-30 percent (Bunch and López, 1999).

These slopes are subject to severe erosion. As a result of land degradation and land shortages, farmers were being forced to cultivate the same plots of land on an increasingly continuous basis. Maize yields were declining and in some case were approximately 400 kg ha⁻¹ (Bunch, 1988). There was much out-migration and many of those who stayed began to work off-farm, hiring themselves out as day labourers to saw mills, sugar cane and coffee plantations.

The philosophy underpinning the project, based on a better land husbandry approach to land management, was succinctly outlined in *Two Ears of Corn* (Bunch, 1982). The focus was on the use of conservation-effective and productivity-enhancing technologies and practices. However, rather than relying on technology transfer, the project emphasised the strengthening of local capacity to innovate and generate solutions to farming problems. Furthermore, it was one of the first efforts in Latin America genuinely to employ villagers as the principal agents of change (Sherwood and Larrea, 2001).

World Neighbors worked in 41 communities in Güinope. World Neighbors recognised from the beginning of the programme that there was a need to link conservation and productivity (Bunch and López, 1999). The use of organic fertiliser (chicken manure) and some chemical fertilisers meant that maize yields increased almost immediately, in some cases threefold (Pretty, 1995b). The programme emphasised the value of organic matter and the importance of soil organisms, such as earthworms, in terms of improving the quality of the land.

Soil conservation technologies were largely unknown in Güinope when World Neighbors started work. World Neighbors promoted a number of technologies including live barriers. The wider objective of World Neighbors' work was, however, human development in terms of the sustainability of the development process via a strengthening of local farmer innovation and leadership capacity (Bunch, 1999; Johnson et al. 2001; Sherwood and Larrea, 2001).

5.4.2   Soil Quality and Increasing Productivity

Active farmer participation in the programme was largely achieved by ensuring that the programme addressed farmers' priorities. A key feature was the recognition that any intervention had to address, almost immediately, the limiting factor in local agricultural production. This was identified as declining soil quality (Bunch, 1988). World Neigbors recognised that it was an increase in yields that would convince farmers of the value of soil protection. Furthermore, it was an increase in yields that would convince farmers to continue to experiment and continue to become actively involved in the development process (Bunch, 1999).

World Neighbors promoted productivity-enhancing and conservation-effective practices such as the use of chicken manure as a natural fertiliser and cover crops, along with technologies such as live barriers. This led to rapid improvements in soil quality and subsequently to higher productivity. Within the first year of participating in the programme, 1,500 farmers in 41 villages tripled maize yields and in some cases increased them by 7-8 fold. Production of beans also increased (see Table 5.4).

Table 5.4 Productivity of maize and beans (*Phaseoulus vulgaris*), in 100 kg ha^{-1}, at three villages in Guinope, at programme initiation in 1981 and in 1994, five years after programme termination (after Bunch, 1999)

Village	Pacayas	Manzaragua	Lavenderos
Maize			
1981	6	6	6
1994	42	20	20
Beans			
1981	2	n.a.	12
1994	8	n.a.	15

Having ascertained that there were available markets, World Neighbors encouraged farmers to diversify production away from the basic grains and towards a combination of basic grains and vegetables. Farmers readily took advantage of this market opportunity.

"In terms of agriculture, this was really a zone where we grew only maize. We hadn't discovered vegetables. It was only when World Neighbors arrived that they opened our eyes to the possibilities, the fact that we could actually grow vegetables in this zone. World Neighbors showed us we had been sleeping on top of this huge potential without realising that it was there" (Farmer in Güinope, Honduras based on a semi-structured interview carried out by the author in 1997).

5.4.3 Farmer First: Active Farmer Participation

Sustaining a productive agriculture requires a constantly changing mix of techniques and inputs. The most effective way forward is to encourage processes by which people can develop their own agriculture (Bunch, 1989).

All too often, resource-poor farmers can fall into a state of fatalism or learned helplessness that engenders defeatist attitudes and leads them to believe that they are not capable of making effective changes in their lives (Sherwood and Larrea, 2001). A key component of the World Neighbors' programme was active farmer participation in the development of resource-conserving practices. The main goal being to work towards a system where farmers had a strong motivation to innovate and work for the common good (Bunch, 1999).

World Neighbors' sought active farmer participation at all stages of the programme. All forms of paternalism were avoided, including giving things away or subsidising farmers' activities. Local people participated in planning and implementation and technologies selected were appropriate to the region. Activities started on a small scale and focused on a handful of technologies and practices that were finely tuned through experimentation by and with farmers (Pretty, 1995:220). The emphasis was practical rather than theoretical (Bunch, 1988).

Farmers' enthusiasm and participation was further encouraged by the practice of extension and training being carried out by local extension agents whose demeanour earned the respect and friendship of local farmers. Farmers' respect for extension agents is not always the case in development projects. Farmers in Güinope complained that extension agents, associated with other projects in the area, were condescending, arrogant, and seemingly disinterested in farmers' plight. Farmers stressed that that this was a disincentive to meaningful collaboration (see Box 5.10).

Box 5.10   Behaviour of extension agents can be a key factor in determining local farmers' participation (from Hellin, 1999b)

While farmers in Güinope, Honduras praised the extension agents working with World Neighbors, they were very critical of extension agents working with other organisations in the area.

"Most of the extension agents (working with other organisations) *seldom come and visit our plots. They usually just come up to the entrance to the house and sometimes they do not even get out of the car, they ought to work with us in our fields like the people from World Neighbors"* (farmer, Güinope).

"If an extension worker says that he is going to arrive at such a time on such a day, he should be there. Otherwise we begin to loose patience and whatever group feeling there was begins to disintegrate" (farmer, Güinope).

Farmers commented that if extension agents were perceived to be arrogant and condescending then farmers were minded to ignore them irrespective of whether the extension message was important or not. Farmers added that they were far more disposed to follow extension advice if the extension agents treated them with respect.

World Neighbors, on the other hand, trained over 50 local farmer extension agents. Local farmers' appreciation of and admiration for World Neighbors' extension agents is evident throughout the area. Furthermore, instead of simply demonstrating the benefits of the technologies and practices, World Neighbors sought to strengthen farmers' capacity to understand agro-ecological principles and to discover for themselves why some practices worked and others did not.

Approximately 80 percent of the work that World Neighbors carried out was practical and only 20 percent was theory. This contrasts with many development projects where the balance is often the other way around. Farmers in Güinope continue to talk favourably of the World Neighbors' work in terms of extension agents working alongside farmers in their fields while together they sought practical solutions to farmer-perceived problems rather that simply adopting prescribed technologies. A farmer who the author interviewed summed up the success of World Neighbors' approach:

"The best way to teach farmers is the way used by World Neighbors. The organisation worked with a few farmers and when others came to ask about the work, World Neighbors said that they would teach the farmers but only in the farmers' fields" (farmer in Güinope Honduras based on a semi-structured interview carried out by the author in 1997).

Farmer confidence grew through active participation in the programme. As we will see below, farmers began to innovate and adapt the promoted technologies so that they better complemented their farming systems (Bunch and López, 1999. Farmer confidence grew as they saw the benefits of their own labour and recognised that they had the ability to manage their land in a more productive and sustainable manner. As one farmer said:

"Once you start receiving training, you find that other doors open. All training is important, some doors that were half closed suddenly open" (farmer in Güinope, Honduras based on a semi-structured interview carried out by the author in 1997).

5.4.4 Farmer Adaptation and Innovation

The goal of the World Neighbors' programme was to build a self-sustaining process of farmer experimentation, adaptation and diffusion of practices and technologies, leading to a sustainable agriculture. Active farmer participation led to farmer empowerment and the confidence to experiment and adapt to changing circumstances. There is much evidence to suggest that farmer experimentation and innovation has continued since the end of the World Neighbors' project in 1989 (Bunch and López, 1999; Hellin and Larrea, 1998). A farmer the author interviewed in 1997 succinctly summed up the positive change:

"While there is no change of mind and heart, a farmer is not going to change his way of farming. And he can only change when he sees that land can be improved. We didn't have this vision that we could improve the land. Once World Neighbors arrived we realised that with the technologies that we had learned from them, the more we farm and work the land, the more it improves, and the yields are better. Basically if you treat the land well, it will go on improving" (farmer in Güinope, Honduras based on a semi-structured interview carried out by the author in 1997).

Throughout Güinope, farmers have adapted the green manure and cover crop technology. They use a variety of species including food crops such as lablab beans (*Dolichos lablab*) and fava beans (*Vicia faba*); animal feed such as sweet clover (*Melilotus albus*) and species such as velvet beans (*Mucuna pruriens*) to control weed problems (Bunch, 1999). The green manure and cover crops are also used to generate large quantities of organic matter, which is incorporated into the soil surface and which leads to an improvement in soil quality and increased yields.

One of the most interesting innovations has been the way that farmers have adapted the live barrier technology originally introduced by World Neighbors. Sergio Larrea, a Bolivian agriculturist studying in Honduras, and the author documented this adaptation process (Hellin and Larrea, 1998). World Neighbors originally promoted two grass species for use in live barriers: *Pennisetum purpureum* (Napier Grass) and *P. purpureum* x *P. typhoides* (King grass). Farmers are now using 19 species in live barriers and many farmers use a mixture of several species in the same live barrier (see Table 5.5).

Table 5.5   Species being used in live barriers in the Güinope region. Species marked* have products that can be both consumed and sold (from Hellin and Larrea, 1998)

Scientific name	Common name (in Spanish)
Grass species	
Vetiveria zizanioides	Zacate valeriana
*Cymbopogon citratus**	Zacate limón
Pennisetum purpureum	Zacate napier
Pennisetum purpureum x *Pennisetum typhoides*	Zacate king grass
Setaria geniculata	Setaria
Panicum maximum	Zacate guinea
*Saccharum officinarum**	Caña de azúcar
Brachiaria mutica	Pasto pará
Paspalum notatum	Pasto bahía
Trees, shrubs and other plants	
*Cajanus cajan**	Gandul
*Manihot esculenta**	Yucca

(Contd.)

(*Contd.*)

*Coffea arabica**	Café
*Citrus limetta**	Lima
*Citrus sinensis**	Naranja
*Prunus persica**	Durazno
Gliricidia sepium	Madreado
Colocasia esculenta	Quíscamo
*Musa acuminata**	Guineo
*Ananas comosus**	Piña

Of the 68 farmers in the municipality of Güinope who Sergio interviewed, 63 had established live barriers between 1980 and 1996 and only three had subsequently rejected them. From 1980-1996, the majority of farmers established live barriers of *P. purpureum* (Napier Grass) or *P. purpureum* x *P. typhoides* (King grass). However by 1988, farmers were establishing more live barriers of other species (Figure 5.1). This trend has continued to the extent that in the period 1992-1996, farmers established five live barriers of *P. purpureum* (Napier Grass) or *P. purpureum* x *P. typhoides* (King grass) compared to 18 live barriers of other species.

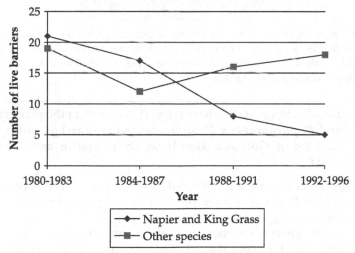

Fig. 5.1 Number and types of live barriers established by 63 farmers 1980-1996 (from Hellin and Larrea, 1998)

Farmers recognised that live barriers of *P. purpureum* (Napier grass) and *P. purpureum* x *P. typhoides* (King grass) retain soil and that over a period of 5-10 years natural terraces are formed. They also pointed out, however, that there are several disadvantages with the live barriers of King and Napier grass (see Box 5.11 and Plate 20).

Plate 20   Live barriers of *Pennisetum purpureum* (Napier grass) and *P. purpureum* x
P. typhoides (King grass) can cause problems for farmers by invading their
fields, if they are not properly managed, and competing with agricultural
crops. Honduras. (Hellin J.).

**Box 5.11   Disadvantages of live barriers of *P. purpureum* (Napier grass)
and *P. purpureum* x *P. typhoides* (King grass) according to
farmers in Güinope, Honduras (from Hellin and Larrea,
1998)**

- Both species are invasive if not regularly managed. This is especially
  the case with *P. purpureum* (Napier grass) and precludes the use of
  both species as a green manure.
- Both species provide excellent fodder but few farmers in Güinope
  have cattle and therefore there is little demand for the amount of
  fodder produced.
- Both species have an extensive root system and therefore compete
  with agricultural crops. There are several examples in the Güinope
  region where the maize plants on either side of the grass barriers
  are stunted compared to the maize between the live barriers.

The live barrier species now being selected by farmers tend to be those
that contribute to the farm household in terms of domestic consumption

and/or the sale of the products of the live barriers. The opportunity cost to the farmer, if land is taken out of production and/or the barrier does little to improve agricultural productivity, is obviously less if the barrier itself is productive i.e. if the products of the barriers could be consumed or sold. Farmers are increasingly using *S. officinarum* (sugar cane) in live barriers and a number of fruit trees including lemon, orange and pear trees (Table 5.5) (Plate 21).

Plate 21   Farmers may show a preference for live barriers that contribute to the farm household in terms of domestic consumption and/or the sale of products. In this case, a farmer has established a barrier of banana plants. Honduras. (Hellin J.).

Farmers are interested in the ability of the live barriers to retain soil but many of the species being used in live barriers are not as effective as Napier and King grass in controlling soil erosion. Local farmers prefer these other species, however, because they make far greater contributions to the farming system than the two grass species because the fruits can be consumed or sold. This type of farmer adaptation has been documented in the Philippines (Fujisaka, 1993), Jamaica (McDonald et al. 1997); Colombia (Ashby et al. 1996) and parts of China, Thailand, Indonesia and Vietnam (Howeler, 1995). More recently, TangYa et al. (2003) have documented the popularity of live barriers of mulberry (*Morus alba*) in China.

Farmers became aware of the importance of soil conservation; they learned useful ideas and practices and developed analytical skills, and they were motivated to take action. If the World Neighbors programme had adopted a transfer-of-technology approach without addressing essential aspects of human development, it is very likely that many farmers would have abandoned the live barrier technology instead of experimenting and improving the technology via the incorporation of additional species (Sherwood and Larrea, 2001). The adaptation of the live barrier technology is a striking contrast to the abandonment of live barriers shown in Figure 2.2 in Chapter 2.

The adaptation of live barriers supports the better land husbandry argument that farmers are primarily concerned with stable and economic production and not with the conservation of soil and water *per se*. It also highlights the importance of collaborative work with farmers in analysing individual farming systems, identifying farmers' needs and selecting appropriate technologies.

5.4.5   Human Development

Sherwood and Larrea (2001) write that it behoves us as researchers and development professionals, concerned about the rural condition, to understand more fully the successful examples development programmes such as World Neighbors' work in Güinope, and to utilise lessons in the design of future interventions. World Neighbors work in Güinope adopted a better land husbandry approach with a focus on active farmer participation and an improvement in soil quality. Farmer innovation continues (Johnson et al. 2001). Güinope, therefore, provides many lessons for designing and implementing better land husbandry programmes.

The Güinope case study exemplifies the degree to which participatory approaches to better land management, where the focus is on conservation-effective and productivity-enhancing practices, can contribute to increased natural, human and social capital (see Box 2.2) and ultimately to on-going farmer experimentation and innovation. Bunch (1999) concludes that one of the most significant outcomes of the programme was that it did not attempt to identify a group of technologies whose use would be somehow sustained. Rather, the objective was to choose a small number of technologies and practices that motivated farmers to become involved in a process of innovation, to search for and experiment with new ideas, adopt those that prove successful and share the results with other farmers.

Natural, human and social capital have undoubtedly increased in Güinope. A majority of men and women in those villages where World Neighbors worked stress that there has been an improvement both economically and in terms of quality of life (Johnson et al. 2001). As such, farmers' priorities and needs have changed. Farmers say that, before the World Neighbors' programme in the 1980s, the greatest obstacle to enhanced and sustained agricultural productivity was 'knowledge' on how to manage land. Farmers now say that they now have that 'knowledge' (and the self-confidence that accompanies the learning process) and that there are now other obstacles to manage. These include recent government 'modernisation' policies that threaten community livelihoods, specifically access to affordable credit and greater access to markets (Hellin, 1999b; Larrea, 1997:vii; Sherwood and Larrea, 2001).

The development process continues in Güinope. Farmer leaders in the area stress that farmers need practical responses to immediate farming demands, but they also need to develop broader learning and action skills that enable them to deal successfully with the multitude of challenges before them. Farmer leaders argue that rural development initiatives should continue to look for ways of strengthening critical thinking, organisation abilities, and other skills needed to participate in modern society, in both agronomic and non-agronomic areas (Sherwood and Larrea, 2001). Better land husbandry does not have all the answers to farmers' problems but it does provide the foundation for future development.

5.5 SOIL MOISTURE AND YIELDS IN AFRICA

5.5.1 Differing Perspectives

As seen in earlier chapters, the conventional soil and water conservation approach has tended to focus on combating soil erosion. The argument is that agricultural yields will rise as erosion is brought under control. The reality, as outlined in Chapter 4, is that agricultural yields are determined more by the quality of soil than the quantity of soil lost through erosion. Improving soil quality, so that the soil provides a better environment for root development, is a fundamental component of the better land husbandry approach. The better the quality of soil the more effectively agriculture crops can utilise soil moisture.

Soil moisture is the hidden and undervalued resource that, by sustaining plant growth, holds the key to the sustainable improvement of an increasingly dire situation of food production and water supplies. All too often poor agronomic practices lead to a deterioration in soil quality and a reduction in available soil moisture. Climatic drought is unavoidable but to worsen it by inducing avoidable runoff and wasting rainfall – as potential soil moisture and possible groundwater – by poor husbandry of the land is inexcusable. Evidence, from parts of Africa, clearly shows the advantages of a better land husbandry as opposed to a conventional soil conservation approach, and what is more demonstrates that such an approach is popular with local farmers.

Between 1968 and 1973 the Sahel was hit by several consecutive drought years. Subsequent soil conservation projects in the 1970s and 1980s focused on the need to control soil erosion. Very little attention was directed at conserving water (Hassane et al. 2000). As a result, farmers showed little interest in participating actively in the projects. Soil conservation practices were established as part of food-for-work programmes but were subsequently abandoned when the projects finished. A new approach was needed, one that changed the focus from soil erosion control to water harvesting and soil fertility management. Encouraging results have come from projects in Niger, Mali and Burkina Faso (Hassane et al. 2000; Wedum et al. 1996 and Ouedraogo and Kaboré, 1996).

5.5.2 Niger

The International Fund for Agricultural Development (IFAD)-funded Soil and Water Conservation Project in Niger's Illéla District ran from 1988 until

1997. It focused on an area with 77 villages and approximately 100,000 inhabitants (Hassane et al. 2000). The area is characterised by low and highly variable rainfall, by land degradation, and by seasonal labour migration to countries to the south. Rainfed agriculture is the dominant mode of production and the main crops are millet, cowpea and sorghum. Rainfall is so low and variable that in three out four years there is insufficient for a good crop.

The project promoted simple and easily replicable technologies and practices. Two of the most popular practices were the use of improved planting pits known as *tassa*, and half moons known locally as *demi-lunes*. These practices are not soil conservation ones *per se*, rather the emphasis is on water harvesting, where the focus is on the collection and concentration of rainfall runoff for improving plant production in arid and semi-arid regions (Critchley et al. 1994). Enlarged planting holes or pits are a feature of certain relatively flat semi-arid regions of West Africa.

The data from the project demonstrated that water harvesting led to a substantial increase in yields in dry years as well as years with good rainfall. The project collected data over a seven- year period (1990-1996) on the impact of the water harvesting practices on agricultural yields. Water harvesting contributed to the rehabilitation of degraded land and also increased household food security in a region considered to be at the northern limit of cultivation.

Farmers reported that without *tassa* it was virtually impossible to grow anything on degraded land. Farmers have voluntarily continued to invest their time and energy in conservation techniques promoted by the project and some farmers have conducted their own experiments with different densities and layout of the water harvesting practices. Families who have invested in water harvesting practices still have a shortage of cereals in crop years but this is much lower than the case prior to establishing *tassa* and *demi-lunes*.

Unimproved *tassa* were very small pits made with a hoe to break the surface crust on existing farm fields before the start of the rains. They were not used to rehabilitate very degraded land and local peoples did not add organic matter. The project introduced a host of improvements. These included increasing the dimension of the pits from 10 cm to 20-30 cm and from a depth of 5 cm to 10-25 cm in order to collect and store more rainfall and runoff. The project also encouraged the adding of organic matter to the pits to improve soil fertility. The organic matter attracted termites that digest organic matter making the nutrients more easily available to the plant roots. The termites also dig small channels and, therefore, increase the water holding capacity of the soil. Farmers often re-dig the *tassa* every three years.

The improved *tassa* builds on indigenous practices and by improving soil quality the focus is on plant production rather then soil conservation *per se*.

Furthermore, as was the case with development efforts in Güinope in Honduras, the extension agents worked with farmers and did not adopt a transfer-of-technology approach where technologies are imposed on farmers. Remarkably farmers' active participation in the project and beyond took place in a macro-economic and macro-political environment that was not very conducive to development: there was a lack of access to credit and other support services and also a degree of political instability.

The yields in *demi-lunes* are similar to those from the *tassa* in average rainfall years but in years with low rainfall, yields are higher because of the greater water storage capacity. The drawback is that the labour costs of constructing and maintaining *demi-lunes* are higher. Table 5.6 illustrates the impact of the water harvesting practices on yields. The data are from demonstration plots established on farmers' fields.

Table 5.6 Impact of *tassa* and *demi-lunes* on millet yields (kg/ha) 1991-1996 in Niger (T0 = no practice, T1 = practice + manure). From Hassane et al. 2000:26.

Rainfall (mm)	1991 654	1992 432	1993 301	1994 597	1995 410	1996 440	Average 1991-1996
Tassa							
T0	n.a	125	144	296	50	11	125
T1	520	297	393	969	347	553	513
Demi-lunes							
T0	n.a.	86	77	206	28	164	112
T1	655	293	416	912	424	511	535

5.5.3 Mali

Planting-pits have also been used in development projects in Mali where they are known as *zaï* (Wedum et al. 1996). As is the case in Niger, *zaï* in Mali is a practice used to improve yields of millet, sorghum and maize. The traditional *zaï* are about 20 cm in diameter and about 10 cm deep. They are dug out with a traditional hoe. *Zaï* are only suitable for gravel soils and clay slopes that generate high levels of runoff.

Improved *zaï* are a means of collecting runoff water and larger than the traditional pits with holes about 30 cm in diameter and 15 cm depth. The excavated earth is used to make a small half moon bund downslope from each pit. As is the case with Niger, data from Mali demonstrate that yields on fields with *zaï* are substantially higher than those from traditionally ploughed fields(see Table 5.7). Although yields are improved, the main drawback of the *zaï* practice is that it requires considerable labour input to dig the holes with a traditional hoe.

Table 5.7   Impact of *zaï* on sorghum yields (kg/ha) 1992-1994 in Mali. *Optimum sowing date. From Wedum et al. 1996.

Season	Yield *zaï*	Yield ploughing
1992-1993	1494	397
1993-1994	620-1288*	280-320*

5.5.4   Burkina Faso

Farmers in Burkino Faso also use *zaï* and *demi-lunes* (Zougmore, 2003). Farmers in the Yatenga region of Burkina Faso have readily adopted *zaï* (Ouedraogo and Kaboré, 1996). With the introduction of the *zaï*, farmers in the Yatenga region have started to intensify land use. The number of *zaï* varies from 12,000 to 25,000 per ha depending on their spacing. They are used to rehabilitate lateritic soil and also sterile, crusted land, with a hard-pan surface. These soils are rock hard and plant roots are unable to penetrate the crust, however, *zaï* are particularly appropriate in these areas because the hard pan promotes runoff into the pits. In Burkina Faso, the use of *zaï* has led to the restoration of approximately 100,000 hectares.

Farmers add manure to the *zaï*. The manure contains grass and tree seeds which have already passed through the digestive tracts of cattle, sheep and goats. As a result the seeds more easily germinate and grow rapidly because of the improved soil moisture and nutrient levels in the planting-pits. Many farmers in the Yatenga region protect the tree species that grow in this way. The area had been characterised by out-migration, predominantly to coffee growing areas in the Ivory Coast. The use of *zaï* has led to the creation of a network of day labourers who have mastered the technique and instead of migrating they find short-term employment in the villages working with local farmers to construct *zaï* (Pretty and Hine, 2001).

In these drier zones, reduced amounts organic matter, rainwater and mineral fertilizer can more used more effectively by concentrating them at planting stations rather than being spread widely and thinly over the whole area. The effects of enriching just the planting pits of the *zaï* shows the benefits of such a concentration, even if maybe half or more of the field does not receive any of the inputs. Furthermore, if this leads to greater plant production from those limited circles then more residues will be available for further soil improvement in the next year (Shaxson and Barber, 2003).

5.5.5   Association for Better Land Husbandry and Double-dug beds in Kenya

The Association for Better Land Husbandry (ABLH) works with resource-poor farmers in Central and Western Kenya and promotes the principles of better land husbandry. The programme began with participatory discussions in which farmers identified their concern for higher cash incomes and stressed that the biggest problem they faced was poor soil fertility. The programme worked with ten self-help groups (SHGs), several of them associated with local churches. Training was offered in several conservation-effective and productivity-enhancing technologies. The most popular technology has been double-dug (to 50 cm depth) composted beds on small areas, generally near to their houses.

Positive results from the double-dug beds derive from the concentration of organic materials along with improved rainwater capture and retention. The improved conditions in the root zone, including better moisture-holding capacity, mean that farmers are able to cultivate a rich diversity of crops well into the dry season each year (Shaxson and Barber, 2003). The plots with double-dug beds are less affected by drought than unimproved plots. The construction of double-dug bed is very labour-intensive but for most farmers this is easily compensated for by the increased agricultural production brought about by the beds' greater water holding capacity and increased levels of organic matter.

Farmers experimented with double-dug beds on small areas of their holdings, and within a few months found that improved fertility and water retention enabled them to grow a new range of crops with good market prospects, such as tomatoes, kale, cabbages, pineapples, carrots, spinach and parsley. Farmers made clear that profit at the earliest opportunity was essential. The programme, therefore, also offered advice and training in the marketing of agricultural products. Collective marketing of fresh produce under the *Conservation Supreme* label has been successful. Producers organised 'pack-house committees' which undertook responsibilities for crop consolidation, collection, grading, bulking, packing, recording, collection of orders and of revenue, and calculation of dues to SHGs and their members (Cheatle and Shaxson, 2001).

Farmers find the advantages of joining a SHG include economics in marketing, the exchange of ideas, and the better ability to call in specialist advice when needed. Overall improvements in food security, nutrition and cash income have encouraged steady spontaneous spread of the agricultural production technologies. More farmers are joining the groups, and many intend to expand the size of the areas they have treated. Hamilton (1997) reported the successes of the project (see Box 5.12). Between 1992 and 1996, the impact of the project included:

- Boosting self-sufficiency in maize from 22 to 48 percent of farmers
- Reducing experience of hunger from 57 to 24 percent of farmers
- Reducing the proportion of farmers buying vegetables from 85 to 11 percent and increasing the number selling to 77 percent

> **Box 5.12   Benefits of double-dug composted beds in Kenya (from Hamilton, 1997)**
>
> - Almost all adopters are hugely satisfied with the improvement in diet that has resulted from the abundance of vegetables that is the most obvious result of the adoption of conservation farming.
> - Adopters are well aware that the new diet is nutritionally better balanced than the old one and that this is important in relation to health, especially of children.
> - Many adopters are very satisfied with the way that the new cash income from the sale of vegetables not only allows purchase of maize and other foods but also essential household needs such as school fees.
> - Adopters have extended organic practices, notably compost, beyond the kitchen garden to the maize fields, even in tea-growing areas.
> - Growing numbers of farmers have adopted the double-dug composted beds and much of this seems to have occurred due to spontaneous adoption by neighbours, who are impressed by the profusion of healthy green vegetables growing on double-dug beds.

5.6   HURRICANE MITCH: FROM SOIL EROSION TO LANDSLIDES

5.6.1   Synoptic History of Hurricane Mitch

Together with many of the positive results outlined above, soil conservation technologies have also been attributed with protecting hillsides from landslide damage caused by extreme rainfall events. This supposition was inadvertently tested by the author at Santa Rosa during Hurricane Mitch. The hurricane struck Central America towards the end of October 1998. The hurricane claimed approximately 11,000 lives and caused billions of dollars of damage. In terms of lives lost and property damaged, it was the most deadly hurricane to strike the Western Hemisphere in the last two centuries (McCown et al. 1998).

Although the hurricane was downgraded to a tropical storm when it moved inland over Honduras, torrential rains caused enormous flooding and triggered numerous landslides throughout Honduras and northern Nicaragua (Perotto-Bladiviezo et al. 2004). In Honduras, the United States Geological Survey estimates that 6,600 persons were killed, 8,052 injured, 1.4 million persons were left homeless and nearly 70 percent of crops were destroyed (Powers, 2001). Hellin et al. (1999b) and Hellin and Haigh (1999) have reported on the synoptic history and rainfall characteristics of Mitch. A summary is given in Box 5.13.

Box 5.13 **Rainfall characteristics during Hurricane Mitch. Times are local time and Honduras is six hours behind Greenwich Mean Time. From Hellin et al. (1999b) and Hellin and Haigh (1999).**

Hurricane Mitch struck Central America towards the end of October 1998. Rainfall was greatest in the north and the south of the country. In the south of Honduras between 1800 on 27 October and 2100 on 31 October 1998, there was 896 mm of rainfall. During this period there were three periods of extreme rainfall intensity:
- 186 mm fell during the six hours 1600 to 2200 on 29 October
- 74 mm fell during the four hours 0100 to 0500 on 30 October
- 254 mm fell during the six hours 1600 to 2200 on 30 October

Maximum rainfall intensities ranged from 138 mm/hr^{-1} (4.6 mm in a 2-minute period) to 58.4 mm/hr^{-1} (60-minute period).

5.6.2 Hurricane Mitch and Landslide Damage in Southern Honduras

In terms of impact, rainfall intensity is associated with landslide activity and erosion flooding especially in small catchments (Hudson, 1995:326). Landslide activity was a major problem and source of sediment pollution during Hurricane Mitch. Observations the author made immediately after Mitch indicate that approximately 5 percent, but in some areas as much as 20 percent of the hillsides, in southern Honduras suffered landslides (Haigh and Hellin, 2001).

Photographic records show the forested hills close to Santa Rosa reduced by a swarm of landslides and debris flows. It is no coincidence that almost all the landslides at Santa Rosa in Southern Honduras and in the surrounding hillsides occurred during the three periods of most intense rainfall between 1600 and 2200 on 29 and 30 October (see Box 5.13) when, according to Martin (2002:19), the maximum daily rainfall in the city of Choluteca was 465 mm against an estimated 50-year daily maximum of 303 mm.

Damage at Santa Rosa is indicative of the frequency and type of landslides that occurred in southern Honduras. At the trial site, five deep-seated landslides originated only on the steep slopes (65-75 percent) where control and live barrier plots were equally affected. The depth of these landslides was approximately 1.5 m - 2 m, far exceeding that of the root-strengthened zone of either vetiver grass or maize crops, although mature tree roots might extend this deep. It is, therefore, no surprise that the landslides destroyed both control and live-barrier plots.

No landslides originated on the more gentle slopes. Here, most of the damage was caused by debris from landslides and surface soil movements originating on steeper slopes (>50 percent) above and outside the research site. These source slopes had been cleared of secondary forest in 1997 and also planted with maize.

It was not just the research plots that were damaged; the disaster brought the scientific study to an abrupt halt (see Plates 22 and CP 8). Recording apparatus, designed to measure sediment losses in terms of a few kilograms, was completely destroyed by landslides that moved several metres of land surface and dumped several tonnes of debris. Only the autographic rain gauge worked perfectly throughout the entire event, a result that impressed even its manufacturers (Hellin and Haigh, 1999).

Plate 22   Hurricane *Mitch* in October 1998, brought the author's research on soil loss and maize production to an abrupt halt. Honduras. (Hellin, J.).

Of course, since Mitch destroyed the catch-pits and barrils, no accurate figures exist for sediment yields during the hurricane. Instead, volumes of soil lost through landslides from the whole area (circa 1 ha) on steeper 65-75 percent slopes were calculated from photographs and measurements made at the trial site some 10 days after the hurricane struck. Following Herweg (1996:50), the volume of soil removed by the five landslides was estimated by measuring the area covered by the landslides (Table 5.8, column 2), this figure was then multiplied by the average depth of the landslide scar from measurements taken at different points. On site measurements suggest an average soil bulk density of just under 1.5 t m^{-3}. This figure can be used to convert the volume of soil removed (m^3) (Table 5.8, column 3), into tonnes of sediment (Table 5.8, column 4) and thence to t ha^{-1} (Table 5.8: column 5).

Table 5.8 Landslides caused by Hurricane Mitch on 65-75 percent slopes on the Honduras test plots (October 1998). From Haigh et al. 2004.

Area of slope	Total area of landslides	Volume soil removed by landslide	Soil loss from landslides	Soil loss from landslides (t ha^{-1})	Average annual loss by soil erosion (t ha^{-1})
1	2	3	4	5	6
10,500 m^2	3027 m^2	6983 m^3	10,460 t	9,961	17

Data from Santa Rosa show that prior to the Hurricane, annual soil loss by erosion for the steep slopes was 17 t ha^{-1}. Comparing Table 5.8, columns 5 and 6, we can see that the sediment yield due to the landslides of October 1998 was approximately 600 times greater than the average annual soil loss caused by water erosion on these same steep slopes. This result raises some very important questions for land use managers, ones that are considered below.

5.6.3   Soil Conservation Technologies and Slope Protection

At the Santa Rosa experimental station live barriers were shown to be ineffectual in terms of protecting hillsides against landslide damage, but what of the evidence from other parts of Honduras? There is some evidence that soil conservation measures worked during Hurricane Mitch. Holt-Giménez (2000 and 2001) reports on a study of 1,804 cultivated plots in Honduras, Nicaragua and Guatemala that confirmed that farms using "sustainable" practices including soil conservation methods appeared to suffer less damage than their "conventional" neighbours.

According to Holt-Giménez (2000), sustainable farms had fewer and smaller gullies and areas of rill erosion. However, on slopes greater than 50 percent and under conditions of high storm intensity the differences between sustainable and conventional farms vanished, indicating that these techniques have thresholds of effectiveness. The report does go on to say that under high rainfall conditions, the occurrence of landslides is more likely to be linked to underlying geological characteristics than the presence or otherwise of vegetation and soil conservation technologies. These conclusions appear to be confirmed by the research at Santa Rosa, where rainfall was some of the heaviest in the country and where the research plots were found on hillsides that span this slope steepness threshold.

Furthermore, Rocha and Christopolos (2001) report that in Nicaragua, where rainfall was also very heavy and intense, many landslides occurred in areas with abundant tree cover and on steep slopes with soil conservation technologies. While reports from northern Honduras indicate that landslides indiscriminately affected hillside farmers (e.g. Sherwood, 1999). Landslides occurred on unprotected fields and also on fields where farmers used cover crops such as *Mucuna* spp. and *Canavalia* spp. It is perhaps no coincidence that the two areas of Honduras that received most rain during Hurricane Mitch were the north and the south of the country (Hellin and Haigh, 1999).

The second question is whether soil conservation technologies targeting the correct problem? A statistical analysis carried out by the Nicaraguan Institute of Territorial Studies (INETER) concluded that the precipitation levels during Mitch are likely to recur about every 150 years (Rocha and Christopolos, 2001). If this were the case then landslides are a four times larger source of sediment than surface wash on steep agricultural hillsides, and defence against landslides rather than soil erosion should be the fundamental concern of all soil conservation activities. Even if Hurricane Mitch were a 1 in 500-year event, on the steepest slopes it would still be the largest factor in sediment production.

The significance of this new problem has been highlighted by the advent of a United States Geological Survey (USGS) led flood hazard mapping project conducted in response to Hurricane Mitch's impacts. These studies are somewhat coy about the possible recurrence interval of the hurricane, which, as the largest event in every hydrological data run – most of which do not achieve a total of 50 years – might be conceived as a <50-year event, or by extrapolation of the local records of historical flood peaks, a >50 or 200 year event (cf. Kresch et al. 2002, Mastin, 2002).

The National Oceanic and Atmospheric Administration (NOAA) in the United States calculates the risk of a hurricane striking Honduras's north coast in any single year as 1–2 percent, and the risk of one reaching Choluteca in the south of the country as very much less (Mastin, 2002:5). However, the 50-year flood estimates of the USGS team share the property that they are all a small fraction of the actual discharge experience during Hurricane Mitch (Mastin, 2002:27) (Table 5.9). Indeed the discharges during Hurricane Mitch cannot be fitted on the exceedance probability graph computed by the study (Kresch et al. 2002:5). According to this, Hurricane Mitch could never happen.

Table 5.9   Flood hazard estimates (m³/s) for the Rio Choluteca at Puente Choluteca gauge close to the Pan American Highway in Choluteca, Honduras (from Kresch et al. 2002)

Peak Discharge during Hurricane Mitch – estimated from indirect measurements	15,500
Next highest measured flood peak on record	2,130
50-year flood discharge estimated from the period of station record (1979-1998)	12,500
50-year flood discharge based on local historical information (56-years)	4,910
50-year flood discharge estimate used for flood hazard mapping purposes	4,613

No doubt, the process of flood hazard mapping could have the benefit of helping local authorities resist the colonisation and development of the most vulnerable flood-plain areas. Unfortunately, it could also have the side-effect of encouraging both the development of the land immediately above the estimated 50-year flood level line and the popular understanding that such development was safe from flooding. It is hard to construct a false sense of security based on the data above. Unfortunately, flood hazard maps exude an air of authority that is hard to overcome and once they are printed, they become 'scientifically established' facts to many users.

5.6.4 Promoting a False Sense of Security?

This book has shown that there are very sound reasons for farmers to practice better land husbandry, not least that by doing so farmers are able to maintain and improve soil quality with a commensurate increase in soil productivity and production. Hurricane Mitch, however, showed that technologies such as cross-slope grass barriers and agronomic practices such as providing soil cover are unlikely to be able to protect the soils of the steepest slopes in hurricane conditions. It also showed that, in the long term, on such slopes, the dominant form of soil loss, and greatest threat to the pollution of runoff and river courses, comes from landslides rather than the wash of surface soils. The landslides induced by one Hurricane Mitch proved to remove as much soil as 600 years of surface erosion, yet landslides had been a minor factor in soil conservation thinking.

While the extent and severity of landslides in 'normal' years are far less than those caused during Mitch, the results of a landslide (however small) can still be catastrophic for a smallholder farmer with limited access to land. Although as Morris et al. (2001) point out, in some respects the poorest households are sometimes less affected by natural disasters such as Hurricane Mitch because they have so little to lose: they are cropping such small plots of land that there is less chance of the plot being affected by a landslide, and they are less likely to lose other assets because they have so few assets to begin with.

The results from Santa Rosa during Hurricane Mitch suggest that mass movements are a far greater threat to sustainable agriculture than soil loss by erosion. It is perhaps no coincidence that farmers who the author interviewed in Honduras expressed much more concern about landslides than soil erosion. A worrying development is that following Hurricane Mitch many organisations that promote soil conservation technologies are now presenting their efforts as being oriented towards disaster mitigation (Rocha and Christopolos, 2001). The danger is that this may well lull farmers and the development community into what can only be a false sense of security. The salutary reality is that soil conservation technologies are unlikely to offer protection during extreme rainfall events. This should come as little surprise:

road engineers who know a thing or two about steep slopes and landslides are well aware that vegetation can be used to protect hillsides from erosion and shallow slope failure but it does not offer protection from landslides (Clark and Hellin, 1996:1).

The danger of soil conservation technologies and other agronomic practices being promoted on the erroneous basis that they can protect hillsides in large rainfall events is even greater in light of the fact that Hellin et al. (1999b) have demonstrated that rainfall totals (one hour – 48 hours) in southern Honduras during Mitch were not exceptional for hurricanes/tropical storms in the Atlantic basin (see Figure 5.2 which is based on data from Chaggar, 1984; Paulhus, 1965; Lawrence et al. 1998; Eyre, 1989 and Rappaport, 1998).

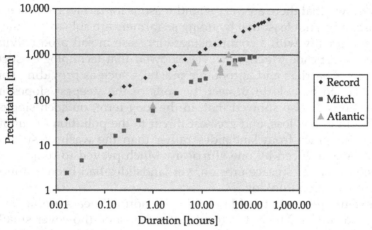

Fig. 5.2 Maximum rainfall amounts (mm) versus duration (hours) during Hurricane Mitch ($y = 51.64 * x^{0.674}$, $R^2 = 0.99$). Also plotted are updated record rainfall events (mm) for different durations that define the curve of maximum potential rainfall ($y = 353.07 * x^{0.519}$, $R^2 = 0.99$) and data from major recent Atlantic hurricanes and tropical storms. From Hellin et al. 1999b.

Countries such as Honduras may well face storms with similar rainfall amounts and intensities to those experienced during Mitch and those farmers who have been encouraged to establish conservation structures and adopt conservation measures could well find that once again these offer little protection against landslide damage associated with high volume and intensity rainfall. The likelihood of countries like Honduras experiencing Mitch-like levels and intensity of rainfall is very difficult to calculate as Honduras is no exception to the phenomenon highlighted by Haigh (2000b) that in most countries only a few meteorological stations have kept records of rainfall intensity for more than 20 years. Hence, relatively little is known about the kind of rare rainfall events that occur infrequently and that cause extensive flooding and landslides.

Even though events like Hurricane Mitch are rare, Thurow and Smith (1998:14) contend that, in hillside farming, gully erosion and mass wasting are more important sources of sediment than sheet erosion. Evidence from other parts of the world also indicates that losses from landslides may be greater than from surface erosion. Blaut et al. (1959) documented that in the Blue Mountains in Jamaica landslides covering a surface area of 19-280 m^2 were common in farming areas. Sillitoe (1993) reports that in Papua New Guinea mass movements are more important with regards to soil loss than sheet erosion. In this context, it would be disingenuous to promote soil conservation technologies and better land husbandry practices as offering substantial protection against landslides.

6

Better Land Husbandry and Policy

"Leaders in the economic and agricultural development communities, as well as environmentalists, must draw the attention of policy-makers to soil degradation concerns and work with them to set priorities for public investment, farmer services and policy. A necessary though not sufficient step is to provide support policies for broad-based agricultural development. Targeted policies and investments will also be needed to address many serious degradation problems" (Scherr, 1999:45).

6.1 A NEW APPROACH TO LAND MANAGEMENT

Land degradation is a social, economic, political, and technical problem requiring multi- and inter-disciplinary solutions. Previous chapters in this book demonstrate that by focusing on soil loss rather than soil quality, cross-slope soil conservation technologies are unlikely to contribute to increased production. Understandably, farmers consider such technologies largely irrelevant. A new approach is needed, which aspires to introduce land management changes that improve soil quality while simultaneously meeting the farmers' needs. Despite the numerous constraints to better land management, farmers can improve soil quality via the use of conservation-effective and productivity-enhancing technologies. An approach best articulated by better land husbandry.

A key component of the better land husbandry approach involves treating farmers as people who can suggest and implement strategies that can help solve the problem of soil degradation (Norman and Douglas, 1994:87). The basis for developing improved land management practices is to identify which of the farmers' existing practices are beneficial from a conservation point of view and which are not, along with the farmers' reasons for practising them (Douglas, 1993). Change tends to be most appropriate when it is generated from within farms and communities and when it emerges from the local agro-ecological cultural, social, and economic context. Add-on technologies that bear no relation to what farmers do at present must be treated with caution (FAO, 1995:18). Rather, improvements to land use and management practices should develop from the knowledge, problem awareness, analysis, ideas, insights, goals, aspirations and priorities of local farmers (Cheatle, 1993).

The better land husbandry message is a persuasive one although as we will see later in this chapter, it is a message that is not always getting through to policy-makers and development practitioners. One of the challenges facing the aficionados of better land husbandry is that the approach to land management cannot be treated in isolation from the policy environment and one of the challenges facing anyone working with smallholder farmers is to persuade policy-makers that smallholder agriculture needs and ought to be supported.

6.2 FUTURE OF SMALLHOLDER FARMING

6.2.1 Why Support Smallholder Agriculture?

Public policy plays a crucial role in supporting and accelerating transitions to sustainable agriculture (Scherr and Yadav, 1996:9) and in this context there are some hard decisions that donors and governments have to make. Despite rapid urbanisation, an estimated 70 to 75 percent of the world's poorest people live in rural areas where their livelihoods are largely dependent on agriculture. Until recently, conventional wisdom was that smallholder agriculture is strategically indispensable to development as a whole, largely because smallholders are efficient users of resources as well as being a relatively equitable means of providing income and food directly to poor people (Kydd, 2002).

There are those, however, who question whether there is really any future for smallholder farming. Maxwell et al. (2001), for example, have pointed out that the liberalisation of markets poses significant new threats to smallholder farmers and there is little hope that smallholders will be able to continue to survive as full-time farmers on tiny patches of land. In a similar vein, Southgate (1992:1) argues that given the market conditions and government policies that smallholder farmers face, the best way to respond to land degradation, declining prices for crops and livestock, policy-induced shortages in rural financial markets, and attenuated property rights is to depreciate farm assets, especially land, as a prelude to exiting the agricultural economy. Despite these pressures, there are three reasons why it may be right to support small-holder agriculture (see Box 6.1).

Box 6.1 Importance of smallholder agriculture

Agriculture is not only an economic activity and source of production and income; it is also an important part of rural people's culture and social organisation. It is partly for this reason that farming enjoys such protection in the developed countries, where the predominant image of farming is the family farm, despite the reality that is one of dominance by agri-business (Tripp, 2001).

Smallholder farming is linked to reductions in rural poverty and inequality. According to the World Bank growth in agricultural incomes is particularly effective at reducing rural poverty, because it has knock-on or multiplier effects on local markets for other goods and services provided by non-farm rural poor, such as construction, manufacturing and repairs (The World Bank, 2001). Strong agricultural growth has also been a feature of countries that have successfully reduced poverty, these countries include Indonesia and China, (DFID, 2002).

Thirdly, a more sustainable agriculture provides environmental services such as the conservation of soil and water, the maintenance of bio-diversity and also contributes to locking up carbon. These services are important to society in both urban and rural areas as well as locally and globally.

It is generally expected that in the future a smaller proportion of the population will be involved in farming and that larger numbers of people will be employed in other parts of the rural and urban economy (Tripp, 2001). However, with the international community committed to the realisation of the Millenium Development Goals (The World Bank Group, 2004), it is highly likely that smallholder agriculture will be supported and will continue to make an important contribution to development as well as being part of the rural livelihoods of many of the world's poorest people, sometimes referred to as the 'Chronic Poor' (Hulme and Shepherd, 2003; Chronic Poverty Research Centre, 2004).

Previous chapters have shown that a better land husbandry approach can contribute to a more sustainable agriculture and that, in this context, this holistic approach to land management should feature more widely in development agendas. There are those, however, who argue that while smallholder farming should be supported, better land husbandry is not necessarily the best way forward. The argument is that there are other approaches, namely by extending the Green Revolution and making fuller use of genetically modified crops.

6.2.2 What About the Green Revolution?

The period since the Second World War has undoubtedly seen remarkable growth in agricultural production and productivity in the developing world. The Green Revolution was indeed a revolution that led to a significant increase in crop production. In other cases, however, and as we saw in Chapter 1, crop production came about due to the clearing of steeplands that are of lower productive potential and higher vulnerability.

As outlined in Chapter 1, the challenge of feeding growing populations in the developing world will require finding sustainable methods of intensive production on soils that have not previously been used for this purpose and by substituting for or rehabilitating degraded soils where there is continuing demand for their use. There are those who argue that we can meet the challenge by driving forward with the Green Revolution. One of the problems with this argument is that the Green Revolution has bypassed millions and millions of farmers who simply cannot afford the expensive inputs such as irrigation, pesticides, herbicides and fertiliser needed to restore and maintain soil productivity (Meinzen-Dick et al. 2004) (see Box 6.2 and Box 6.3)

Box 6.2   The Green Revolution: bypassing smallholder farmers

When the author was working in Honduras in the 1990s, he made numerous visits to a farming area in the south of the country. On one occasion, he visited a small community in the hills. The valley below was criss-crossed with neat green lines of honey melons destined for export to Europe and the United States. Standing on the hillside above the melon farm, the author was surrounded by a mosaic of maize plots belonging to a local farmer and remnants of forest.

The author was witness to two contrasting realities. The melon farm in the valley below exemplifies the type of agriculture that emerged during the Green Revolution, an agricultural system that relies on expensive inputs such as irrigation, pesticides, herbicides and fertiliser to restore and maintain soil productivity. The maize plots perched precariously on the hillsides represented another reality, one faced by millions of smallholder farmers worldwide.

While, the Green Revolution has undoubtedly led to a substantial increase in production, it has also bypassed millions and millions of farmers who simply cannot afford the costly inputs and are highly unlikely to be able to do so in the future.

Box 6.3   The Green Revolution and irrigation

The success of the Green Revolution agriculture depends on copious amounts of pesticides and fertilizer and also water. For example, between 1950 and 1995 grainland productivity increased 240 percent while water use for irrigation increased 220 percent (Leslie, 2000). The scarcity of water is becoming one the most important factors in limiting agricultural production.

There is therefore a need to increase water productivity. One very effective way is via drip irrigation that can reduce water use by 30 to 70 percent while increasing yields by 20 to 90 percent (Leslie, 2000). The drawback is that at US$500 to $1,000 per acre it is an expensive technology that is well beyond the reach of most resource-poor farmers. Much of the food needed by those in the developing world will, therefore, have to come from rainfed rather than irrigated land.

The policy environment that enabled the Green Revolution is in sharp contrast to the policy environment of today. Some development experts, therefore, question whether it is realistic to pin our hopes on a new Green Revolution: *"the Asian Green Revolution was predicated on comprehensive agricultural support policies that have been dismantled and disallowed in the post-liberalisation dispensation originating in the structural adjustment policies of the 1980s and 1990s. In those days there were fixed prices, floor prices, buffer stocks, fertilizer subsidies, credit subsidies and public irrigation schemes, all paid for by the state or by donors, and none of these policy instruments are available in the current lexicon of acceptable public sector interventions in rural areas"* (Ellis and Harris, 2004).

Many of the high-input agricultural systems that emerged during the Green Revolution treat soils as though they are an inanimate collection of minerals rather than dynamic and living systems. These agricultural systems are only sustainable as long as someone is prepared and able to pay for the inputs. It is, therefore, unrealistic to rely on Green Revolution techniques being made available to anything but a small minority of resource-poor farmers. Meanwhile it is still unclear the extent to which genetically modified foods can boost agricultural production in smallholder farming communities.

There is a very strong argument for promoting a better land husbandry approach that seeks to work with nature by focusing on achieving a self-sustainable system based on improving soil quality via soil protection, incorporation of organic matter, and the use of soil organisms. This does not necessarily entail a purist approach whereby farmers stop using all artificial fertilisers and pesticides. What it entails is reduced reliance on these inputs. This approach is more likely to complement the needs and resources of millions of resource-poor farmers worldwide (Pretty, 1995; Fowler and Rockstrom, 2001).

6.3   A NEW PROFESSIONALISM

6.3.1   There is Still Much to Do

Previous chapters have demonstrated that rethinking the concepts and approaches for helping farmers adopt improved land use and husbandry on steep lands has many implications for extension, training, research, policy formulation, institutional organisation, project formulation and legislation.

We saw at the end of Chapter 4 that inter-disciplinary approaches, such as better land husbandry, have taken time to evolve because, above all, their implementation requires a new professionalism for agricultural scientists and extension agents. This new professionalism involves new concepts, values, methods and behaviour (Pretty and Chambers, 1994; Pretty, 1995:203). As Bunch (1982:vi) has written:

> *"Agricultural improvement will always be more of an art than a science... Though general guidelines for program design can be established, the final*

outcome will depend more upon good judgement and understanding than strict adherence to a set of guidelines. The success of programs depends on an understanding of people's needs, motivations, values and viewpoints and of the possible consequences of the social pressures that programs are setting in motion" (Bunch, 1992:vi).

A number of inter-related changes are needed. Firstly, scientists need to break out of the single disciplinary research paradigm and communicate more effectively with farmers (Doran et al. 1996). Secondly, professionals need to encourage and enhance farmers' active participation in development initiatives. Thirdly, there is a need to focus down from watershed management to a more community-based level of planning.

6.3.2 Effective Communication

An approach, such as land husbandry, has the potential to draw social and natural scientists into a productive debate leading to practical and realistic development initiatives. Land husbandry is not a panacea but it does offer a viable alternative to the legacy of failed soil conservation initiatives. Farmer enthusiasm and receptivity to new ideas will come from seeing beneficial changes in the short-term and at an affordable cost. This, in turn, depends on a partnership between social and natural scientists and farmers that can address the complexity of issues surrounding soil and land degradation.

Chapters 2 and 4 show that much of the success of better land husbandry initiatives rests on effective communication between different actors. Firstly, there is the relationship between outsiders and farmers. Secondly, because of the complexity of farmers' needs, there is a need for outsiders to communicate effectively amongst themselves as part of interdisciplinary teams (Norman and Douglas, 1994:110; Bellon et al. 2003).

Farmers' priority problems range from pests and diseases to a lack of rural credit. These problems cannot be addressed by single disciplinary paradigms. For example, giving farmers more secure access to land or providing sources of credit *per se* will not directly address the issue of declining productivity due to a deterioration of soil quality. Röling (2000) has referred to the difficulties that natural and social scientists have in communicating and working effectively together as well as the imperative of them doing so (see Box 6.4). While Sillitoe (1998), has noted there is a tendency on the part of anti-positivist social scientists to undermine natural scientists: a tendency that does not bode well for the better functioning of interdisciplinary teams.

Box 6.4 Gateway to the global garden: beta/gamma science for dealing with ecological rationality (from Röling, 2000)

Röling refers to the role of natural and social science, and especially their inter-disciplinary encounter that he calls beta/gamma science, in

---

dealing with today's ecological challenges. According to Röling, natural scientists typically consider social science as an oxymoron. For the natural scientist, reality exists independently from the observer: through scientific methods it is possible to build objectively true knowledge about that reality. Science seeks to predict natural phenomena, such as soil erosion, based on discovering natural laws governing cause and effect. Applied scientists then use that knowledge to develop technical solutions to deal with these problems.

Social science also tries to predict, or at least explain human behaviour, but the causal factors involved are invisible moving targets. They include desires (emotion, motivation, objectives, attitudes, norms and values), beliefs (knowledge, sense making, influence and interpretation) and action. Beta/gamma approaches have developed as scientists have recognised that the delivery of science-based technologies to groups, such as smallholder farmers, does not work. The beta/gamma approach is, essentially, a new area of science professionalism in which people are not objects that can be instrumentally or even strategically manipulated. On the contrary, they must participate in this process.

Existing scientists may also be unable to accept the prospects of a sustainable agriculture that is built on a new professionalism because of the threat it represents to established career patterns (Pretty, 1995:202). As Chambers (1997:180) writes, scientists on research stations seek peer approval and promotion while resource-poor farmers on their farms seek livelihood security. Farmers and scientists often live in different worlds. It is this dichotomy that partly explains both why better land husbandry took so long to conceptualise and also why the land husbandry message is not spreading as rapidly as might have been anticipated and expected.

6.3.3   Encouraging Active Farmer Participation

A key component of the better land husbandry approach is active farmer participation. There is a need to spread among professionals, effective means to enhance farmers' participation, competence and choice (Chambers, 1993). The changes are radical because they go beyond putting the last (farmers) first, and require a process of disempowerment, whereby researchers and extension agents are put last (Chambers, 1997:211). Instead of teachers, outsiders become facilitators, initiating, supporting and maintaining a process of change (Enters and Hagmann, 1996; Pretty, 1998b:24).

Extension agents and scientists may find it difficult to assume the role of facilitator when they have traditionally been trained as advisers. As we saw in earlier chapters, programmes of assistance that try to expand rapidly, bringing new skills and opportunities to many farmers, tend to do so prescriptively. Technical assistance to combat soil and land degradation is offered as a defined package; farmers are told what to do in order to reduce soil loss. This is not empowerment and once the external assistance is withdrawn, there is a danger that the development initiatives will peter out. Although there has already been a paradigm shift away from the transfer of technology model towards a more participatory approach there is still some way to go (Röling and de Jong, 1998).

Experience in Latin America with a range of participatory extension and research models such as Farmer Field Schools, Local Agricultural Research Committees and Farmer-to-Farmer extension models demonstrate that these may be effective in empowering farmers and encouraging them to seek solutions to their own production problems (Williamson, 2002; Meinzen-Dick et al. 2004) (see Box 6.5). By doing so these models provide farmers with some of the skills needed to manage better their land (Humphries et al. 2000; Johnson et al. 2003).

Box 6.5 The *Kamayoq*: Farmer-to-farmer extension services in the Andes (from Hellin et al. 2004)

The Andean region is beset by low endowments of 'geographic capital' (natural, social, human and physical capital) and rural poverty is endemic. Farmers often need technical advice and training in order to achieve some degree of livelihood security. The Peruvian government used to be largely responsible for the provision of agricultural extension services. During the 1990s, however, structural adjustments and cuts in fiscal deficits led to a reduction in the provision of government extension services. Private extension provision has grown but few resource-poor farmers are able to pay for this private extension. As a result it has generally been directed at larger commercial farmers (Chapman and Tripp, 2003).

Since 1997, ITDG, a non-governmental development organisation, has been working in farming communities in the Peruvian Andes and exploring approaches to extension service provision that complement smallholder farmers' needs and ability to pay. The farming communities are located at over 3,500 metres above sea level. Farm households have one or two head of cattle, some sheep and a number of guinea pigs (a food staple in the Andes). The most common crops are maize, potatoes and beans. The defining characteristic of the extension initiatives is the training of local farmer-to-farmer extension agents known as *Kamayoq*. ITDG established a *Kamayoq* School and to date over 140 *Kamayoq* have been trained of whom 20 percent are women. Training at the *Kamayoq* School is characterised as follows:

- Trainees come from and are selected by the communities. One of the key criteria in the selection process is the willingness of the trained farmers to work subsequently in their communities.

- Training is provided in Quechua, the local language. Instructors include ITDG staff, long-serving *Kamayoq* and experts from regional universities.
- The course lasts eight months and involves attendance for one day per week. The emphasis is on practical learning and training occurs at different field locations and only periodically in a classroom.
- The course focuses on local farmers' veterinary and agricultural needs including: identification and treatment of pests and diseases of agricultural crops and farm animals; improved irrigation via the use of a network of drainage channels; and improved breeding of farm animals such as guinea pigs.

Kamayoq who have received training at the school are subsequently able to address farmers' veterinary and agricultural needs. Farmers pay the *Kamayoq* for their services in cash or in kind. They are able and willing to do so because the advice and technical assistance they receive has led to an increase in family income of 10-40 percent. This has come about through increased production and sales of animals and crops. For example, the technical advice that farmers have received on irrigation and improved pasture has led to increases in milk production of up to 50 percent.

One of the most sought after services is the diagnosis and treatment of various animal diseases. In each of the 33 communities where the *Kamayoq* are active, mortality rates among cattle have fallen dramatically. This has partly come about through the *Kamayoq* working with local farmers to find solutions to veterinary problems. A good example of this process of Participatory Technology Development has been the discovery of a natural medicine to treat the parasitic disease *Fasciola hepatica*. Over a three-year period, the *Kamayoq* and local villagers experimented with a range of natural medicines until they discovered a particularly effective treatment that is also cheaper than conventional medicines.

The *Kamayoq* is largely an unsubsidised farmer-to-farmer extension service with ITDG only covering the cost of the training provided at the *Kamayoq* School. The success of the *Kamayoq*, demonstrate that in the context of debilitated or non-existent government extension programmes it is possible to provide extension advice in a way that complements smallholder farmers' needs and resources.

Having worked principally with farming communities in highland valleys, ITDG is now extending the *Kamayoq* scheme to communities in the higher reaches of the Andes where livelihood security is linked to the husbandry of alpacas. A group of 32 *Kamayoq* has recently been trained to address the veterinary needs of alpaca herders. There is also a need to broaden the focus of the *Kamayoqs'* work. Like most conventional agricultural extension provision, the *Kamayoq* have worked predominantly on improving and increasing production at the farm level. ITDG is investigating whether the *Kamayoq* model is one that can be adapted so as to provide farmers with the business development services they need in order to benefit from emerging market opportunities (see Section 6.4).

Farmer Field Schools is a training approach that was originally developed for helping farmers understand integrated pest management as an alternative to chemical control. The format is now being extended to help farmers learn about land management, market demand and product requirements as well as how to negotiate in new markets. Local Agricultural Research Committees develop farmers' research and learning capacities; they aim to encourage farmers to learn by doing, to criticise their own and others' work, and to adapt their processes to changing conditions.

Local Agricultural Research Committees are essentially a 'little school for learning' and are building block for a more sustainable land use and empowerment because *'agricultural research is but one element of a broader process of social change leading to empowerment'* (Humphries et al. 2000). These participatory methods can stimulate local innovation, because the emphasis is on principles and processes rather than recipes or technology packages. In some cases, farmers who participate in Local Agricultural Research Committees, go onto to learn how to manage funds, plan time, launch micro-credit schemes, prepare proposals to access external resources, and deal with outside agronomists and professionals on a more equal basis (Sherwood et al. 2000; Humphries et al. 2000).

Suitably empowered, farmers are better able to influence formal research and extension systems to their own benefit and to gain access to potentially useful skills, information and research products. The Farmer Field Schools, Local Agricultural Research Committees and Farmer-to-Farmer approaches to extension seem to offer encouraging results. However, it is important to acknowledge that these approaches do take time and resources.

6.3.4 Watershed Catchment to Community Level

Another challenge facing development practitioners is the need to focus down: the topographic catchment or watershed with the people it contains, has generally been seen as the optimum unit for programme planning, and for implementing improvements in land use and management (including demonstrating the effects of technical recommendations). The reasons are related to the focus on streamflow and water and sediment movements within and out of the catchment. If, however, the focus is on better management of the soil in and around each farm in a village, the primacy of the catchment basis diminishes. In its place, a community, and the land it occupies and uses, becomes the optimum focus for village-planning and implementing improvements in land use and management. Sometimes, the focus should even be the individual farm unit.

6.4   FROM PRODUCTION TO MARKETS

As part of securing their livelihoods, farmers need adequate food and money. In some cases market opportunities can act as a strong incentives to improved better land management, including the use of conventional soil conservation technologies (Tiffen et al. 1994). For example, in some steeplands where market conditions have generated a demand for crops from which farmers can gain acceptable levels of income, much effort and money has been invested by farmers and governments on bench terracing to provide a sustainable physical basis for commercial production. The crops is question include fruit trees on terraces. On the other hand, markets can also, as we well know, contribute to very poor management.

Land management and market opportunities can complement each other. In many situations more available soil moisture, and attractive market prices, will do much to enable and encourage rural families to achieve better livelihoods through greater food security and higher income from selling crops and livestock. Well-functioning markets are important in generating growth and expanding opportunities for poor people (Hellin and Higman, 2003). Managed well, trade can lift millions of poor people out of poverty; managed badly it will leave whole economies even more marginalized (Oxfam, 2002). Developing these markets and the institutions to support them is difficult and takes time and even when international and national markets do work properly, poor people need help to overcome obstacles that prevent them from participating on a fair footing.

There is little point in farmers growing surpluses if there are no markets for their produce. Conversely, efforts to improve farmers' access to markets will be wasted if farmers' natural resource base becomes so degraded that they are unable to farm. Better land husbandry, although a holistic approach to better land management, has tended not to give enough attention to the challenge of facilitating smallholder farmers' access to markets (see Box 6.6).

Box 6.6   Moving beyond agricultural production

In 2000, Sophie, my wife and I decided to carry out some research on the problems and opportunities facing smallholder farmers in South America (Hellin and Higman, 2003). We spent twelve months travelling from the southern tip of Tierra del Fuego to the northern border of Ecuador. The farmers we met mentioned a number of problems including some of those mentioned in Table 2.2 such as irregular rains and pests and diseases. Overwhelmingly, however, smallholder farmers and those working with them talked about the obstacles they face in accessing local, national and global markets. The growth of these markets has accompanied the process of globalisation, the economic integration of countries and national economies into a single world system.

While many farmers still produce food for themselves and their families, they also sell increasing amounts in their produce. The farmers we met need and want to sell their produce in the market place. They need money to buy agricultural inputs such as fertilisers and pesticides, to pay for their children's education, to cover medical costs and to purchase consumer goods. Opting out of the market place is not an option for the majority of farmers and they have no wish to return to subsistence farming. Time and time again farmers described the difficulty of meeting market demands for quality, quantity and continuity of production. Donors and development practitioners agreed that meeting these market requirements represents an enormous challenge for smallholder farmers who seldom have the technical and financial resources to understand and adapt quickly to these market demands.

As we travelled slowly northwards we began to question whether as a natural resource managers we have been focusing too much on the production side of agriculture and too little on how and where farmers can sell their agricultural produce. After all, for every additional US$1 generated through agricultural production in developing countries, economic linkages can add another US$3 to the rural economy (Watkins and Von Braun, 2003). We were not questioning the validity and usefulness of our work, as much as recognising that we had missed the big picture. Better land management is critical to farmers' livelihoods and, as earlier chapters have demonstrated, without it soil and land degradation will continue to undermine farmers' livelihood security. Better land management on its own, however, is often not enough. As one farmer told us in Bolivia "*I know how to farm better but what is the point of improving my farming methods if I have nowhere to sell the extra production*"?

Pretty and Hine (2001) document 208 cases of sustainable agriculture practices in 52 countries worldwide. In these projects and initiatives, approximately 9 million farmers have adopted sustainable agriculture practices and technologies on almost 29 million hectares (equivalent to 3 percent of the 960 million hectares of arable and permanent crops in Africa, Asia and Latin America.) But only a minority of these projects and initiatives have given much attention to adding value (through activities such as agro-processing) or marketing. Pretty and Hine (2001) calculate that only about 12-15 percent of the 208 cases addressed these issues.

In order to have more control over their lives and to benefit from global markets, farmers need to be aware of market requirements and to sell their produce accordingly (Hellin and Higman, 2001; Hellin and Higman, 2002). Farmers' participation in local, national and international markets involves a complex inter-locking system of agricultural inputs, technical extension, packing, processing and marketing activities. Aside from the challenge of meeting strict quality criteria and ensuring consistent supply, farmers need leadership qualities: contacts need to be forged, negotiations carried out and capital is needed to improve farm infrastructure, purchase agricultural inputs and, in some cases, pay for certification. Farmers also need to be adept at financial planning and control, and forecasting. In other words, they must become more business-like (Hellin and Higman, 2003).

These demands are being placed on farmers at precisely the same time that structural adjustment and cuts in fiscal deficits have led to a dismemberment of classical agricultural extension and research services to the extent that these services are unable to serve the needs of farmers living in complex, diverse and risk prone environments. Facilitating smallholder farmers' access to markets is a complex challenge that is being met by relatively few better land husbandry initiatives. One example, though, is the Association for Better Land Husbandry in Kenya (see Box 6.7).

Box 6.7   Association for Better Land Husbandry in Kenya (from http://www.landhusbandry.cwc.net/abkenya.htm)

In the first phase of direct action (1994-1999) the Association for better land husbandry (ABLH) worked with self-help groups. There were four components to the strategy: group organisation and management, group capitalisation and planning, marketing, and conservation farming. Initial technical support focused on working with the self-help groups to build up organisational and management skills so that they could develop a savings scheme that in turn could fund the business plan. Once the self-help groups were organised, ABLH started to look at better land husbandry approaches and practices including:

- Composting and other forms of organic recycling using on farm resources
- Use of natural pesticides and cultural methods of pest control
- Contoured double dug beds and pits with compost or manure added

ABLH has positioned itself as an objective and accountable management services provider aiming to facilitate community base organisations, non-governmental organisations and public agencies with the skills and knowledge that enable them to develop the business skills of smallholder farmers.

To reduce transaction costs ABLH is experimenting with farmer-to-farmer advisory and training services. This entails the establishment of a cadre of farmer specialists who can train other farmers in better land husbandry in exchange for a very small fee. Some farmer groups are selling produce to Farmers Own Ltd. This is a not-for-profit trading company and subsidiary of ABLH. The trading company is intended to grow and provide an overarching manufacturing and marketing service, not just for ABLH partners but also for other farmer groups. The vision includes numerous producer groups selling to the trading company, some of them involved in partial processing and, possibly, cottage factory manufacture of processed goods.

It is a real challenge to find the judicious balance between better land management *per se* and improved market opportunities. For example, there are those who argue that on-farm diversification will not only generate additional income for farmers but will also reduce their vulnerability to market and price fluctuations and to the impacts of extreme climatic events. Diversification and better marketing opportunities are often key factors that can enable farmers to break through the poverty barrier that surrounds them (Shaxson, 1999:14). There are, however, those who argue that there is a contradiction between exploiting market opportunities and maintaining on-farm diversity. Box 6.8 includes examples from both sides of the argument. What is clear is that the improved land management and greater access to markets for smallholder farmers must be taken more into account by policy-makers.

Box 6.8   Economic sustainability and on-farm diversity

There are economies of scale in accumulating transactions knowledge relevant to a particular product (Kydd, 2002). Banana producers in the Chapare region in Bolivia are being encouraged by an international development project to take advantage of market opportunities by following the example of more advanced banana associations in the region and move away from atomised farm plots of a few hectares towards larger consolidated holdings, essentially monocultures (Hellin and Higman, 2003). By consolidating the agricultural activities of dispersed farmers into production and marketing centres, the farmers can reduce their costs through shared equipment and volume buying of agricultural inputs, and justify the expense of a permanent staff of administrators and marketers. Development practitioners refer to these large blocks as potentially "economically sustainable units".

While large blocks of single species may, in some circumstances, be "economically sustainable", are they " environmentally sustainable"? Other development efforts in the Chapare stress the importance of agricultural diversification at the farm level. For example, a project funded by the United Nations Drug Control Program (UNDCP) promotes, at the farm level, diverse agroforestry systems incorporating licit agricultural crops. Although these may prove to be environmentally sustainable, can an agriculture based on diverse farm plots survive in the globablized economy? Evidence from Guatemala and the Andean region suggests that in some case the answer is 'yes'.

Hamilton and Fischer (2003) show how Kaqchikel Maya farmers in the central Guatemalan highlands are involved in non-traditional export agriculture while also maintaining cultural values. In the 1970s, the smallholder farmers began growing snow peas, cauliflower and broccoli and have expanded to include beans. In contrast to large-scale producers who may plant all their land to export crops, the Maya farmers, usually with less than 4 hectares of land, plant only a relatively small proportion to export crops. Instead, they continue to grow traditional maize and beans. Farmers' decision to plant small areas of export crops is partly due to the lack of accessible credit and the need to reduce their vulnerability to the risks of crop loss or price drops. The Guatemalan example contrasts with other export production schemes in Central America where the small producers have been forced out. Reasons for the success of the Kaqchikel Maya example hinge on the communities' social and cultural capital: traditional and non-traditional cultural relations, family labour, women's marketing experience, and indigenous knowledge of integrated pest management.

The Papa Andina project, co-ordinated by the Centro Internacional de la Papa (CIP) in Lima, is looking for ways to capitalize on small farmers' knowledge, abilities and the diversity of their potato heritage. A specific focus of the project is to identify the market niches where small farmers actually have a competitive advantage (Hellin and Higman, 2005). The project aims to identify markets where smallholder farmers have a long-term competitive advantage, because of their location, local knowledge, access to local varieties or crop management practices. For example, some market niches require small tubers grown at high planting densities and manually harvested, requirements that are much more difficult for mechanised farmers to achieve. The project facilitates contacts between smallholder potato farmers and the processing companies that are becoming increasingly important buyers of potatoes. Farmers learn more about the processors' demands in terms of preferred potato varieties, volumes required, quality and timing of production. The processors, in turn, learn about the varieties of potatoes that farmers grow and how they grow them. With a greater understanding of the reality faced by both parties, the processors can utilise potato varieties that have previously been ignored and, hence, encourage on-farm crop diversity.

6.5 BROADENING THE POLICY AGENDA

6.5.1 The Bigger Picture

According to Pretty (1995:19-25) there are four main conditions for sustainable agriculture:

- Development and use of resource-conserving technologies
- Full participation and collective action by groups and communities at a local level
- Supportive and enabling external institutions that facilitate farmer participation
- Favourable policy environment

The better land husbandry approach, with its focus on conservation-effective and productivity-enhancing technologies and active farmer participation, addresses the first two conditions. Better land husbandry as outlined in this book, shows what can be achieved at the production end. However, Pretty's (1995:19-25) third and fourth conditions, outlined above, are critical because trade, macroeconomic and sectoral policies affect the income, production, and investment choices of households and communities (Vosti and Reardon, 1997b). Such policies, therefore, have an impact on the natural resource base.

In the past, governments have often penalised agriculture through a variety of mechanisms including import and export taxes, foreign exchange controls, export licensing requirements and controls, and bureaucratic marketing boards. Food subsidies have allowed governments to keep food prices low, often to appease vocal urban constituents, but at the expense of rural producers. Such policies and practices have reduced farmers' incentives to look after their land, increase local food grain production and use modern inputs to improve productivity (Gruhn et al. 2000).

6.5.2 A Juggling Act

Scherr (1999:48) has written that economists and policy makers need to focus greater attention on soil quality management as a central natural resource issue for sustainable agriculture in the developing world. Norman and Douglas (1994:137) argue that it makes no sense pursuing a research and development agenda on improving land management through the promotion of soil quality, without addressing (in a serious way) the external factors that ultimately shape agricultural practice via farm families' decision-making. Box 6.9 gives an example of the impact of these 'extraneous factors' on land management decisions.

Box 6.9   Policies affecting land management on the north coast of Honduras

The example of the maize-*Mucuna* spp. (velvetbean) cropping system in northern Honduras demonstrates the impact that policy can have on the costs and benefits of resource use and investments at the farm level. Smallholder farmers' practice of intercropping *Mucuna* spp. with maize has been well documented (e.g. Buckles et al. 1998). The system enabled farmers to double their average yields of maize while maintaining soil nitrogen and increasing organic matter content. The system was able to support smallholder production in the 1970s and 1980s.

As external conditions have changed, however, the practice is no longer so attractive and is being abandoned. Since the early 1990s, economic restructuring policies adopted by the Honduran government have contributed to a greater concentration of land ownership and control over markets. The practice of *mucuna*/maize intercrops is declining due to the competition from higher value palm oil and dairy cattle industries (Humphries et al. 2000). On selling their land, farmers are either clearing more land in the remaining forested regions (and, hence, extending the agricultural frontier) or migrating to the cities.

A key first step is to provide support policies for broad-based agricultural development. Farmers' willingness to invest in soil improvement is closely associated with the overall economic profitability of farming and an economic and policy environment that facilitates commercialisation, reduces price risks, increases access to infrastructure, increases security of land access and encourages technical innovation (Scherr, 1999:10).

In some situations, the supportive policy environment may be such that simply promoting information dissemination about good land husbandry practices and supporting research on technologies to reduce conservation costs may be sufficient for addressing degradation concerns (Scherr, 1999; Douglas, 1993). This is the case where recommended practices can be adopted within the existing resources of the farmers and extension support services.

In other cases, it may first be necessary to overcome local constraints such as credit services and/or larger constraints. Farmer investment in better land husbandry practices, for example, should increase where agricultural markets perform more effectively, reducing the costs of inputs and increasing farmgate output prices; where profitable farming opportunities raise the value of agricultural land; where technological change makes higher, sustainable yields possible; and where land tenure is secure. A conscientious effort on the part of government is needed to bring about policies that support this change.

Even assuming that governments are committed to supporting the agriculture sector, it is not easy finding the correct blend of policies. Vosti and Reardon (1997) identify the critical triangle of links between growth, poverty alleviation and sustainable natural resource use. A major challenge is to find policies, institutions and technologies that make the three goals more compatible. The goals are complementary and linked but it may take generations to achieve the complementarities.

Vosti and Reardon (1997) contend that, as countries strive to find the balance within the critical triangle there may, in the short-term, be trade-offs. For example, reducing poverty may involve cultivating more steeplands, which in turn may undermine efforts to alleviate poverty by causing more land degradation. This may partly explain why efforts to stimulate agricultural growth, such as trade liberalisation and the promotion of agricultural export crops, have exacerbated land degradation problems rather than led to the type of sustainable land use and poverty alleviation envisaged by organisations such as the World Bank.

6.5.3 Rural Non-agricultural Employment

For policy-makers, it is important to locate support to agriculture in the wider context of rural development and so adopt a more balanced set of policies. There is no case for exclusive reliance on agricultural development to improve the quality of life in rural areas. In the developing world, farming's capacity to provide the sole means of survival for rural populations is diminishing fast. Whether because of declining crop prices, competition for land and access to markets, or declining productivity due to soil and land degradation, smallholder farmers are diversifying their livelihood option and are moving into rural non-agricultural work (defined as work that excludes primary production such as agriculture, fisheries, livestock) or migrating to the towns and cities (Berdegué et al. 2000).

Resource-poor farmers worldwide, therefore, juggle the labour demands imposed by farm and off-farm economic activities. Any strategy aimed at reducing rural poverty should, therefore, augment access to productive resources involving both land and enhanced opportunities for off-farm employment (Stonich, 1993:167; Enters, 1996). In this context, rural non-agricultural employment (RNAE), in agro-processing, manufacturing and transport sectors, is re-emerging as a critical issue in sustaining viable rural economies and reducing rural poverty.

Whether through self-employment or wage-earning, RNAE enables resource-poor farmers to increase total income as well as offset the effects of fluctuations in income flows during the year (Berdegué et al. 2000). The mix of agricultural and non-agricultural work is, therefore, a strategy to reduce vulnerability because farmers are better able to substitute between opportunities that are in decline and those that are expanding. The contribution of RNAE to rural people's livelihoods should not be underestimated. In Latin America and the Caribbean, for example, it has been calculated that on average rural non-agricultural income accounts for over 40 percent of total rural income

(Derdegué et al. 2000). The figure for South Asia is approximately 60 percent (Ellis, 1999).

The importance of RNAE is likely to increase because agriculture today requires improved linkages with input supply systems, agricultural processing chains, and systems for the distribution of fresh and processed products. Modern agriculture requires cooperation with agro-industry in order to meet successfully the demanding quality and safety norms and standards of international markets. It also needs access to management, administrative and advisory services.

Strengthening the rural non-agricultural enterprise sector is, therefore, one way to achieve both value addition in the rural economy and poverty reduction. Efforts to bolster RNAE are essential to sustain the local economy and reduce the pressure on the rural farming population to migrate to urban areas. The development of RNAE offers a different trajectory for modernising the rural environment, through the *in-situ* development of industry and services and as part of a more general process of 'rural-urbanisation'

The rural enterprise sector, however, faces a variety of hurdles that must be resolved before a significant impact on rural poverty can be achieved. These hurdles include: identification of market opportunities; greater inclusion and empowerment of women; better access to appropriate processing technologies; implementation of effective business organization practices; more efficient farm to market channels; and the timely access to affordable financial and business services.

Economic benefits are unlikely to reach the poorest if the enterprises they work in, the value-chains they belong to, and the business environments within which they operate, lack the quality, reliability, market sensitivity and innovative capabilities needed to compete in a globalising world. These limitations to pro-poor economic growth can be mitigated through the development of skills, services and alliances between local and external actors and agencies (see Box 6.10).

Box 6.10 Development of rural enterprises in Bangladesh (from Hellin et al. 2004b)

Bangladesh is an exceptionally densely populated country and agriculture is the mainstay of the economy. The vast majority of Bangladesh's farmers live on the floodplains of the three major rivers that make up the Bengal Delta. During the height of the monsoon, from June to September, the rivers burst their banks and the floodplains become a vast expanse of brown muddy and fast-flowing water. The outflow is the third highest in the world after the Congo and Amazon rivers.

By December, the seasonal floodwaters have receded leaving behind rich deposits of silt that provide some of the best farmland in Bangladesh. During the dry season the floodplains have become a mosaic of greens and oranges, a patchwork of agricultural plots where local farmers grow a diversity of crops: rice, onions, garlic, mustard, egg-plant and pumpkin. The flatness of the floodplain is criss-crossed by elevated dirt roads and scattered villages.

The scene conveys a rural idyll but the reality is somewhat different. As in other parts of the developing world, farming's capacity to provide the sole means of survival for rural populations is diminishing fast. Whether because of declining crop prices, competition for land and access to markets, or declining productivity due to land degradation, smallholder farmers in Bangladesh are being forced to diversify into rural non-agricultural work or migrate to the towns and cities.

In Bangladesh, ITDG, a non-governmental development organisation, is exploring how to establish these alliances and ensure that the re-source-poor in rural areas have the skills and capability to participate in the rural non-agricultural enterprise sector.

ITDG Bangladesh has been researching and promoting services for rural non-agricultural enterprises since 1996. Through its Small Enterprise Unit, ITDG is supporting six large local development NGOs and around 120 small village-level organisations that they work with, to deliver business services to thousands of their members, mainly women. These services include training in technical and business skills, brokering of market linkages, coordination of input procurement and new product development.

ITDG Bangladesh has sought to promote access to business services for rural enterprises through this network of Bangladeshi NGOs, for three reasons:

- Village-level NGOs in Bangladesh have established a key role in the delivery of services related to health, nutrition, education and social empowerment of women. They, therefore, have unparalleled access to those most affected by poverty and in some cases a highly motivated 'socially responsible' workforce of staff and volunteers.
- Many village-level NGOs have active group-based savings and credit schemes. There are many advantages to delivering business services in conjunction with these schemes.
- The absence of any significant commercial market for business services in remote and poorly connected rural areas, means that village-level NGOs are often the sole agencies capable of reaching the poorest rural entrepreneurs.

ITDG's approach is to promote sustainable and market-orientated service delivery for RNAE. It does this by encouraging NGOs at district and village level to test the value of their services through fee-recovery; by promoting the use of private trainers to delivery business services; and by brokering links among NGOs, local authorities and private companies in each local region.

- ITDG's practical support to the partner NGOs takes the form of:
- Publishing a series of highly popular income-generating activity profiles that provide basic information in Bangla, the local language, about the technicalities, skills, equipment and investment required for a wide range of different types of enterprises.
- Business skills training and orientation for NGO staff, including advice on establishing revenue-generating activities in areas such as agro-processing, batik-making and small-scale manufacturing.
- Technical skills training for NGO field officers and volunteers.
- Training in organisational planning and administration as well as market analysis, including sub-sector studies.

It is also important to note that change does not have to come all at once, relatively small advances such as introducing electricity into communities can lead to economic diversification such as carpentry shops. Diversification can provide alternative livelihood sources, reduce dependence on marginal farms and/or generate resources for investment in land improvements (Scherr and Yadav, 1996:23; Vosti and Reardon, 1997b). However, Bunch (1982:45) warns that the difficulties of finding markets and teaching people how to manage a business should not be underestimated.

Policies that encourage investment in support services for RNAE could benefit resource-poor farmers in two ways: by strengthening the markets for local agricultural production, and by providing farmers with the vulnerability-reducing opportunity to diversify their livelihoods. Supporting RNAE complements the better land husbandry approach, an approach that on its own is unlikely to resolve the enormous problem of rural poverty. However, there are cases where the incentive to invest in improved land management is reduced as farm households gain more access to non-farm incomes, Holden et al. (2004) report such a phenomenon in the Ethiopian Highlands.

6.6   BETTER LAND HUSBANDRY AND FUTURE RESEARCH

6.6.1   What can Researchers and Development Practitioners Do?

Researchers and development practitioners can do much to influence policy although it is not always clear how. Research by the London-based Overseas Development Institute (Court et al. 2005) suggests that research is more likely to contribute to evidence-based policy if:

- It fits within the political and institutional limits and pressures of policy makers and resonates with their ideological assumptions or sufficient pressure is exerted to challenge those limits.
- The evidence is credible and convincing, provides practical solutions to current policy problems and is packaged to attract policy-makers interest.

- Researchers and policy makers share common networks, trust each other, honestly and openly represent the interests of all stakeholders and communicate effectively.

The reality, of course, is that these three conditions are rarely met in practice. Although researchers can control the credibility of their evidence and ensure that they interact with and communicate well with policy-makers, they often have limited capacity to influence the political context within which they work, especially in less democratic countries. Despite this, researchers and development practitioners can still do much if they want to achieve policy impact for example in land management (see Table 6.1).

Table 6.1 Influencing policy (from Court et al. 2005)

What researchers and development practitioners need to know	What researchers and development practitioners need to do	How to do it
Political context:	**Political context:**	**Political context:**
• Who are the policy makers?	• Get to know the policy makers, their agendas and the constraints they operate under	• Work with policy makers
• Are there policy maker demands for new ideas?	• Identify potential supporters and opponents	• Seek commissions
• What are the sources and strengths of resistance?	• Keep an eye on the horizon and prepare for opportunities in regular policy processes	• Line up research programmes with high-profile policy events
• What is the policy-making process?	• Look out for and react to unexpected policy windows	• Reserve resources to be able to move quickly to respond to policy makers
• What are the opportunities and timing for input into formal processes?		• Allow sufficient time and resources
Evidence:	**Evidence:**	**Evidence:**
• What is the current theory?	• Establish credibility over the long-term	• Build up respected programmes of high-quality work
• What are the prevailing narratives?	• Provide practical solutions to problems	• Action-research and pilot projects to demonstrate benefits of new approaches
• How divergent is the new evidence?	• Establish legitimacy	• Use participatory approaches to help with legitimacy and implementation
• What sort of evidence will convince policy makers	• Build a convincing case and present clear policy options	
	• Package new ideas in familiar theory or narratives	• Clear strategy and resources for
	• Communicate effectively	

(Contd.)

(Contd.)

		communication from start
		• Real communication "seeing is believing"
Links:	**Links:**	**Links:**
• Who are the key stakeholders in the policy discourse	• Get to know the other stakeholders	• Partnerships between researchers, policy makers and communities
• What links and networks existe between them?	• Establish a presence in existing networks	• Identify key networkers and salesmen
• Who are the intermediaries and what influence do they have?	• Build coalitions with like-minded stakeholders	• Use informal contacts
• Whose side are they on?	• Build new policy networks	

6.6.2   Need for More Research

In the context of growing soil and land degradation problems, there are increasing demands, from various quarters, for more applied research on soil and water loss and changes in productivity (on steep slopes) even though this type of research is both capital-intensive and time-consuming (see Box 6.11). The demand stems from the recognition of the problems associated with farming on steep slopes, the increasing number of farmers cultivating these marginal areas, and the dearth of data on the erosion processes on steep slopes and consequences of these erosion processes on agricultural productivity.

Box 6.11   Demands for more research

The World Bank, the Food and Agriculture Organisation, United Nations Development Programme, United Nations Environment Programme and other international and national organisations have emphasised the need for better field measurements of soil erosion and its effects upon production. They highlight steeplands as one of the areas with highly distinctive problems of land resource management (Pieri et al.1995:15). Thereby highlighting the unjustified and unproven assumption that there is a direct link between soil loss and yield loss.

Wu and Wang (1998) and IFPRI (Scherr and Yadav, 1996:21) have also identified the need for more research on soil loss from steeplands. Although Lal (1994) recognises that it is difficult to relate soil erosion to the erosion-caused effects on soil properties and crop yields, he identifies

---

the need for more research to establish quantitative relationships between soil loss and crop yield for a range of management systems (Lal, 1982 and 1988). Lutz et al. (1994) and Tengberg et al. (1998) have also identified the need for more research on erosion-induced loss in soil productivity.

Meanwhile, Rosegrant and Cline (2003) point out that the research agenda also needs to encompass participatory plant breeding for yield increases in rainfed agricultural systems along with innovations in agroecological approaches.

Future research will need to be largely funded from public and non-profit funds (Runge et al. 2003:144). While it is true that the private sector is spending more on agricultural research and development, the vast majority of this research is conducted in developed countries (Pardey and Beintema, 2001). Furthermore, much of the research carried out by the private sector involves food-processing and crop technologies that are more suited to capital-intensive forms of commercial agriculture. Pardey and Beintema (2001) stress that *"it is folly to think that private research will substantially replace public science in developing countries anytime soon."*

Altieri (2002), however, has argued that natural resource problems experienced by resource-poor farmers are not amenable to the research approaches previously used by the international research community even when public funds are being used. He argues that in most research organisations, including the centres associated with the Consultative Group for International Agricultural Research (CGIAR), research has been commodity-oriented with the goal of improving yields of specific food crops or livestock and far too little attention has been directed at understanding the needs and opportunities of the poor.

This criticism is less valid than it used to be: many of the international research centres' work is now guided by a livelihoods and pro-poverty perspective. For example, the preface to the International Maize and Wheat Improvement Center's (CIMMYT) *Strategy for helping to reduce poverty and hunger by 2020* states that *"the Center ...... has a new structure to bridge the disciplinary divides that occur within most research centers, The strategy focuses CIMMYT research more sharply on reducing people's vulnerability to poverty by looking at the entire context in which poor households operate and not exclusively at maize or wheat."* (CIMMYT, 2004:v). Furthermore, there is a growing body of evidence that demonstrates that agricultural research conducted by the CGIAR and other institutions has had a beneficial impact on the livelihoods of smallholder farmers, largely through increased agricultural production and subsequently higher rural incomes for the producers (The World Bank, 2003; Thirtle et al. 2003; Hazell and Haddad, 2001; Runge et al. 2003:142).

The focus of current research and extension activity linked to hillside agriculture is still, however, too focused on the conventional soil conservation approach, where the concepts of transfer of technology and soil erosion control are favoured in place of farmer first and soil quality. Segerros and Kerr (1996) contend that many manuals on soil conservation continue to focus on the control of soil erosion and ignore issues of soil quality and the complexity of farmers' realities and decision-making processes.

One such technology that continues to be widely promoted is live barriers of vetiver grass (Grimshaw, 1995). According to Young (1997), farmer-adoption rates of the live barrier technology are variable but no worse than for earlier conservation projects. If some of the data in Table 3.2 are reliable, however, higher adoption rates might have been expected. Young (1997b:71) observes that there is some evidence that contour hedgerows' ability to control soil and water loss diminishes as slope angle increases. Young (1997b:76) states that more research is needed specifically on whether live barriers of vetiver grass are effective in controlling runoff and erosion on steep slopes.

Bearing in mind some of the fundamental problems with the conventional soil conservation approach is this really where research efforts should be directed? Should we note be directing research efforts more towards improving soil quality? Over ten years ago, Hudson (1995:204) wrote: *"If the present trend continues 'soil conservation' may become a pejorative term"* (Hudson, 1995:204). Despite a growing appreciation of the virtues of a better land husbandry approach, it is clear that soil conservation still has some way to go before it becomes a pejorative term. The aficionados of the better land husbandry approach still face a challenge in getting their message across even though there is much evidence that progress is being made.

It is important, though, that the better land husbandry message remains a coherent one. Soil conservation and land management problems are multi-dimensional and will require cross-disciplinary approaches to resolve them. The danger is that in the context of poor and falling productivity and in the search for alternatives to the conventional soil conservation approach, the pendulum has swung proportionally too far towards the social sciences.

Gardner and Lewis (1996:67) correctly point out that the success of all projects depends upon whether or not they are socially and culturally appropriate. However, these same projects also have to be technically appropriate. Christiansson et al. (1993) report that *"the trend seems to be one of ignoring quantitative biophysical data and processes to the extent that we might start losing more than we gain."* Chapters 3 and 4 and the above discussion demonstrate that the agro-ecological components of better land husbandry are critical and any programme that ignores these components will almost certainly fail.

Whilst it is not a panacea to land degradation, an approach that encourages better land husbandry has the potential to draw social and natural scientists into productive dialogue with the land users, leading to practical and realistically sustainable land management development initiatives that can play a critical role in sustainable agriculture intensification and the sound management of natural resources. It is very important, though, to be able to assess the impact of better land management initiatives. Appendix 1 outlines some approaches and tools for monitoring and assessing the environmental and livelihood impact of better land husbandry.

APPENDIX 1

Impact Monitoring and Assessment

IMPORTANCE OF IMPACT MONITORING AND ASSESSMENT

Better land husbandry is a cross-disciplinary approach to improved land management that involves social, economic and agro-ecological aspects. Any impact monitoring and assessment system needs to measure or assess impact in all these areas. In this Appendix, some basic approaches and tools are provided for environmental impact (covering the agro-ecological aspects of better land husbandry) and livelihood impact (covering the social and economic aspects of better land husbandry). The environmental impact section comes from Douglas (1997) and the livelihood impact section comes from Hellin et al. 2003.

Impact can be defined as the mid- and long-term implications that a project has for the context and its population. Impact is closely related to the effectiveness of a project, although the reality is that a project can only establish and show plausible relations between its actions and changes in the context (Herweg and Steiner, 2002). Herweg and Steiner (2002) give details on how an impact monitoring and assessment procedure can be integrated into project cycle management for sustainable land management projects.

ENVIRONMENTAL IMPACT

One of the challenges facing development practitioners who wish to work with farmers to achieve better land management is how you can best characterise the baseline condition of the land and the improved land use changes that can be attributed to the project. The following guidelines cover conservation effectiveness, the use of visual indicators of erosion, and criteria for determining the better land husbandry rating.

Conservation-effectiveness

The overall conservation-effectiveness of farmers' land use and management practices should be used as the basic criterion for determining the environmental impact of any project intervention. In this context any project intervention can be assessed qualitatively to determine whether it is:

- **Conservation-effective**—the intervention has directly or indirectly contributed to the maintenance and enhancement of the soils' productivity and has prevented further degradation

- **Conservation-neutral**—the intervention has had no direct or indirect beneficial or negative impact on the soils' productivity or land degradation
- **Conservation-negative**—the intervention is believed to have directly or indirectly contributed to a decline in soil productivity and has exacerbated land degradation

Clearly, the aim is to use interventions, such as the use of cover crops, which are conservation-effective. In order to be so, the intervention needs to play a positive role in one or more of the following:

- **Improved crop management**—adopted practice(s) should lead to an increase in the protective ground cover provided by the growing crops (e.g. intercropping, early planting, use of improved seed and fertiliser). See Table A 1.1.
- **Improved soil management**—adopted practice(s) should lead to an increase in organic matter levels, biological activity and topsoil erosion resistance (e.g. use of animal manure, compost, mulches, incorporation of crop residues and green manure crops, improved tillage techniques). See Table A 1.2.
- **Improved rainwater management** - adopted practice(s) should lead to a reduction in surface runoff and increase infiltration (e.g. contour strip cropping, tied ridges, hedgerows and other cross-slope barriers). See Table A 1.3.

Some of the key factors that should be considered in assessing the conservation-effectiveness of a project intervention and its associated land use and management practices are shown in Box 1. More details are provided in Tables A1.1, A1.2 and A1.3. The use of some or all of these factors would facilitate the monitoring and evaluation of the environmental impact of any project intervention.

Box 1   Three broad types of indicators to determine soil quality

1. **Above the soil surface as related to yields, for example:**
   Cover close to the ground—its density, distribution, duration and timing etc.
   Stress in plants—growth rates, timing and frequency of wilting; visible nutrient deficiencies or imbalances
2. **On the soil surface as affecting soil moisture and runoff and erosion, for example:**
   Porosity of first few millimetres of topsoil layers—proportions of incident rainfall becoming infiltrated
3. **Below the soil surface, for example:**
   Organic matter content and biological activity—as affecting soil architecture (structural stability, gas exchange, water movement and retention/release) and cation exchange capacity (nutrient capture and retention, pH buffering; nutrient availability, source of small amounts of recycled nutrients)

Table A1.1 Crop management considerations (from Douglas, 1997)

Improved crop management indicators	Conservation-effective	Conservation-neutral	Conservation-negative
Change in percentage ground cover provided by growing crop(s)	Net increase in ground cover provided by annual crops–at least 40% cover achieved within 30 days of the start of rainy season	No change in percentage ground cover provided by annual crops during cropping season	Decrease in percentage ground cover provided by annual crops - ground cover remains below 40 % for most of cropping season
Intercropping/relay cropping	Change in existing intercropping/relay cropping practices leading to improved ground cover and/or increase in ratio of legumes (N-fixing) to N-demanding crops	No change	Change in existing intercropping/relay cropping practices leading to a reduction in ground cover and/or a decrease in ratio of legumes (N-fixing) to N-demanding crops
Spacing/plant density	Improvement in ground cover through closer crop spacing/increased plant density per unit area	No change	Reduction in ground cover due to wider crop spacing/decreased plant density per unit area
Improved seed/planting material	Adoption of improved seed/planting material results in production of more biomass and better ground cover than farmers' traditional varieties	No change in crop biomass and ground cover	Adoption of improved seed/planting material results in production of less biomass and inferior ground cover than farmers' traditional varieties
Fertiliser and/or organic manure	Increase in quantity of fertiliser and/or organic manure used resulting in production of more crop biomass and better ground cover	No change in quantity of fertiliser and/or organic manure used for crop production	Reduction in quantity of fertiliser and/or organic manure used resulting in production of less crop biomass and poorer ground cover
Crop residues	Crop residues incorporated into soil or retained on soil surface as a protective mulch	Not applicable	Crop residues burnt or fed to livestock

Table A1.2 Soil management considerations (from Douglas, 1997)

Improved crop management indicators	Conservation-effective	Conservation-neutral	Conservation-negative
Soil organic matter	Project interventions and good land husbandry practices that enhance soil organic matter levels, e.g. • Incorporation of all crop residues • Application of at least 3 tons/hectare/year of compost and/or animal manure • Application of at least 5 tons/hectare/year of fresh green manure	Project interventions and other land husbandry practices that maintain soil organic matter levels, e.g. • Grazing livestock on crop residues *in-situ* • Application of compost and/or animal manure at a rate below 3 tons/hectare/year • Application of fresh green manure at a rate below 5 tons/hectare/year	Poor land husbandry practices associated with specific interventions that continue the reduction in organic matter levels, e.g. • Removal or burning of all crop residues • No application of compost and/or animal manure • No application of green manure (i.e. all hedgerow biomass removed as fuel or fodder)
Soil chemical properties	Project interventions and good land husbandry practices that replace soil nutrients lost by leaching, topsoil erosion and removed in harvested products, e.g. • Application of compost and/or animal manure • Use of nitrogen-fixing species (crop rotation and intercropping with legumes, nitrogen-rich green manures and hedgerows) • Enriched fallows	Traditional low input fertility management practices capable of achieving low levels of nutrient replenishment, e.g. • Short bush fallow • Tethered grazing of livestock within farm plots on crop residues and weeds • Retention of a few scattered trees on the croplands	Poor land husbandry practices associated with specific interventions that continue the depletion of soil nutrients, e.g. • Continuous cultivation of cereal and root crops • Burning of crop residues • Little if any use of compost, organic manures or chemical fertiliser

(Contd.)

(*Contd.*)

	• Application of chemical fertiliser as a supplement to but not a substitute for organic manure		Poor land husbandry practices associated with specific interventions that continue the physical degradation of the soil, e.g. • Excessive tillage • Continuous cultivation • No incorporation of any organic matter • Trampling by people and livestock
Soil physical properties	Project interventions and good land husbandry practices that maintain and enhance topsoil structures, e.g. • Minimum tillage • Planted pasture and enriched fallows • Incorporation of crop residues, compost, animal manure, green manure and tree litter	Traditional low input land husbandry practices that neither combat nor promote the physical degradation of the soil, e.g. • Partial tillage • Short bush fallow • Retention of a few scattered trees on the croplands	

Table A1.3 Rainwater management considerations (from Douglas, 1997)

Improved crop management indicators	Conservation-effective	Conservation-neutral	Conservation-negative
Decrease large rain-drops' power to damage and seal soil surface	See aspects of <u>cover</u> in Table A 1.1. Crop management considerations		
Infiltration	Project interventions and good land husbandry practices that increase infiltration, e.g. • Maintenance of an open structure on the soil surface through appropriate tillage and organic materials' management practice • *In-situ* entrapment of rainwater (tied crop ridges, pits and micro-basins) • Permeable cross-slope barriers to slow down but not totally arrest runoff (hedgerows, grass strips, trash barriers, rock walls) • Impermeable cross-slope barriers to check all runoff (contour ditches, earth banks.)	Project interventions and associated land husbandry practices that neither reduce nor increase infiltration, e.g. • Tree planting in anything other than closely-spaced hedgerows • Bio-intensive gardens	Poor land husbandry practices associated with specific interventions that reduce infiltration, e.g. • Up and down slope cultivation • Poor construction and maintenance of infiltration structures leading to filling-in of ditches and breaching of cross-slope barriers • No incorporation of any organic matter • Surface compaction due to trampling by people and livestock

(Contd.)

(Contd.)

Reduction of runoff volume and velocity	Project interventions and good land husbandry practices that significantly reduce surface runoff volume and velocity, e.g.	Project interventions and associated land husbandry practices that neither reduce nor increase runoff, e.g.	Poor land husbandry practices associated with specific interventions that concentrate and speed up runoff, e.g.
	• Contour cultivation	• Tree planting in anything other than closely-spaced hedgerows	• Up and down slope cultivation
	• Increased surface roughness (litter, stone mulch, soil clods)	• Bio-intensive gardens	• Poor alignment and maintenance of cross-slope barriers
	• *In-situ* entrapment of rainwater (tied crop ridges, pits and mico-basins)		• Extensive hillside cultivation with no soil and water conservation measures to reduce effective slope length
	• Permeable cross-slope barriers to slow down but not totally arrest runoff (hedgerows, grass strips, trash barriers, rock walls)		
	• Impermeable cross-slope barriers to check all runoff (contour ditches, earth banks)		

Qualitative Visual Indicators to Estimate the Status, Type and Severity of Soil Erosion

Monitoring changes in the status, type and severity of erosion within the project area will provide a means for evaluating the impact of the project interventions on the biophysical environment. Such changes can be monitored at:

- Individual plot/micro level
- Farm household level (total farm holding)
- Macro geographic area level (based on the topographic boundaries of a mini watershed or the socio/cultural boundaries of a participating community)

It is recommended that erosion changes be monitored, for each level, at the end of each cropping or rainy season. The monitoring visits should be combined with those for collecting data for the socio-economic impact monitoring. Qualitative estimates would be made during these visits by the use of direct observation in individual farmers' fields and when undertaking participatory transects (i.e. walking through an area with a group of farmers and discussing with them the observable land use changes and visual indicators of land degradation). Note that data on changes should be recorded on a Land Management Unit basis rather than just for the total project area.

The following notes provide examples of some visual parameters that could be used for assessing qualitatively the state and severity of erosion. Such visual parameters call for the observer to make a subjective visual assessment based on his/her past experience and local knowledge. It is not possible to give quantitative or precise definitions of what constitutes slight, moderate or severe erosion. There is therefore an element of imprecision in this approach, in that it is possible for different observers to arrive at different classifications for the same area. However, it is believed that different observers can achieve a degree of uniformity through shared training and field experience. Herweg (1996) has produced a very useful booklet on field tools for the assessment of erosion damage.

State of Erosion

A	Active	One or both of the following conditions apply: evidence of sediment movement; sides and/or floors of erosion forms (e.g. rills, gullies) are relatively bare of vegetation.
B	Partly stabilized	Evidence of some active erosion but also some evidence of stabilization.
C	Stabilized	One or both of the following conditions apply: no evidence of sediment movement; sides and/or floors of erosion forms are vegetated.

Sheet Erosion

Sheet erosion is the relatively uniform removal of soil from an area without the development of conspicuous channels. Indicators of sheet erosion include 'pedestalling'[1]; root exposure; exposure of subsoils; soil deposits against field boundaries, hedgerows and conservation structure downslope.

X	Not apparent	No obvious signs of sheet erosion but evidence of minor sheet erosion may have been masked by for instance tillage.
0	No sheet erosion	No visual indicators of sheet erosion.
1	Slight	Some visual evidence of the movement of topsoil particles downslope through surface wash; no evidence of pedestal development; only a few superficial roots exposed.
2	Moderate	Clear signs of the transportation and deposition of topsoil particles downslope through surface wash; some pedestalling but individuals pedestals no more than 5 cms high; some tree and crop roots exposed within the topsoil; evidence of topsoil removal but no subsoil horizons exposed.
3	Severe	Clear evidence of the wholesale transportation and deposition of topsoil particles downslope through surface wash; individual pedestals over 5 cms high; extensive exposure of tree and crop roots; subsoil horizons exposed at or close to the soil surface.

Rill Erosion

A rill is a small channel less than 300 mm deep. Rills can be completely smoothed out by cultivation with animal- or machine-drawn implements, although traces (depression lines within the field) may remain where all cultivation is done by hand.

0	No rill erosion	No rills present within the field.
1	Slight	A few shallow (less than 100 mm depth) rills affecting no more than 5 percent of the surface area.
2	Moderate	Presence of shallow to moderately deep rills (<200 mm depth) and/or rills affectinq up to 25 percent of the surface area.
3	Severe	Presence of deep rills (up to 300 mm depth) and/or rills affecting more than 25 percent of the surface area.

---

[1]Pedestalling occurs when an easily eroded soil is protected from raindrop impact by a stone or tree root, isolated pedestals capped by the resistant material are left standing up from the surrounding ground. Providing there is little or no undercutting at the base of the pedestal then the removal of the surrounding soil is the result of splash erosion rather than by surface flow.

Gully Erosion

A gully is a channel 300 mm or more deep. It will provide a physical impediment to the movement, across the slope, of animal- or machine-drawn implements. It cannot be smoothed out in the course of normal cultivation.

0	No gully erosion	No gullies present within the field.
1	Slight	A few shallow (<0.5 m) gullies affecting no more than 5 percent of the surface area.
2	Moderate	Presence of shallow to moderately deep gullies (0.5-1.0 m depth) and/or gullies affecting 5-25 percent of the surface area.
3	Severe	Presence of deep gullies (>1 m depth) and/or gullies affecting more than 25 percent of the surface area.

Mass Movement

This includes all relatively large downslope movements of soil and/or rock e.g. landslides, slumps, earth flows and debris avalanches. This category of land degradation would be described for relatively large land units, such as watersheds, rather than for individual fields.

Status

A	Active	Landslide scars clearly visible with sharp boundaries and less than 10 percent cover within the landslide area.
B	Partly stabilised	Landslide scars early visible with vegetation cover between 10-50 percent of the area of the landslide.
C	Stabilised/inactive	Landslide scars still detectable but no longer with sharp boundaries and with greater than 50 percent vegetation cover within the landslide area.

Severity

0	No mass movement			No visible evidence of mass movement
P	Present	1	Slight	Isolated examples of mass movement. Individual events small in size and/or affecting less than 0.1 percent of total area.
		2	Moderate	A moderate number of mass movement events. Individual events small to moderate in size and/or affecting 0.1-1.0 percent of total area.
		3	Severe	Significant number of mass movement events. Individual events may be large in size and/or affect over 1 percent of its total area.

Better Land Husbandry Rating-rating the Overall Care of the Land

The assessments of, firstly, the conservation effectiveness of particular interventions and, secondly the status, type and severity of soil erosion can be used as the basis for arriving at an overall 'better land husbandry' rating. Such a rating would of necessity be subjective but it would enable a qualitative assessment to be made of the overall environmental impact of the project. Its main purpose with regard to the project's monitoring and evaluation system would be to provide a clear indication to the extent to which the land use management practices on a particular plot, individual farm holding or over a wider geographic area conform to the principles and practices of better land husbandry. If they do, then they would be in line with the project's goal and purpose and could be used as an indicator of success. The degree to which the land use management practices conform to the principles and practice of better land husbandry would correspond to the following ratings:

Rating	Criteria	Score
Excellent	The land husbandry practices are exemplary	4
Good	The land husbandry practices are of acceptable quality	3
Fair	The land husbandry practices give some cause for concern and require minor corrective actions	2
Poor	The land husbandry practices give major cause for concern and require considerable corrective action	1
Very poor	Conforms to none of the requirements for better land husbandry	0

The following tables provide indicative guidelines for determining the specific better land husbandry rating. Note that, in arriving at an individual rating, it would not be necessary for the land use management practices within the area being assessed, to conform to everyone of the factors listed but obviously, the more the better. Likewise the assessment can be applied to areas in which there are no project soil conservation interventions. For example an area of gently sloping land with deep well-drained soils used for arable farming could still be rated as excellent even if no soil conservation structures had been installed, providing all the crop husbandry practices being followed conform to the requirements for better land husbandry.

Excellent Land Husbandry Rating

Description	Score
A No evidence of active erosion. Gullies completely stabilised and healed.	4
B Ground cover provides the best protection against splash erosion that could be expected, given the prevailing climate.	
1 **Croplands**: annual crops achieve at least 40 percent ground cover within 20 days of the start of the rainy season.	

(Contd.)

(Contd.)

2 **Pasture**: grasses are evenly and closely spaced with no bare areas.

3 **Woodland**: mature trees, closed canopy and continuous litter layer.

C **Land management exemplary**

1 **Crop husbandry**

Contour cultivation and minimum tillage.

Use of improved varieties.

Optimum crop spacing/plant density.

High ratio of legumes (N-fixing) to N-demanding crops.

Optimum plant nutrition (minimum 5 tons/ha/yr of compost/ animal manure or minimum 10 tons/ha/yr green manure from hedgerows, supplemented as needed with chemical fertilizer.)

All crop residues returned to the soil.

Crop rotation incorporating a 2-3 year partially enriched fallow.

No annual crop production on land with slope greater than 30 percent.

2 **Pasture**

Use of improved pasture management practices (e.g. controlled grazing).

Use of improved pasture species (grasses and herbaceous legumes).

On-farm forage production with contour hedgerows and grass strips used on a cut-and-carry basis (zero grazing).

3 **Trees**

Closed-canopy multi-storey home garden with a good ground-level herbaceous and litter layer.

Minimum of 30 mature trees per hectare as scattered or boundary plantings within the cropped lands.

Well-managed woodlots/orchards with retention of litter below the trees.

4 **Water**

Streams run clear during the rains.

Springs flow for 12 months of the year.

D **Project soil conservation interventions of exemplary quality**.

1 **Initial construction and establishment**

Follow the contour.

All the stems of the shrubs (in the hedgerows) and/or the perennial crops close enough together to function as a cross-slope runoff control barrier.

Rock walls and gully plugs well constructed and stable.

No gaps or low points.

2 **Maintenance**

Gaps filled and storm damage speedily repaired.

(Contd.)

(Contd.)

Rock walls, gully plugs and earth banks raised in line with which soil accumulates behind them.

No signs of rill or sheet erosion immediately below individual hedgerows, rock walls and earth banks, no active gullying within the plugged gullies.

Good Land Husbandry Rating

Description	Score
A Evidence of slight sheet or rill erosion. No active gullying.	3

B   Ground cover still provides reasonable protection against splash erosion, but with some scope for improvement.

1 **Croplands**: annual crops achieve at least 50 percent ground cover within 30 days of the start of the rainy season.

2 **Pasture**: occasional bare spots in an otherwise continuous grass sward.

3 **Woodland**: mature trees, open canopy with an almost continuous ground layer of herbs and leaf litter.

C   **Land management good**

1 **Crop husbandry**

Contour cultivation and minimum tillage.

Use of improved varieties.

Slightly below optimum crop spacing/plant density.

Ratio of legumes (N-fixing) to N-demanding crops is still good but scope for improvement.

Good plant nutrition (minimum 3 tons/ha/yr of compost / animal manure or minimum 5 tons/ha/yr green manure from hedgerows, supplemented as needed with chemical fertilizer.)

All crop residues returned to the soil.

Crop rotation incorporating a partially enriched fallow of no more than two years.

No annual crop production on land with slope greater than 30 percent.

2 **Pasture**

Use of improved pasture management practices (e.g. controlled grazing).

Limited use of improved pasture species (grasses/herbaceous legumes).

On-farm forage production with contour hedgerows and grass strips used on a cut-and-carry basis (zero grazing).

3 **Trees**

Partially closed-canopy multi-storey home garden with a moderately good ground-level herbaceous and litter layer.

Minimum of 20 mature trees per hectare as scattered or boundary plantings within the cropped lands.

(Contd.)

(Contd.)

Moderately well-managed woodlots/orchards with retention of most of the litter below the trees.

4 **Water**

Streams run clear in the rainy season except during severe storm events.

Springs flow for all but 1-2 months of the year.

D **Project soil conservation interventions of acceptable quality.**

1 **Initial construction and establishment**

Follow the contour.

Over 90 percent of the stems of the shrubs (in the hedgerows) and/or the perennial crops close enough together to function as a cross-slope runoff control barrier.

Rock walls and gully plugs acceptably constructed and stable

No major gaps or low points.

2 **Maintenance**

Gaps filled and storm damage repaired within a season.

Rock walls, gully plugs and earth banks raised in line with the rate at which soil accumulates behind them.

Only slight signs of rill or sheet erosion immediately below individual hedgerows, rock walls and earth banks, no active gullying within the plugged gullies.

Fair Land Husbandry Rating

Description	Score
A Evidence of moderate sheet or rill erosion. Some slight gully erosion.	2
B Ground cover provides only fair protection against splash erosion, with considerable scope for improvement.	
1 **Croplands**: annual crops achieve at least 40 percent ground cover but takes more than 30 days from the start of the rainy season.	
2 **Pasture**: frequent bare spots in patchy grass sward.	
3 **Woodland**: scrubby regrowth with a fair ground layer of herbs and leaf litter.	
C **Land management fair**	
1 **Crop husbandry**	
Cultivation approximately on the contour.	
Limited use of improved varieties.	
Below optimum crop spacing/plant density.	
Ratio of legumes (N-fixing) to N-demanding crops is still good but scope for improvement.	
Below optimum plant nutrition (animal manure from livestock tethered and grazing in the field, some N from the roots of leguminous hedgerow plants but no use of prunings or chemical fertilizer.)	

(Contd.)

(Contd.)

Crop residues burnt with the ashes returned to the soil.

Crop rotation incorporating a short bush fallow lasting at most two years.

Annual crop production may occur on land with slopes up to 40 percent.

2 Pasture

Generally uncontrolled grazing on unimproved natural pasture.

Very limited on-farm forage production from a few contour hedgerows that are generally grazed *in-situ*.

3 Trees

Open canopy multi-storey home garden with a patchy ground-level herbaceous and litter layer.

Between 15-20 trees per hectare as scattered or boundary plantings within the cropped lands.

No consolidated tree planting in woodlots or orchards.

4 Water

Streams are frequently discoloured with silt-laden runoff during the rainy season.

Springs flow for up to eight months of the year.

D **Project soil conservation interventions found with minor problems needing correction to improve their conservation effectiveness.**

1 Initial construction and establishment

Minor problems requiring some corrective action.

75-90 percent of the stems of the shrubs (in the hedgerows) and/or the perennial crops close enough together to function as a cross-slope runoff control barrier.

Minor problems requiring some corrective action to with regards to the construction of the rock walls and gully plugs.

Obvious gaps and low points requiring corrective action.

2 Maintenance

Delay in filling gaps and repairing storm damage.

Rock walls, gully plugs and earth banks rarely raised to allow for further soil to accumulate behind them.

Moderate signs of rill and sheet erosion immediately below individual hedgerows, rock walls and earth banks, some slight gully erosion still taking place within the plugged gullies.

Poor Land Husbandry Rating

Description	Score

A   Evidence of moderate sheet or rill erosion with moderate gully erosion.   1

B   Ground cover thin, provides little protection against splash erosion and requires improvement.
1   **Croplands**: annual crops provide less than 40 percent ground cover for most of the rainy season.
2   **Pasture**: less than 40 percent cover from a very patchy grass sward.
3   **Woodland**: severely degraded with only some scrubby regrowth, poor ground layer of herbs and leaf litter.

C   Land management poor.
1   **Crop husbandry**
Cultivation does not adhere to the contour.
No use of improved varieties.
Wide crop spacing and low plant density.
Low ratio of legumes (N-fixing) to N-demanding crops
No use of animal manure, green manure, or chemical fertilizer.
All crop residues burnt or removed from the field for fuel and fodder.
Continuous cultivation with only infrequent periods of short bush fallow lasting at most two years.
Annual crop production commonly on land with slopes up to 45 percent.
2   **Pasture**
Uncontrolled grazing on unimproved natural pasture.
No on-farm forage production.
3   **Trees**
Multi-storey home garden comprises only a few trees and shrubs with predominantly annual crops below.
Between 1-5 trees and shrubs per hectare as scattered or boundary plantings within the cropped lands.
No other tree planting.
4   **Water**
Streams discoloured with silt-laden runoff during the rainy season.
Springs dry up shortly after the end of each rainy.

D   Project soil conservation interventions not adopted or found with serious problems needing correction to improve their conservation effectiveness.
1   **Initial construction and establishment**
Serious problems requiring drastic corrective action to improve the contour alignment.
Only 50-75 percent of the stems of the shrubs (in the hedgerows) and/or the perennial crops close enough together to function as a cross-slope runoff control barrier.

(Contd.)

(Contd.)

Serious problems requiring major corrective action with regard
to the construction of the rock walls and gully plugs.

Frequent gaps and low points requiring major corrective action.

2 **Maintenance**

No attempt to fill gaps and repair storm damage.

Rock walls, gully plugs and earth banks very rarely if ever
raised to allow for further soil to accumulate behind them.

Moderate signs of rill and sheet erosion immediately below
individual hedgerows, rock walls and earth banks,
moderate gully erosion still taking place within the plugged
gullies.

Very Poor Land Husbandry Rating

Description		Score
A	Evidence of severe sheet and/or rill erosion with moderate to severe gully erosion.	0
B	Ground cover very thin or absent, providing little if any protection against splash erosion.	

    1 **Croplands**: annual crops provide less than 40 percent ground
cover for all of the rainy season.

    2 **Pasture**: less than 30 percent cover from a very patchy grass
sward.

    3 **Woodland**: severely degraded with only a few stumps and
some scrubby regrowth with little in the way of grass and
litter and much bare ground.

C   **Land management very poor**

    1 **Crop husbandry**

Cultivation up and down slope.

No use of improved varieties.

Very wide crop spacing and very low plant density.

Very few legumes (N-fixing) compared to quantity of N-
demanding crops.

No use of animal manure, green manure, or chemical fertilizer.

All crop residues burnt or removed from the field for fuel
and fodder.

Continuous cultivation with only infrequent periods of short
bush fallow lasting at most one year.

Annual crop production commonly on land with slopes over
45 percent.

    2 **Pasture**

Uncontrolled grazing on unimproved natural pasture.

No on-farm forage production.

    3 **Trees**

Multi-storey home garden comprises only a few trees and
shrubs with predominantly annual crops below.

(Contd.)

(Contd.)

No trees occurring as scattered or boundary plantings within the cropped lands.

No other tree planting.

4 **Water**

Streams flow intermittently and highly discoloured with silt-laden runoff during the rainy season.

Springs flow for only short periods during each rainy.

D Project soil conservation interventions not adopted or found with serious problems needing correction to improve their conservation effectiveness.

1 **Initial construction and establishment**

Very serious problems requiring replanting or reconstruction to achieve the required contour alignment.

Less than 50 percent of the stems of the shrubs (in the hedgerows) and/or the perennial crops close enough together to function as a cross-slope runoff control barrier.

Very serious problems requiring the reconstruction of the rock walls and gully plugs.

Large gaps and low points requiring major corrective action.

2 **Maintenance**

No attempt to fill gaps and repair storm damage.

Rock walls, gully plugs and earth banks have never been raised to allow for further soil to accumulate behind them.

Signs of severe rill and sheet erosion immediately below individual hedgerows, rock walls and earth banks, moderate to severe gully erosion still taking place within the plugged gullies.

LIVELIHOOD IMPACT

The Importance of and Obstacles to Assessing Livelihood Impact

There is a tendency in better land husbandry work to report impacts in terms of increased soil cover, reduced soil loss and improved agricultural production. While these results in terms of increased natural capital may be impressive they do not comprehensively capture the true impact of a project or programme's activities on the livelihoods of smallholder farmers. The problem is that impact in terms of reduced soil loss and/or increased agricultural yields conceal other benefits. These include farm income from the sale of surplus crops and also farmers' assessment of the qualitative changes in their attitude, self-esteem and confidence.

As outlined in Box 2.2, the five capitals of the sustainable livelihood framework (natural, social, human, physical and financial) can be seen as livelihood building blocks. Sustainable livelihood development depends on achieving an appropriate balance between all five of these building blocks rather than advances in access to one of them, for example natural capital. Despite this, many development projects and programmes fail to measure the livelihood impact of their work. There are three main reasons for this:

- **Time**—Project staff are often under great pressure to implement activities and ensure that outputs are achieved within the lifetime of a project. They have little time to monitor the livelihood impact of the project's activities.
- **How?**—Firstly, there is confusion over what is meant by a 'livelihood'. Secondly, even if there is broad agreement on the definition of a livelihood, it is seldom clear what ought to be measured or assessed. Thirdly there are few guidelines on what practical tools project staff should use to assess the impact.
- **Why bother?**—While donors are now demanding evidence of impact (see below), to date the emphasis has been on demonstrating that activities are being carried out as envisaged in the logical framework, even though the carrying out of these activities *per se* seldom reveals much about the impact on the livelihoods of the beneficiaries.

Despite these obstacles, an assessment of the livelihood impact of development work is becoming increasingly important. Firstly, the conventional approach of many non-governmental organisations and research organisations is to secure funding from different organisations in order to implement projects and programmes. Donor organisations are demanding more evidence that work funded by them is indeed having a positive impact. This perhaps stems from the realisation that the objectively verifiable indicators used in logical frameworks seldom measure the livelihood impact of a project. Secondly, organisations themselves are making strenuous efforts to improve the quality of their work; an assessment of the impact of their work on the livelihoods of the rural and urban poor is part of the quality assurance process. As assessment of this impact also enables organisations to amend field activities in order to make them more effective.

A Three-stepped Approach to Assessing Livelihood Impact

There is clearly a need for a user-friendly and not too costly approach, as part of a monitoring and evaluation system, to assess the livelihood impact of better land husbandry work. Guided by the sustainable livelihoods framework outlined in Section 2.2, project staff and local farmers can easily and relatively cheaply follow a three-stepped approach to measuring the impact of a better land husbandry initiative on local people's livelihoods.

Step one—compares achievements to the work plan and logical framework.

Step two—interprets expected achievements in terms of the Sustainable Livelihoods Framework (SLF) and the five assets (natural, financial, social, human and physical impact) of the framework.

Step three—outlines qualitative and quantitative research tools that can be used to measure/assess livelihood impact in the context of the five assets of the SLF.

Step One—The Work Plan and Logical Framework

Almost all project proposals now require a logical framework otherwise known as a logframe. The second column of the logical framework is for the *Objectively Verifiable Indicators.* Indicators are used to show the extent to which the objectives of the project are being met. They are quantitative and qualitative criteria that are used to check whether proposed changes have taken place. For each output there are also a number of activities that the project is supposed to carry out in order for the output to be achieved. These activities often have indicators attached to them (either in the logical framework and/ or in a work plan) to make it easier to determine if the project is progressing as planned.

There are, in turn, different types of indicators. Type 1 indicators demonstrate that a particular activity has been carried out or completed, for example that a training course on the use of cover crops has been held by a predetermined time. Type 2 indicators go beyond whether an activity has been carried out and start to address the consequences of a particular activity, for example that as a result of the training course resource-poor farmers are now aware of the benefits of vegetative soil cover and have access to planting material.

Clearly an organisation needs to know the extent to which work is taking place as envisaged in project documents such as the logical framework. This progress is often included in monthly, quarterly and annual reports. More often than not these reports tend to focus on factual information, on whether the activities leading to a particular output have been carried out and, to a lesser degree, on the extent to which the outputs of a project are contributing to the project purpose (and goal).

In the absence of any clear alternatives and guided by project documents, such as the logical framework, there has been a tendency to use Type 1 and Type 2 indicators to measure the impact of project work on the livelihoods of the rural poor. This is step one of the three-stepped approach and is what we normally measure as part of a monitoring and evaluation system. The problem is that the assessment process does not go far enough; the indicators used to measure progress very seldom measure livelihood impact *per se.* This is especially the case with Type 1 logical framework indicators. While Type 2 indicators reveal more information on impact, they tend to focus on only one component of livelihoods such as the build up of natural capital rather than exploring the other facets of livelihoods (see below).

A hypothetical example is a project in which the activities include an increase in production via better land management as well as the construc-tion of grain banks to store surplus grain. Understandably, and as required by the project documents, staff will report crop yields and the fact that the grain bank has been constructed. These are typical Type 1 indicators that actually reveal very little about the impact on people's livelihoods. It may be the case that the subsequent impact on peoples' lives is a positive one, but there is also a danger that the impact will be negative (see Box 2).

Box 2   Hypothetical livelihood impact of increased agricultural production and the construction of a grain bank in a farming community

Positive livelihood impact Farmers invest time in improving land management practices. Agricultural yields increase and farmers are able to store grain in the grain bank. They are subsequently able to sell surplus grain when the post-harvest glut has ended and the market price for grain has increased. Storage also enables farmers to have a reserve of food for lean months.

Negative livelihood impact Agricultural yields increase as a result of farmers' better land husbandry. Farmers store the harvest in the new grain bank in the hope of selling it later at a higher price. A few weeks later a beetle appears and starts eating the stored grain. Farmers do not realise immediately what is happening and by the time they do, they have lost almost all of the stored grain. They have little to sell and few supplies to eat in the lean months.

Any project or programme intents to create positive impacts but development actions also often generate unforeseen negative impacts. The problem is that Type 1 and to a lesser degree Type 2 indicators do not themselves tell us about the real livelihood impact of the project. In the case of the example in Box 2, the realisation of the targets of increased yields and the construction of grain banks do not themselves tell us about the real livelihood impact of the project: an impact that could be either positive or negative.

Step Two—Expected Achievements and the Sustainable Livelihoods Framework

The sustainable livelihoods framework is explained in Section 2.2. All too often the emphasis in soil and water conservation and better land husbandry work is on increasing natural and to a lesser degree financial capital (from the sale of agricultural produce). While these are important, development practitioners should not lose sight of the need to work with local people to increase their other assets (physical, social and human). These other assets support the accumulation of natural and financial capital. In fact for many resource-poor people, the reality is that they may be unable to increase their natural and financial capital without these assets.

The sustainable livelihood framework provides us with a useful tool for measuring the livelihood impact of our work. The second step of the three-stepped approach involves field staff identifying the likely impact of their work in terms of the five capitals (see Box 3).

Box 3   Identifying the likely livelihood impact of a better land husbandry project

- **Natural capital** through reduced soil loss, more efficient use of water resources, and increased agricultural production.
- **Financial capital** via the sale of agricultural products from the increases in production.
- **Social capital** by virtue of the farming community having participated together in training workshops and having subsequently shared their experiences with improved land management practices.
- **Physical capital** from farmers using the income from the sale of agricultural crops to improve their homes, buy a radio, television or bicycle.
- **Human capital** in terms of the skills and knowledge farmers develop from attending the training sessions and interacting with extension agents.

It is also important to remember that they may be 'unexpected' negative livelihood impacts such as the hypothetical example of the grain bank (see Box 2). Furthermore, the identification of livelihood impacts should not necessarily be confined to the immediate beneficiaries. There are likely to be knock-on effects of any development intervention. These can be referred to as 'indirect impacts' and in any impact assessment they should be considered along with the direct impact.

The sustainable livelihoods framework provides an understanding of what a livelihood means in terms of the five assets. It facilitates the identification of the likely livelihood impact of a project or programme (both positive and negative). The third and final step is to measure/assess the impact in terms of the five assets.

Step Three—Measuring Livelihood Impact

Intuitively, it makes sense to ask farmers themselves what they believe the livelihood impact of a better land husbandry project has been. Farm families' comments are likely to provide information on many different aspects of change that they perceive and these same families are likely to use their own indicators and their views of the explanations of observed changes. Indeed it is essential to hear their voice because their views of the apparent advantages and disadvantages arising from changes will essentially determine whether they will sustain any improvements in the future.

Hence, often the full diversity and richness of livelihoods can best be understood only by qualitative and participatory analysis at a local level. There is a range of commonplace qualitative and quantitative tools available and a plethora of literature on the strength and weaknesses of these tools and the most effective ways of using them (see section 2.4.2).

In the case of the better land husbandry project outlined in Box 3, having identified in Step Two the likely livelihood impact of work on local farmers' livelihoods, project managers have a much better idea what sort of information to gather with the active participation of local farmers. They can subsequently use one or more qualitative and/or quantitative research tools to assess the actual impact. Although there are no hard and fast rules on which tools to use in different situations, details are provided in Section 2.4.2 on when it might be appropriate to use a certain tool and when it might not.

It is at the project staffs' discretion to decide which communities to focus on, how many farmers to interview in each community, and the extent of farmer participation in the monitoring and evaluation process. Project staff should also be encouraged to document and report negative impacts. The reporting of negative impacts is essential because often these 'failures' can be turned into a 'successes' if projects learn from the experience and subsequently either rectify the situation and/or ensure that the similar mistakes are avoided in the future.

An Iterative Process

The three-stepped approach to measuring the impact of development initiatives, such as better land husbandry work, on local people's livelihoods is unashamedly 'quick and dirty' as opposed to one involving much in-depth research. In the context of many rural development projects such an approach is needed because an impact assessment entails costs and, therefore, diverts resources from other project activities.

The three-stepped approach is designed to be easily replicable but in all cases decisions will have to be made on what sort of information to record, how often it will be recorded, and how frequently and in what format it will be reported. Attention also has to be focused on how the results will be analysed and fed back into project management.

What is clear from the author's own experiences is that there is often a huge amount of information on livelihood impact that development practitioners are exposed to on a regular basis even though they do not always 'see' this information for what it is (see Box 4). If these practitioners know what to look for–in this case changes in social, natural, financial, physical and human capital–then assessing the livelihood impact of development work need not be an onerous task.

Box 4   Seeing but not seeing

In early 2003, and with the birth of our first child, my wife suggested that we buy a bigger car. I freely admit to knowing very little about cars and having no interest in them: they are a means of getting from A to B. After some research my wife concluded that we really ought to get a Volkswagen Passat Estate. I had no idea what sort of car this was but the literature indicated that it was reliable and had enough space for the paraphernalia associated with a growing family. We duly bought a second hand Volkswagen Passat Estate.

I remember very clearly the first day that we took the car out for a drive. I suddenly noticed numerous other Volkswagen Passat Estates on the road. One interpretation of this is that many of these other drivers had like me decided to go and buy the same model of car at the same time that I purchased mine. This is clearly nonsense. What really happened is that I had never noticed the other Volkswagen Passat Estate cars because I was not looking for them. Development practitioners are no different to other human beings in terms seeing but not seeing evidence of change in project areas.

REFERENCES

Abrams, E.M. and Rue, D.J. 1988. The causes and consequences of deforestation among the prehistoric Maya. *Human Ecology*, 16 (4): 377-395.

Adato, M. and Meinzen-Dick, R. 2002. *Assessing the Impact of Agricultural Research on Poverty Using the Livelihoods Framework*. Food Consumption and Nutrition Division Discussion Paper Number 128. International Food Policy Research Institute, Washington D.C., USA.

Ahmed, N. and Breckner, E. 1973. Soil erosion on three Tobago soils. Paper presented at the conference on soils of the Caribbean and Tropical America, University of the West Indies, Trinidad.

Alegre, J.C. and Rao, M.R. 1996. Soil and water conservation by contour hedging in the humid tropics of Peru. *Agriculture, Ecosystem and Environment*, 57 (1): 17-25.

Altieri, M. 1999. Enhancing the productivity of Latin American traditional peasant farming systems through an agroecological approach. Paper prepared for Bellagio Conference on Sustainable Agriculture. http:// www.rodaleinstitute.org/global/6 00 99.html (Accessed December 2004).

Altieri, M.A. 1990. Agroecology and rural development in Latin America. In: Altieri, M.A. and Hecht, S.B. (eds.). *Agroecology and Small Farm Development*, pp. 113-118. CRC Press, Boca Raton, Ann Arbor, Boston, USA.

Altieri, M.A., 2002. Agroecology: the science of natural resource management for poor farmers in marginal environments. *Agriculture, Ecosystems & Environment*, 93 (1-3): 1-24.

Anderson S., Gündel S., Pound B. and Triomphe B. 2001. *Cover Crops in Smallholder Agriculture: Lessons from Latin America*. Intermediate Technology Development Group Publishing, London, UK.

Anderson, J. M. and Ingram, J. S. I. 1993. *Tropical soil biology and fertility: A handbook of methods*. CAB International, Wallingford, Oxon, UK.

Arévalo-Méndez, I. 1994. *The Assessment and Development of Sustainable Hillside Conservation Technology for Small Farms in Central America*. Report on 1994 field work. Department of Geography, Loughbrough University of Technology, UK.

Arnold, M. and Dewees, P. 1997. Rethinking approaches to tree management by farmers. *Natural Resource Perspectives* 26, Overseas Development Institute, London, U

Ashby, J.A. 1985. The social ecology of soil erosion in a Colombian farming system. *Rural Sociology*, 50 (3): 377-396.

Ashby, J.A., Beltrán, J.A., Guerrero, M. and Ramos, H.F. 1996. Improving the acceptability to farmers of soil conservation practices. *Journal of Soil and Water Conservation*, 51 (4): 309-312.

Ashish, M. 1979. Agricultural economy of Kuamaun hills: threat of ecological disaster. *Economic and Political Weekly*, 14 (25): 1058-1064.

Astatke, A., Jabbar, M. and Tanner, D. 2003. Participatory conservation tillage: an experience with minimum tillage on an Ethiopian highland Vertisol. *Agriculture, Ecosystems & Environment*, 95 (2-3): 401-415.

Banda, A.Z., Maghembe, J.A., Ngugi, D.N. and Chome, V.A. 1994. Effect of intercropping maize and closely spaced *Leucaena* hedgerows on soil conservation and maize yield on a steep slope at Ntcheu, Malawi. *Agroforestry Systems*, 27: 17-22.

Barber, R.G. 2001. Evaluación de los impactos de las tecnologías agroforestales y silvopastoriles sobre el manejo y recuperación de suelos y agua en fincas privadas. Draft consultancy report for the Inter-American Development Bank.

Barghouti, S. and Hazell, P. 2000. *The Role of Agricultural Science*. Focus 4, Brief 9 of 9. International Food Policy Research Institute, Washington D.C., USA.

Barker, D.H. (ed.). 1995. *Vegetation and Slopes: Stabilisation, Protection and Ecology*. Institute of Civil Engineers and Thomas Telford, London, UK.

Barrios, E. and Trejo, M.T. 2003. Implications of local soil knowledge for integrated soil management in Latin America. *Geoderma*, 111 (3-4): 217-231.

Beach, T. and Dunning, N.P. 1995. Ancient Maya terracing and modern conservation in the Petén rainforest of Guatemala. *Journal of Soil and Water Conservation*, 50 (2): 138-145.

Bellon, M., Adato, M., Becerril, J. and Mindek, D. 2003. *The Impact of Improved Maize Germplasm on Poverty Alleviation: the Case of Tuxpeño-Derived Material in Mexico*. Food Consumption and Nutrition Division Discussion Paper Number 162. International Food Policy Research Institute, Washington D.C., USA.

Bellon, M.R. 2001. *Participatory Methods for Technology Evaluation: A Manual for Scientists Working with Farmers*. CIMMYT, Mexico, D.F., Mexico.

Bentley, J.W. 1993. What farmers don't know. *Ceres*, 14: 42-45.

Berdegué, J., Reardon, T., Escobar, G. and Echeverría, R. (2000). Policies to promote non-farm rural employment in Latin America. *Natural Resource Perspectives* 55, Overseas Development Institute, London, UK.

Bergsma, Ir.E. 1997. Erosion hazard evaluation from soil micro-topographic features – an application on soil loss plots, northern Thailand. *Land Husbandry*, 2 (1): 45-58.

Bernstein, H. 1977. Notes on capital and peasantry. *Review of African Political Economy*, 10: 60-73.

Bewket, W. and Sterk, G. 2002. Farmers' participation in soil and water conservation activities in the Chemoga watershed, Blue Nile basin, Ethiopia. *Land Degradation & Development*, 13 (3): 189-200.

Bharad, G.M. 1995. Vetiver hedges on contour for *in-situ* rain water management on arable, non-arable and wet lands. *Newsletter of the Vetiver Network*, 13: 32-34.

Bharad, G.M. and Krishnappa, A.M. 1990. Research Update. *Newsletter of the Vetiver Network*, 4: 7-8.

Bhardwaj, S.P. 1996. Comparative study of hedgerow of vetiver and other grasses with mechanical measures on erosion losses at 4 % runoff plots. *Newsletter of the Vetiver Network*, 15: 20.

Blaikie, P. 1989. Explanation and policy in land degradation and rehabilitation for developing countries. *Land Degradation and Rehabilitation*, 1: 23-37.

Blaikie, P. 1993. What glasses are we wearing? Different views of environmental management. In: Christiansson, C., Dahlberg, A., Loiske, V-M and Östberg, W (eds.). *A Discussion on Natural Resources Research in the Third World*, pp. 32-35. School of Geography, Stockholm University, Sweden.

Blaikie, P. and Brookfield, H. 1987. Approaches to the study of land degradation. In: Blaikie, P. and Brookfield, H. (eds.). *Land Degradation and Society*, pp. 27-48. Methuen, London, UK.

Blaut, J.M., Blaut, R.P., Harman, N. and Moerman, M. 1959. A study of cultural determinants of soil erosion and convservation in the Blue Mountains of Jamaica. *Social and Economic Studies*, 8 (4): 403-420.

Bleek, W. 1987. Lying informants: A fieldwork experience from Ghana. *Population and Development Review* 13 (2): 314-322.

Blustain, H. 1982. Social issues in technology choice: Soil conservation in Jamaica. *Journal of Soil and Water Conservation*, 37 (6): 323-325.

Bocco, G. 1991. Traditional knowledge for soil conservation in central Mexico. *Journal of Soil and Water Conservation*, 46 (5): 346 – 348.

Bodnar, F. and De Graaff, J. 2003. Factors influencing adoption of soil and water conservation measures in southern Mali. *Land Degradation & Development*, 14 (6): 515-525.

Boehringer, A. and Caldwell, R. 1989. *Cajanus cajan* (L.) Millsp. as a potential agroforestry component in the Eastern Province of Zambia. *Agroforestry Systems*, 9: 157-140.

Bohlen, P.J. and Edwards, C.A. 1994. The response of nematode trophic groups to organic and inorganic nutirent inputs in agroecosystems. In: Doran, J.W., Coleman, D.C., Bezdicek, D.F. and Stewart, B.A. (eds.). *Definining Soil Quality For a Sustainable Environment*, pp. 235-244. Soil Science Society of America Special Publication Number 35, Madison, Wisconsin, USA.

Bonnard, P. 1995. *Land Tenure, Land Titling, and the Adoption of Improved Soil Management Practices in Honduras*. PhD dissertation, Department of Agricultural Economics, Michigan State University, USA.

Boserup, E. 1965. *The Conditions of Agricultural Growth: the Economics of Agrarian Change under Population Pressure.* George Allen and Unwin, London, UK.

Boubakari, M. and Morgan, R.P.C. 1999. Contour grass strips for soil erosion control on steep slopes: a laboratory evaluation. *Soil Use and Management,* 15: 21-26.

Bryan, R.B. 1979. The influence of slope angle on soil entrainment by sheetwash and rainsplash. *Earth Surface Processes,* 4: 43-58.

Bryan, R.B. and Luk, S.H. 1981. Laboratory experiments on the variation of soil erosion under simulated rainfall. *Geoderma,* 26 (4): 245-265.

Buckles, D., Triomphe, B. and Sain, G. 1998. *Cover Crops in Hillside Agriculture: Farmer Innovation with Mucuna.* International Development Research Centre, Ottawa, Canada and International Maize and Wheat Improvement Center, Mexico D.F., Mexico.

Budowski, G. 1987. The development of agroforestry in Central America. In: Steppler, H.A. and Nair, P.K.R. (eds.). *Agroforestry a Decade of Development,* pp. 69-88. International Institute for Research in Agroforestry, Nairobi, Kenya.

Bunch, R. 1982. Two Ears of Corn: A Guide to People-Centered Agriculture. World Neighbors, Oklahoma, U.S.A.

Bunch, R. 1988. Guinope integrated development program, Honduras. In: Conroy, C. and Litvinoff, M. (eds.). *The Greening of Aid,* pp. 40-44. Earthscan Publications Ltd in association with The International Institute for Environment and Development, London, UK.

Bunch, R. 1989. Encouraging farmers' experiments. In: Chambers, R., Pacey, A. and Thrup, L.A. (eds.). *Farmer First: Farmer Innovation and Agricultural Research,* pp. 55-61. Intermediate Technology Publications, London, UK.

Bunch, R. 1997. Achieving sustainability in the use of green manures. *LEISA,* 13 (3): 12-13.

Bunch, R. 1999. Learning how to 'make the soil grow': three case studies on soil recuperation adoption and adaptation from Honduras and Guatemala. In: McDonald, M.A. and Brown, K. (eds.). *Issues and Options in the Design of Soil and Water Conservation Projects,* pp. 23-37. School of Agricultural and Forest Sciences Publication Number 17, University of Wales, Bangor, UK.

Bunch, R. 1999b. Personal communication (e-mail dated 22 October 1999).

Bunch, R. 2003. Adoption of green manure and cover crops. *LEISA,* 19 (4): 16-18.

Bunch, R. and López, G. 1999. Soil recuperation in Central America: How innovation was sustained after project intervention. In: Hinchcliffe, F., Thompson, J., Pretty, J., Guijt, I. and Shah, P. (eds.). *Fertile Ground: The Impacts of Participatory Watershed Management,* pp. 32-41. Intermediate Technology Publications, London, UK.

Busscher, W.J., Reeves, D.W., Kochhann, R.A., Bauer, P.J., Mullins, G.L., Clapham, W.M., Kemper, W.D. and Galerani, P.R. 1996. Conservation farming in southern Brazil: Using cover crops to decrease erosion and increase infiltration. *Journal of Soil and Water Conservation*, 51 (3): 188-192.

Carrasco, D.A. and Witter, S.G. 1993. Constraints to sustainable soil and water conservation: A Dominican Republic example. *Ambio*, 22 (6): 347-350.

Carter, J.C. 1995. Alley Farming: have resource-poor farmers benefited? *Natural Resource Perspectives* 3, Overseas Development Institute, London, UK

Cartier van Dissel, S. and de Graff, J. 1999. Differences between farmers and scientists in the perception of soil erosion: a South African case study. *Indigenous Knowledge and Development Monitor*, 6(3).

Cassel, D.K. and Lal, R. 1992. Soil physical properties of the tropics: Common beliefs and management restraints. In: Lal, R. and Sanchez, P.A. (eds.). *Myths and Science of Soils in the Tropics, pp. 61-89.* Soil Science Society of America Special Publication Number 29, Madison, Wisconsin, USA.

Chaggar, T.S. 1984. Reunion sets new rainfall records. *Weather*, 39: 12-14.

Chambers, R. 1993. Sustainable small farm development – frontiers in participation. In: Hudson, N.W. and Cheatle, R. (eds.). *Working with Farmers for Better Land Husbandry*, pp. 96-101. Intermediate Technology Publications, London, UK.

Chambers, R. 1997. *Whose Reality Counts? Putting the First Last.* Intermediate Technology Publications, London, UK.

Chambers, R. and Conway, R. 1992. Sustainable Rural Livelihoods: Practical Concepts for the 21st Century'. *IDS Discussion Paper* No. 296, IDS, Brighton, UK.

Chan, S.K., Tan, S.L., Ghulam Mohammed, H. and Howeler, R H. 1994. Soil erosion control in cassava cultivation using tillage and cropping techniques. *Journal of the Malaysian Agricultural Research and Development Institute*, 22 (1): 55-66.

Chapman, R. and Tripp, R. 2003. Changing incentives for agricultural extension - a review of provatised extension in practice. *Agriculture Research & Extension Network Paper* 132, Overseas Development Institute, London, UK.

Cheatle, R.J. 1993. Next steps towards better land husbandry. In: Hudson, N.W. and Cheatle, R. (eds.). *Working with Farmers for Better Land Husbandry*, pp. 223-230. Intermediate Technology Publications, London, UK.

Cheatle, R.J. and Shaxson, T.F. 2001. Conservation is for business – self help groups in Kenya. In: Stott, D.E., Mohtar, R.H. and Steinhardt, G.C. (eds). *Sustaining The Global Farm.* Selected papers from the 10th International Soil Conservation Organisation Meeting, Purdue University and the USDA-ARS National Erosion Research Laboratory.

Chinene, V.R.N., Shaxson, T.F., Molumeli, P. and Segerros, K.H.M. 1996. *Guidelines to Better Land Husbandry in the SADC Region*. Environment and Land Management Sector, Southern African Development Community, Maseru and Agriculture Development Division, Commonwealth Secretariat, London, UK.

Chirwa, P.W., Nair, P.K.R. and Nkedi-Kizza, P. 1994. Pattern of soil moisture depletion in alley-cropping under semiarid conditions in Zambia. *Agroforestry Systems*, 26: 89-99.

Christiansson, C., Dahlberg, A., Loiske, V-M and Östberg, W. 1993. Environments and their inhabitants: Debating research experiences and prospects. In: Christiansson, C., Dahlberg, A., Loiske, V-M and Östberg, W (eds.). *A Discussion on Natural Resources Research in the Third World*, pp. 8-19. School of Geography, Stockholm University, Sweden.

Chronic Poverty Research Centre. 2004. *The Chronic Poverty Report 2004-05*. Institute of Development Policy & Management, University of Manchester, UK.

Clark, J. and Hellin, J. 1996. *Bio-engineering for Effective Road Maintenance in the Caribbean*. Natural Resources Institute, Chatham, UK.

Clark, R. 1996. *Methodologies for the Economic Analysis of Soil Erosion and Conservation*. Centre for Social and Economic Research on the Global Environment, Working Paper GEC 96-13, University of East Anglia, UK.

Clark, R., Durón, G., Quispe, G. and Stocking, M. 1999. Boundary bunds or piles of stones? Using farmers' practices in Bolivia to aid soil conservation. *Mountain Research and Development*, 19 (3): 235-240.

Collins, J. L. 1987. Labor scarcity and ecological change. In Little, P.D., Horowitz, M.H. and Nyerges, A.E. (eds.). *Lands at Risk in the Third World: Local perspectives*, pp. 19-37. Westview Press, Boulder, Colorado, USA.

Collinson, M. 1981. A low-cost approach to understanding small farmers. *Agricultural Administration*, 8 (6): 433-450.

Comia, R.A., Paningbatan, E.P. and Håkansson, I. 1994. Erosion and crop yield response to soil conditions under alley cropping systems in the Philippines. *Soil & Tillage Research*, 31: 249-261.

Conroy, M.E., Murray, D.L. and Rosset, P.M. 1996. *A Cautionary Tale: Failed U.S. Development Policy in Central America*. Food First Development Studies, Lynne Rienner Publishers, Boulder, USA and London, UK.

Cook, M.G. 1988. Soil conservation on steep lands in the tropics. In: Moldenhauer, W. C. and Hudson, N. W. (eds.). *Conservation Farming on Steeplands*, pp. 18-22. Soil and Water Conservation Society and World Association of Soil and Water Conservation, Ankeny, Iowa, USA.

Cools, N., De Pauw, E. and Deckers, J. 2003. Towards an integration of conventional land evaluation methods and farmers' soil suitability assessment: a case study in northwestern Syria. *Agriculture, Ecosystems & Environment*, 95 (1): 327-342.

Cornwall, A., Guijt, I. and Welbourn, A. 1994. Acknowledging process: challenges for agricultural research and extension methodology. In: Scoones, I. and Thompson, J. (eds.). *Beyond Farmer First: Rural People's Knowledge, Agricultural Research and Extension Practice*, pp. 98-117. Intermediate Technology Publications, London, UK.

Court, J., Hovland, I. and Young, J. (eds.). 2005. *Bridging Research and Policy in Development: Evidence and the Change Process*. ITDG Publishing, London, UK.

Critchley, W.R.S. and Bruijnzeel, L.A. 1996. *Environmental Impacts of Converting Moist Tropical Forest to Agriculture and Plantation*. IHP Humid Tropics Programme Series No. 10, UNESCO, Paris, France.

Critchley, W.R.S., Reij, C. and Willcocks, T.J. 1994. Indigenous soil and water conservation: a review of the state of knowledge and prospects for building on traditions. *Land Degradation and Rehabilitation*, 5: 293-314.

Current, D., Lutz, E. and Scherr, S.J. 1995. Costs, benefits, and farmer adoption of agroforestry. In: Current, D., Lutz, E. and Scherr, S.J. (eds.). *Costs, Benefits, and Farmer Adoption of Agroforestry: Project Experience in Central America and the Caribbean*, pp. 1-27. World Bank Environment Paper 14, The World Bank, Washington, D.C., USA.

Dalton, P.A. and Truong, P.N.V 1999. Soil moisture competition between vetiver hedges and sorghum under irrigated and dryland conditions. *Newsletter of the Vetiver Network*, 20: 22-25.

Dangler, E.W. and El-Swaify, S.A. 1976. Erosion of selected Hawaii soils by simulated rainfall. *Soil Science Society of America Journal*, 40: 769-773.

Daniels, R.B., Gilliam, J.W., Cassel, D.K. and Nelson, L.A. 1985. Soil erosion class and landscape position in the North Carolina Piedmont. *Soil Science Society of America Journal*, 49: 991-995.

David, S. 1995. What do farmers think? Farmer evaluations of hedgerow intercropping under semi-arid conditions. *Agroforestry Systems*, 32: 15-28.

de Freitas, V.H. 2000. *Soil Management and Conservation for Small Farms: Strategies and Methods of Introduction, Technologies and Equipment*. FAO Soils Bulletin 77, Food and Agriculture Organization of the United Nations, Rome, Italy.

de Janvry, A. and Helfand, S. 1990. The dynamics of peasant agriculture in Latin America: Implications for rural development and agroecology. In: Altieri, M.A. and Hecht, S.B. (eds.). *Agroecology and Small Farm Development*, pp. 61-69. CRC Press, Boca Raton, Ann Arbor, Boston, USA.

de Janvry, A., Sadoulet, E. and Young, L.W. 1989. Land and labour in Latin American Agriculture from the 1950s to the 1980s. *Journal of Peasant Studies*, 16 (3): 396-424.

De Ploey, J.; Savat, J. and Moeyersons, J. 1976. The differential impact of some creep loss factors on flow, runoff creep and rainwash. *Earth Surface Processes*, 1: 151-161.

de Vaus, D.A. 1996. *Surveys in Social Research.* University College London Press, London, UK.

Deevey, E.S., Rice, D.S., Rice, P.M., Vaughan, H.H., Brenner, M. and Flannery, M.S. 1979. Mayan Urbanism: Impact on a tropical karst environment. *Science,* 206: 298-306.

Dehn, M. 1995. An evaluation of soil conservation techniques in the Ecuadorian Andes. *Mountain Research and Development,* 15 (2): 175-182.

Department for International Development. 2002. *Better Livelihoods for Poor People: the Role of Agriculture.* Department for International Development, London, UK.

Derpsch, R., Roth, C.H., Sidiras, N. and Kopke, U. 1991. *Controle da Erosao no Paraná, Brasil: Sistemas de Cobertura do Solo, Plantio Direto e Preparo Conservacionista do Solo.* GTZ, Eschborn, Germany.

Dewalt, B. 1985. Microcosmic and macrocosmic processes of agrarian change in southern Honduras: The cattle are eating the forest. In: DeWalt, B. and Pelto, P.J. (eds.). *Micro and Macro Levels of Analysis in Anthropology: Issues in Theory and Practice,* pp. 165-186. Westivew Press, Boulder, Colorado, USA.

DeWalt, B. 1994. Using indigenous knowledge to improve agriculture and natural resource management. *Human Organization,* 53(2): 123-131.

Dick, R.P. 1994. Soil enzyme activities as indicators of soil quality. In: Doran, J.W., Coleman, D.C., Bezdicek, D.F. and Stewart, B.A. (eds.). *Defining Soil Quality for a Sustainable Environment,* pp. 107-124. Soil Science Society of America Special Publication Number 35, Madison, Wisconsin, USA.

Dickson, D. 2003. Let's not get too romantic about traditional knowledge. SciDevNet. Available at http://www.scidev.net/editorials. Accessed December 2004.

Dillaha, T.A., Sherrard, J.H., Lee, D., Shanholtz, V.O., Mostaghimi, S. and Magette, W.L. 1986. *Use of Vegetative Filter Strips to Minimize Sediment and Phosphorous Losses From Feedlots: Phase 1, Experimental Plot Studies.* Virginia Water Resources Research Center, Virginia Polytechnic Institute and State University Bulletin 151, USA.

Dominguez, J., Bohlen, P.J. and Parmelee, R.W. 2004. Earthworms increase nitrogen leaching to greater soil depths in row crop agroecosystems. *Ecosystems,* 7 (6): 672-685.

Doran, J.W. and Jones, A.J. 1996. *Methods for Assessing Soil Quality.* Soil Science Society of America Special Publication Number 49, Madison, Wisconsin, USA.

Doran, J.W. and Parkin, T.B. 1994. Defining and assessing soil quality. In: Doran, J.W., Coleman, D.C., Bezdicek, D.F. and Stewart, B.A. (eds.). *Defining Soil Quality for a Sustainable Environment,* pp. 3-21. Soil Science Society of America Special Publication Number 35, Madison, Wisconsin, USA.

Doran, J.W., Coleman, D.C., Bezdicek, D.F. and Stewart, B.A. (eds.). 1994. *Definining Soil Quality for a Sustainable Environment*. Soil Science Society of America Special Publication Number 35, Madison, Wisconsin, USA.

Doran, J.W., Sarrantonio, M. and Liebig, M.A. 1996. Soil health and sustainability. *Advances in Agronomy*, 52: 1-54.

Douglas, M.G. 1988. Integrating conservation into farming systems: The Malawi experience. In: Moldenhauer, W. C. and Hudson, N. W. (eds.). *Conservation Farming on Steeplands*, pp. 215-227. Soil and Water Conservation Society and World Association of Soil and Water Conservation, Ankeny, Iowa, USA.

Douglas, M.G. 1993. Making conservation farmer-friendly. In: Hudson, N.W. and Cheatle, R. (eds.). *Working with Farmers for Better Land Husbandry*, pp. 4-15. Intermediate Technology Publications, London, UK.

Douglas, M.G. 1997. *Guidelines for the Monitoring and Evaluation of Better Land Husbandry*. The Association for Better Land Husbandry, UK.

Douglas, T.D., Kirkby, S.J., Critchley, R.W. and Park, G.J. 1994. Agricultural terrace abandonment in the Alpujjara, Andalucia, Spain. *Land Degradation & Rehabilitation*, 5: 281-291.

Dumanski, J., Gameda, S. and Pieri, C. 1998. *Indicators of Land Quality and Sustainable Land Management: An Annotated Bibliography*. The World Bank, Washington, D.C., USA.

Dunne, T. and Dietrich, W. 1982. Sediment sources in tropical drainage basins. In: Kussow, W., El-Swaify, S.A and Mannering, J. (eds.). *Soil Erosion and Conservation in the Tropics* pp. 41-55. American Society of Agronomy Special Publication Number 43, Soil Science Society of America, Madison, Wisconsin, USA.

Durham, W.H. 1979. *Scarcity and Survival in Central America: Ecological Origins of the Soccer War*. Stanford University Press, Stanford, California, USA.

Durham, W.H. 1995. Political ecology and environmental destruction in Latin America. In: Painter, M. and Durham, W.H. (eds.). *The Social Causes of Environmental Destruction in Latin America*, pp. 249-264. The University of Michigan Press, USA.

Dvorak, K.A. 1991. Methods of on-farm, diagnostic research on adoption potential of alley cropping. *Agroforestry Systems*, 15: 167-181.

Eash, N.S., Karelen, D.L. and Parkin, T.B. 1994. Fungal contributions to soil aggregation and soil quality. In: Doran, J.W., Coleman, D.C., Bezdicek, D.F. and Stewart, B.A. (eds.). *Definining Soil Quality for a Sustainable Environment*, pp. 221-228. Soil Science Society of America Special Publication Number 35, Madison, Wisconsin, USA.

Edwards, M. 1989. The irrelevance of development studies. *Third World Quarterly* 11 (1): 116-135.

Eilittä, M., Sollenberger, L.E., Little, R.C. and Harrington, L.W. 2003. On-farm experiments with maize-*mucuna* systems in the Los Tuxtlas region of Veracruz, Mexico. I. *Mucuna* biomass and maize grain yield. *Experimental Agriculture*, 39: 5-17.

El-Ashry, M.T. 1988. Foreword. In: Moldenhauer, W. C. and Hudson, N. W. (eds.). *Conservation Farming on Steeplands*, pp. ix-xi. Soil and Water Conservation Society and World Association of Soil and Water Conservation, Ankeny, Iowa, USA.

Ellis F. and Seeley J. 2001. *Globalisation and Sustainable Livelihoods: An Initial Note*. Department for International Development, London, UK.

Ellis, F. 1999. Rural livelihood diversity in developing countries: evidence and policy implications. *Natural Resource Perspectives* 40, Overseas Development Institute, London, UK.

Ellis, F. and Harris, N. 2004. *New Thinking About Urban and Rural Development*. Keynote paper for DFID Sustainable Development Retreat, University of Surrey, UK.

Ellis-Jones, J. and Sims, B. 1995. An appraisal of soil conservation technologies on hillside farms in Honduras, Mexico and Nicaragua. *Project Appraisal*, 10(2): 125-134.

El-Swaify, S.A. 1994. State of the art for assessing soil and water conservation needs and technologies: a global perspective. In: Napier, T.L., Camboni, S.M. and El-Swaify, S.A. (eds.). *Adopting Conservation on the Farm: An International Perspective on the Socioeconomics of Soil and Water Conservation*, pp. 13-27. Soil and Water Conservation Society, Ankeny, Iowa, USA.

El-Swaify, S.A. 1997. Factors affecting soil erosion hazards and conservation needs for tropical steeplands. *Soil Technology*, 11: pp. 3-16.

El-Swaify, S.A. and Dangler, E.W. 1982. Rainfall erosion in the tropics: a state of the art. In: Kussow, W., El-Swaify, S.A and Mannering, J. (eds.). *Soil Erosion and Conservation in the Tropics* pp. 1-25. American Society of Agronomy Special Publication Number 43, Soil Science Society of America, Madison, Wisconsin, USA.

El-Swaify, S.A., Dangler, E.W. and Armstrong, C.L. 1982. *Soil Erosion by Water in the Tropics*. Research and Extension Service 024, Hawaii Institute of Tropical Agriculture, University of Hawaii Press, Honolulu. USA.

Elwell, H.A. and Stocking, M.A. 1982. Developing a simple yet practical method of soil-loss estimation. *Tropical Agriculture* (Trinidad), 59(1): 43-48.

Enters, T. 1994. Now you see it, now you don't: the effects of the ecocrisis theory on research. Paper presented at the IUFRO, FORSPA, CIFOR, FAO/RAPA workshop on *The barriers to the application of forestry research results*, Bangkok, Thailand.

Enters, T. 1996. The token line: adoption and non-adoption of soil conservation practices in highlands of northern Thailand. In: Sombatpanit, S., Zöbisch, M.A., Sanders, D.W. and Cook, M.G. (eds.). *Soil Conservation Extension: From Concepts to Adoption*, pp. 417-427. Soil and Water Conservation Society of Thailand, Bangkok.

Enters, T. 1998. *Methods for the Economic Assessment of the On- and Off-Site Impacts of Soil Erosion*. International Board for Soil Research and

Management, Issues in Sustainable Land Management, No. 2, Bangkok, Thailand.

Enters, T. and Hagmann, J. 1996. One-way, two-way, which way? Extension workers: from messengers to facilitators. *Unasylva*, 46 (184):1-9.

Enters, T., 1999. Incentives as policy instruments – key concept and definitions. In: Sanders, D.W., Huszar, P.C., Sombatpanit, S., Enters, T. (eds). *Incentives in Soil Conservation: from Theory to Practice*, pp. 25-40. Oxford & IBH Publishing, New Delhi, India.

Erenstein, O. 2003. Smallholder conservation farming in the tropics and sub-tropics: a guide to the development and dissemination of mulching with crop residues and cover crops. *Agriculture, Ecosystems & Environment* 100 (1): 17-37.

Ericksen, P.J. 1998. *An Evaluation of Sustainable Land Use in a Hillside Agroecosystem in Central Honduras.* PhD dissertation, University of Wisconsin-Madison, USA

Ericksen, P.J. and Ardon, M. 2003. Similarities and differences between farmer and scientist views on soil quality issues in central Honduras. *Geoderma*, 111(3-4): 233-248.

Evans, R. 1995. Some methods of directly assessing water erosion of cultivated land – a comparison of measurements made on plots and in fields. *Progress in Physical Geography*, 19 (1): 115-129.

Eyre, L.A. 1989. Hurricane *Gilbert*: Caribbean record-breaker. *Weather*, 44(4): 160-164.

Fahlen, A. 2002. Mixed tree-vegetative barrier designs: experiences from project works in northern Vietnam. *Land Degradation & Development*, 13 (4): 307-329.

Feder, G. and Onchan, T. 1987. Land ownership security and farm investment in Thailand. *American Journal of Agricultural. Economics*, 69 (2): 311-320.

Floyd, S. 1999. When is quantitative data collection appropriate in farmer participatory research and development? Who should analyse the data and how? *Agricultural Research & Extension Network Paper* 92b, Overseas Development Institute, London, UK.

Food and Agriculture Organization of the United Nations. 1994. Documento del Proyecto. GCP/HON/016/Net *Desarrollo Rural del Sur de Lempira*, Tegucigalpa, Honduras.

Food and Agriculture Organization of the United Nations. 1995. *Agricultural Investment to Promote Improved Capture and Use of Rainfall in Dryland Farming*. FAO Investment Centre Technical Paper 10, Rome, Italy.

Food and Agriculture Organization of the United Nations. 2001. *The Impact of HIV/AIDS on Food Security*. Twenty-seventh session of the Committee on Food Security, Rome 28 May - 1 June 2001. Available at http://www.fao.org/docrep/meeting/003/Y0310E.htm. Accessed September 2004.

Food and Agriculture Organization of the United Nations. 2001b. *No-Till Agriculture for Sustainable Land Management: Lessons Learnt From the 2000 Brazil Study Tour*. Draft for TCI Occasional Paper Series, FAO Investment Centre Division, Rome, Italy.

Foster, G.R., Moldenhauer, W.C. and Wischmeier, W.H. 1982. Transferability of U.S. technology for prediction and control of erosion in the tropics. In: Kussow, W., El-Swaify, S.A and Mannering, J. (eds.). *Soil Erosion and Conservation in the Tropics*, pp. 135-149. American Society of Agronomy Special Publication Number 43, Soil Science Society of America, Madison, Wisconsin,USA.

Fowler, R. and Rockstrom, J. 2001. Conservation tillage for sustainable agriculture – an agrarian revolution gathers momentum in Africa. *Soil & Tillage Research*, 61 (1-2): 93-107.

Frye, W.W. and Thomas, G.W. 1991. Management of long-term field experiments. *Agronomy Jounral*, 83: 38-44.

Fujisaka, S. 1989. A method for farmer-participatory research and technology transfer: upland soil conservation in the Philippines. *Experimental Agriculture*, 25: 423-433.

Fujisaka, S. 1991. Thirteen reasons why farmers do not adopt innovations intended to improve the sustainability of upland agriculture. In: *Evaluation for Sustainable Land Management in the Developing World*, IBSRAM Proceedings, 12(2): 509-522, Bangkok, Thailand.

Fujisaka, S. 1991b. Improving productivity of an upland rice and maize system: farmer cropping choices or researcher cropping pattern trapeziods. *Experimental Agriculture*, 27: 253-261.

Fujisaka, S. 1993. A case of farmer adaptation and adoption of contour hedgerows for soil conservation. *Experimental Agriculture*, 29: 97-105.

Fujisaka, S. 1994. Learning from six reasons why farmers do not adopt innovations intended to improve sustainability of upland agriculture. *Agricultural Systems*, 46: 409-425.

Fujisaka, S. 1997. Research: help or hindrance to good farmers in high risk systems? *Agricultural Systems*, 54 (2): 137-152.

Fujisaka, S. and Cenas, P. 1993. Contour hedgerow technology in the Philippines: not yet sustainable. *Indigenous Knowledge and Development Monitor*, 1 (1):14-16.

Fujisaka, S., Jayson, E. and Dapusala, A. 1994. Trees, grasses and weeds: species choices in farmer-developed contour hedgerows. *Agroforestry Systems*, 25: 13-22.

Gardner, K. and Lewis, D. 1996. *Anthropology, Development and the Post-Modern Challenge*. Pluto Press, London, UK and Chicago, Illinois, UK.

Gardner. R. and Mawdesley, K. 1997. Soil erosion in the Middle Mountains of Nepal. *Agroforestry Forum*, 8(4): 25-29.

Garlynd, M.J., Romig, D.E., Harris, R.F. and Kurakov, A.V. 1994. Descriptive and analytical characterization of quality/health. In: Doran, J.W.,

Coleman, D.C., Bezdicek, D.F. and Stewart, B.A. (eds.). *Definining Soil Quality for a Sustainable Environment*, pp. 159-168. Soil Science Society of America Special Publication Number 35, Madison, Wisconsin, USA

Garrity, D.P and Mercado, A.R. 1994. Nitrogen fixation capacity in the component species of contour hedgerows: how important? *Agroforestry Systems*, 27: 241-258.

Garrity, D.P. 1996. Tree-soil-crop interactions on slopes. In: Ong, C.K. and Huxley, P. (eds.). *Tree-Crop interactions: A Physiological Approach*, pp. 299-318. CAB International, Wallingford, Oxon., UK.

Garrity, D.P., Stark, M. and Mercado, A. 1997. Rapid soil distribution within alleys: why simple extension models for contour hedgerows may not be appropriate. *Agroforestry Forum*, 8 (4): 5-7.

Gicheru, P., Gachene, C., Mbuvi, J. and Mare, E. 2004. Effects of soil management practices and tillage systems on surfaced soil water conservation and crust formation on a sandy loam in semi-arid Kenya. *Soil & Tillage Research*, 75(2): 173-184.

Gill, G.J. 1993. *O.K., the Data's Lousy, But It's All We've Got (Being a Critique of Conventional Methods)*. IIED Gatekeeper Series 38, International Institute for Environment and Development, London, UK.

Govers, G., Vandaele, K., Desmet, P., Poesen, J. and Bunte, K. 1994. The role of tillage in soil redistribution on hillslopes. *European Journal of Soil Science*, 45: 469-478.

Greenberg, J.B. and Park, T.K. 1994. Political Ecology. *Journal of Political Ecology*, 1: 1-12.

Greenfield, J.C. 1989. *Vetiver Grass (Vetiveria spp.) the Ideal Plant for Vegetative Soil and Moisture Conservation*. The World Bank, Washington D.C., USA.

Grimshaw, R.G. 1995. Vetiver grass – its use for slope ans structure stabilisation under tropical and semi tropical conditions. In Barker, D.H. (ed.). 1995. *Vegetation and Slopes: Stabilisation, Protection and Ecology*, pp. 26-35. Institute of Civil Engineers, Thomas Telford, London, UK.

Grossman, J.M. 2003. Exploring farmer knowledge of soil processes in organic coffee systems of Chiapas, Mexico. *Geoderma*, 111 (3-4): 267-287.

Gruhn, P., Goletti, F. and Yudelman, M. 2000. Integrated Nutrient Management, Soil Fertility and Sustainable Agriculture: Current Issues and Future Challenges. Food Agriculture and the Environment Discussion Paper 32, International Food Policy Research Institute, Washington D.C., USA.

Guba, E.G. 1981. Criteria for assessing the trustworthiness of naturalistic inquiries. *Educational Communication of Technology Journal*, 29 (2): 75-91.

Guillet, D.W., Furbee, L., Sandor, J. and Benfer, R. 1995. The Lari soils project in Peru - a methodology for combining cognitive and behavioural research. In: Warren, D.M., Slikkerveer, L.J. and Brokensha, D. (eds.). *The Cultural Dimension of Development: Indigenous Knowledge Systems*, pp. 71-81. Intermediate Technology Publications, London, UK.

Haggar, J.P and Beer, J.W. 1993. Effect on maize growth of the interaction between increased nitrogen availability and competition with trees in alley cropping. *Agroforestry Systems*, 21: 239-249.

Haigh, M.J, Jansky, L. and Hellin, J. 2004. Headwater deforestation: a challenge for environmental management. *Global Environmental Change*, 14(1): 51-62.

Haigh, M.J. 1977. The use of erosion pins in the study of slope evolution. In: *Shorter Technical Methods* (II), *Technical Bulletin* 18, British Geomorphological Research Group, Geo Books, Norwich, Norfolk, UK.

Haigh, M.J. 1999. Deforestation. In: Paccione, M. (ed.). *Applied Geography: Principles and Practice*, pp. 200-221. Routledge, London, UK.

Haigh, M.J. 2000. Soil stewardship on reclaimed lands In: Haigh, M.J. (eds.). *Land Reconstruction and Management* 1, pp. 165-274. A.A. Balkema, Rotterdam, The Netherlands.

Haigh, M.J. 2000b. Erosion control: Principles and some technical options. In: Haigh, M.J. (ed.). *Land Reconstruction and Management* 1, pp. 75-109. A.A. Balkema, Rotterdam, The Netherlands.

Haigh, M.J. 2004. Sustainable Management of Headwater Resources: The Nairobi 'Headwater' Declaration (2002) and Beyond. *Asian Journal of Water, Environment and Pollution*, 1 (1-2): 17-28.

Haigh, M.J. and Hellin, J. 2001. Extreme events and headwater management: reflections on the magnitude-frequency issue. In: Subramanian, V.S. and Ramanathan, A.L. (eds). *Ecohydrology, Proceedings of the International Workshop*. UNESCO, JNU & Capital Books, New Delhi, India.

Hallsworth, E.G. 1987. *Anatomy, Physiology and Psychology of Erosion*. The International Federation of Institutes of Advanced Study. John Wiley & Sons Ltd., Chichester, UK.

Hamilton, P. 1997. *Goodbye to Hunger! The Adoption, Diffusion and Impact of Conservation Farming Practices in Rural Kenya*. The Association for Better Land Husbandry, Nairobi, Kenya.

Hamilton, S. and Fischer, E.F. 2003. Non-traditional agricultural exports in highland Guatemala: Understandings of risk and perceptions of change. *Latin American Research Review*, 38 (3): 82-110.

Harden, C.P. 1993. Upland erosion and sediment yield in a large Andean drainage basin. *Physical Geography*, 14 (3): 254-271.

Harris, R.F. and Bezdicek, D.F. 1994. Descriptive aspects of soil quality/health. In: Doran, J.W., Coleman, D.C., Bezdicek, D.F. and Stewart, B.A. (eds.). *Defining Soil Quality for a Sustainable Environment*, pp. 23-35. Soil Science Society of America Special Publication Number 35, Madison, Wisconsin, USA

Harrison, P. 1992. *The Third Revolution: Environment, Population and a Sustainable World*. I.B. Tauris & Co. Ltd., UK.

Hassane, A., Martin, P. and Reij, C. 2000. *Water harvesting, land rehabilitation and household food security in Niger: IFAD's soil and water conservation project in Illéla District.* International Fund for Agricultural Development (IFAD), Rome, Italy.

Haug, G.H., Günther, D., Peterson, L.C., Sigman, D.M., Hughen, K.A. and Aeschlimann, B. 2003. Climate and collapse of Maya civilization. *Science,* 299 (5613): 1731-1735.

Hauser, S. 1993. Distribution and activity of earthworms and contribution to nutrient recycling in alley cropping. *Biology and Fertility of Soils,*15: 16-20.

Hawkesworth S. and Perez, J.D.G. 2003. Potentials and constraints of the farmer-to-farmer programme for environmental protection in Nicaragua. *Land Degradation & Development,* 14(2): 175-188.

Hazell, P. and Haddad, L. 2001. *Agricultural Research and Poverty Reduction.* 20/20 Brief 70, International Food Policy Research Institute, Washington D.C., USA

Hecht, S.B. 1990. Indigenous soil management in the Latin American Tropics: neglected knowledge of native peoples. In: Altieri, M.A. and Hecht, S.B. (eds.). *Agroecology and Small Farm Development,* pp. 151-158. CRC Press, Boca Raton, Ann Arbor, Boston, USA.

Heissenhuber, A., Kantelhardt, J. and Pahl, H., 1998. *Economic Considerations Concerning the Possibilities of Environmental Protection in Agriculture.* Medit 3/98, Edagricole, Bologna, Italy.

Hellin, J. 1998. *El Sistema Quezungual: Un Sistema Agroforestal Indígena de Lempira Sur, Honduras.* Consultant's report to the *Proyecto Lempira Sur* (Government of Honduras and Food and Agriculture Organization of the United Nations), Tegucigalpa, Honduras.

Hellin, J. 1999. Land degradation in Honduras: A challenge to an Ecological Marxism? *Capitalism, Nature, Socialism,* 10(3): 105-125.

Hellin, J. 1999b. *Soil and Water Conservation in Honduras: a Land Husbandry Approach.* PhD dissertation, Oxford Brookes University, UK

Hellin, J. 1999c. Soil and water conservation in Honduras: Addressing whose reality? Research note in the Newsletter of the *Agriculture Research & Extension Network* 40: 12-16. Overseas Development Institute, London, UK.

Hellin, J. and Haigh, M.J. 1999. Rainfall in Honduras during Hurricane *Mitch. Weather* 50 (11): 350-359.

Hellin, J. and Haigh, M.J. 2002. Better land husbandry in Honduras: towards the new paradigm in conserving soil, water and productivity. *Land Degradation & Development,* 13 (3): 233-250.

Hellin, J. and Haigh, M.J. 2002b. Impact of *Vetiveria zizanioides* (vetiver grass) live barriers on maize production in Honduras. In: Wang, L., Wu, D., Tu, X. and Nie, J. (eds.). *Sustainable utilization of global soil and water resources.* Proceedings of the 12th International Soil Conservation Organization Conference, Vol. 3, pp. 277-281. Beijing, China.

Hellin, J. and Higman, S. 2001. Competing in the market: Farmers need new skills. *Appropriate Technology* 28(2): 5-7.

Hellin, J. and Higman, S. 2002. Smallholders and niche markets: Lessons from the Andes. *Agriculture Research & Extension Network Paper* 118, Overseas Development Institute, London, UK.

Hellin, J. and Higman, S. 2003. *Feeding the Market: South American Farmers, Trade and Globalization*. ITDG Publishing and Latin American Bureau, UK.

Hellin, J. and Higman, S. 2005. Crop diversity and livelihood security in the Andes. *Development in Practice*, 15(2): 165-174.

Hellin, J. and Larrea, S. 1997. Live barriers on hillside farms: are we really addressing farmers' needs? *Agroforestry Forum*, 8(4): 17-20.

Hellin, J. and Larrea, S. 1998. Ecological and socio-economic reasons for the adoption and adaptation of live barriers in Güinope, Honduras. In: Blume, H.-P., Eger, H., Fleischhauer, E., Hebel, A., Reij, C., and Stenier, K.G. (eds.). *Towards Sustainable Land Use: Furthering Cooperation Between People and Institutions*. Selected papers of the 9th conference of the International Soil Conservation Organisation, Bonn, Germany. *Advances in Geoecology*, 31: 1383-1388

Hellin, J. and Schrader, K. 2003. The case against direct incentives and the search for alternative approaches to better land management in Central America. *Agriculture, Ecosystems & Environment*, 99 (1-3): 61-81.

Hellin, J., Albu, M. and Abdur, R. 2004b. Rural populations need non-agricultural employment. *Appropriate Technology*, 31(2): 17-20.

Hellin, J., Alvárez, L. and Cherrett, I. 1999. The Quezungual System: an indigenous agrofrestry system from western Honduras. *Agroforestry Systems*, 46(3): 229-237.

Hellin, J., Haigh, M.J. and Marks, F. 1999b. Rainfall characteristics of Hurricane Mitch. *Nature*, 399: 316.

Hellin, J., Rodriguez, D. and Coello, J. 2003. Measuring the livelihood impact of farmer-to-farmer extension services in the Andes. Paper presented at the Conference on *New Directions in impact assessment for development: methods and practice*, University of Manchester, 24-25 November, 2003. Enterprise Development Impact Assessment Information Service (EDIAIS), Institute for Development Policy and Management (IDPM) University of Manchester, UK.

Hellin, J., Rodriguez, D. and Coello, J. 2004. Sustainable farmer-to-farmer extension. *Appropriate Technology*, 31 (1): 12-14.

Herweg, K. 1996. *Field Manual for Assessment of Current Erosion Damage*. Soil conservation Research Programme, Ethiopia and Centre for Development and Environment, University of Berne, Switzerland.

Herweg, K. and Ludi, E. 1999. The performance of selected soil and water conservation measures – case studies from Ethiopia and Eritrea. *Catena*, 36: 99-114.

Herweg, K. and Ostrowski, M.W. 1997. *The Influence of Errors on Erosion Process Analysis.* Soil Conservation Research Programme, Ethiopia, Research Report 33. University of Berne, Switzerland in association with The Ministry of Agriculture, Ethiopia.

Herweg, K. and Steiner, K. 2002. *Instruments for Use in Rural Development Projects with a Focus On sustainable Land Management.* Volumes 1 (Procedure) and Volume 2 (Toolbox). CDE and GTZ, Wabern, Switzerland.

Hill, J. and Woodland, W. 2003. Contrasting water management techniques in Tunisia: towards sustainable agricultural use. *The Geographical Journal,* 169 (4): 342-357.

Hillel, D. 1991. *Out of the Earth: Civilization and the Life of the Soil.* University of California Press, Berkeley, Los Angeles, USA.

Hinchclifffe, F., Guijt, I., Pretty, J.N. and Shah, P. 1995. *New Horizons: the Economic, Social and Environmental Impacts of Participatory Watershed Development.* IIED Gatekeeper Series 50, International Institute for Environment and Development, London, UK.

Hobbes. T. 1651. *Leviathan.* Collins Fount Paperbacks 1983. Plamenatz, J. (ed.).

Holden, S., Shiferaw, B. and Pender, J. 2004. Non-farm income, household welfare and sustainable land management in a less-favoured area in the Ethiopian Highlands. *Food Policy,* 29 (4): 369-392.

Holt-Gimenez, E. 2000. *Midiendo la Resistencia Agroecológica Campesina ante El Huracán Mitch en Centroamérica .* World Neighbors, Tegucigalpa, Honduras [http://www.agroecology.org/people/eric/resist.htm] (Accessed April 2003)

Holt-Gimenez, E. 2001. Measuring farmers' agroecological resistance to hurricane *Mitch. LEISA,* 17 (1): 18-21.

Horton, D. 1990. *Tips for Planning Formal Farm Surveys in Developing Countries.* International Institute of Tropical Agriculture Research Guide 31, Ibadan, Nigeria.

Howeler, R .H., Nguyen The Dang and Vongkasem, W. 1996. Farmer participatory selection of vetiver grass as the most effective way to control erosion in cassava-based cropping systems in Vietnam and Thailand. Paper presented at the International Conference on Vetiver, held in Chiang Rai, Thailand, February 1996.

Howeler, R. H. 1987. Soil conservation practices in cassava-based cropping systems. In: Tay, T.H., Mokhtaruddin, A.M. and Zahari, A.B. (eds.). *Proceeding of an International Conference on Steepland Agriculture in the Humid Tropics,* held in Kuala Lumpur, Malaysia, August 1987, pp. 490-517.

Howeler, R. H. 1994. Integrated soil and crop management to prevent environmental degradation in cassava-based cropping systems. In: Bottema, J.W.T. and Stolz, D.R. (eds.) *Upland Agriculture in Asia,* pp. 195-224. Proceedings of Workshops held in Bogor, Indonesia. April, 1993.

Howeler, R. H. 1995. The use of farmer participatory research methodologies to enhance the adoption of soil conservation practices in cassava-based cropping systems in Asia. Paper presented at the International Workshop on Soil Conservation Extension held in Chiangmai, Thailand, June 1995.

Hsieh, Y-P. 1992. A mesh-bag method for field assessment of soil erosion. *Journal of Soil and Water Conservation*, 47 (6): 495-499.

Hudson, N.W. 1982. Soil conservation research and training requirements in developing tropical countries. In: Kussow, W., El-Swaify, S.A and Mannering, J. (eds.). *Soil Erosion and Conservation in the Tropics*, pp. 121-133. American Society of Agronomy Special Publication Number 43, Soil Science Society of America, Madison, Wisconsin, USA.

Hudson, N.W. 1983. Soil conservation strategies in the Third World. *Journal of Soil and Water Conservation*, 38 (6): 446-450.

Hudson, N.W. 1988. Tilting at windmills or fighting real battles. In: Moldenhauer, W.C. and Hudson, N.W. (eds.). *Conservation Farming on Steeplands*, pp.3-8. Soil and Water Conservation Society and World Association of Soil and Water Conservation, Ankeny, Iowa. USA.

Hudson, N.W. 1991. *A Study of the Reasons for Success and Failure of Soil Conservation Projects*. Soils Bulletin 64, Food and Agriculture Organization of the United Nations, Rome, Italy.

Hudson, N.W. 1992. Success and failure of soil conservation programmes. In: Tato, K and Hurni, H. (eds.). *Soil Conservation for Survival*, pp. 129-142. Soil and Water Conservation Society in cooperation with the International Soil Conservation Organisation and World Association of Soil and Water Conservation, Ankeny, USA.

Hudson, N.W. 1992b. *Land Husbandry*. B.T. Batsford Limited, London, UK.

Hudson, N.W. 1993. *Field Measurement of Soil Erosion and Runoff*. Soils Bulletin 68, Food and Agriculture Organization of the United Nations, Rome, Italy

Hudson, N.W. 1993b. Soil and water management for the nineties – New pressures, new objectives. In: Hudson, N.W. and Cheatle, R. (eds.). *Working with Farmers for Better Land Husbandry*, pp.15-20. Intermediate Technology Publications, in association with World Association of Soil and Water Conservation, London, UK.

Hudson, N.W. 1995. *Soil Conservation*. B.T. Batsford Limited, London, UK.

Hudson, N.W. and Cheatle, R. (eds.). 1993. *Working with Farmers for Better Land Husbandry*. Intermediate Technology Publications in association with World Association of Soil and Water Conservation, London, UK.

Hughes, J.D. 1999. The classic Maya collapse. *Capitalism, Nature, Socialism*, 10(1): 81-89.

Hulme, D. and Shepherd, A. 2003. Conceptualizing chronic poverty. *World Development*, 3 (3): 403-423.

Humphries, S., Gonzales, J., Jimenez, J. and Sierra, F. 2000. Searching for sustainable land use practices in Honduras: Lessons from a programme

of participatory research with hillside farmers. *Agricultural Research &
Extension Network Paper* 104. Overseas Development Institute, London,
UK.

Hurni, H., with the assitance of an international group of contributors. 1996.
*Precious Earth: From Soil and Water Conservation to Sustainable Land
Management*. International Soil Conservation Organisation (ISCO), and
Centre of Development and Environment (CDE), Berne, Switerland.

Huszar, P.C. 1999. Justification for using soil conservation incentives. In:
Sanders, D.W., Huszar, P.C., Sombatpanit, S. and Enters, T. (eds.).
Incentives in Soil Conservation: from Theory to Practice, pp. 57-68. Oxford
& IBH Publishing, New Delhi, India.

Hwang, S.W., Alwang, J. and Norton, G.W. 1994. Soil conservation practices
and farm income in the Dominican Republic. *Agricultural Systems*, 46:
59-77.

Instituto Interamericano de Cooperación para la Agricultura (IICA) and La
Dirección de Planeamiento, Programación, Proyectos y Auditoría Técnica
(DIPRAT). 1995. *Honduras - Diagnóstico del Sector Agropecuario*. IICA, San
José, Costa Rica.

International Maize and Wheat Improvement Center (CIMMYT). 2004. *Seeds
of Innovation: CIMMYT's Strategy for Helping to Reduce Poverty and Hunger
by 2020*. CIMMYT, Mexico, D.F., Mexico.

Jama, B.A., Nair, P.K.R. and Rao, M.R. 1995. Productivity of hedgerow shrubs
and maize under alleycropping and block planting systems in semiarid
Kenya. *Agroforestry Systems*, 31: 257-274.

Johnson, N., Lilja, N. Ashby, J.A.. 2001. *Characterizing and Measuring the Effects
of Incorporating Stakeholder Participation in Natural Resource Management
Research: Analysis of Research Benefits and Costs in Three Case Studies*.
Working Document No. 17, PRGA Program. Cali, Colombia.

Johnson, N.L., Lilja, N. and Ashby, J.A. 2003. Measuring the impact of user
participation in agricultural and natural resource management research.
Agricultural Systems, 78 (2): 287-306.

Karim, A.B., Savill, P.S. and Rhodes, E.R. 1993. The effects of between-row
(alley widths) and within-row spacings of *Gliricidia sepium* on alley-
cropped maize in Sierra Leone. *Agroforestry Systems*, 24: 81-93.

Karlen, D.L. and Stott, D.E. 1994. A framework for evaluating physical and
chemical indicators of soil quality. In: Doran, J.W., Coleman, D.C.,
Bezdicek, D.F. and Stewart, B.A. (eds.). *Definining Soil Quality for a
Sustainable Environment*, pp. 53-72. Soil Science Society of America Special
Publication Number 35, Madison, Wisconsin, USA.

Kass, D. 1999. Personal communication (e-mail dated 21 October 1999).

Kass, D.C.L., Foletti, C., Szott, L.T., Landaverde, R. and Nolasco, R. 1993.
Traditional fallow systems of the Americas. *Agroforestry Systems*, 23:
207-218.

Kellman, M. and Tackaberry, R. 1997. *Tropical Environments: The Functioning and Management of Tropical Ecosystems.* Routledge, London and New York.

Kerven, C., Dolva, H. and Renna, R. 1995. Indigenous soil classification systems in northern Zambia. In: Warren, D.M., Slikkerveer, L.J. and Brokensha, D. (eds.). *The Cultural Dimension of Development: Indigenous Knowledge Systems,* pp. 82-87. Intermediate Technology Publications, London, UK.

Kiepe, P. 1995. Effect of *Cassia siamea* hedgerow barriers on soil physical properties. *Geoderma,* 66: 113-120.

Kiepe, P. 1996. Cover and barrier effect of *Cassia siamea* hedgerows on soil conservation in semi-arid Kenya. *Soil Technology,* 9: 161-171.

Kiome, R.M. and Stocking, M. 1995. Rationality of farmer perception of soil erosion: The effectiveness of soil conservation in semi-arid Kenya. *Global Environmental Change,* 5 (4): 281-295.

Kloosterboer, E.H. and Eppink, L.A.A.J. 1989. Soil and water conservation in very steep areas – a case study of Santo Antão Island, Cape Verde. In: Baum, E,E., Wolff, P. and Zöbisch, M.A. (eds.). *The Extent of Soil Erosion – Regional Comparisons.* Volume 1 in *Topics in Applied Resource Management in the Tropics,* pp. 111-142. German Institute for Tropical and Subtropical Agriculture, Witzenhausen, Germany.

Kloppenburg, J. 1991. Social theory and the de/reconstruction of agricultural science: local knowledge for an alternative agriculture. *Rural Sociology,* 56: 519-548.

Kon, K.F. and Lim, F.W. 1991. Vetiver research in Malaysia – some preliminary results on soil loss, runoff and yield. *Newsletter of the Vetiver Network,* 5: 4-5.

Kresch, D.L., Mastin, M.C. and Olsen-Tacoma, T.D. 2002. *Fifty-Year Flood-Inundation Maps for Choluteca,* U.S. Geological Survey Open-File Report 02-250, 1-19. http://water.usgs.gov/pubs/of/ofr02250/pdf/OFR02250.pdf. (Accessed April 2003).

Kydd, J. 2002. Agriculture and rural livelihoods: is globalisation opening or blocking paths out of rural poverty? *Agricultural Research & Extension Network Paper* 121, Overseas Development Institute, London, UK.

Laing, D. 1992. Correspondence in *Newsletter of the Vetiver Network,* 8: 13-14.

Lal, R. 1982. Effective conservation farming systems for the humid tropics. In: Kussow, W., El-Swaify, S.A and Mannering, J. (eds.). *Soil Erosion and Conservation in the Tropics* pp. 57-76. American Society of Agronomy Special Publication Number 43, Soil Science Society of America, Madison, Wisconsin, USA.

Lal, R. 1988. Soil erosion research on steeplands. In: Moldenhauer, W.C. and Hudson, N.W. (eds.). *Conservation Farming on Steeplands,* pp. 45-53. Soil and Water Conservation Society and World Association of Soil and Water Conservation, Ankeny, Iowa. USA.

Lal, R. 1989. Agroforestry systems and soil surface management of a tropical alfisol. *Agroforestry Systems*, 8: 1-6.

Lal, R. 1989b. Agroforestry systems and soil surface management of a tropical alfisol: I. Soil moisture and crop yields. *Agroforestry Systems*, 8: 7-29.

Lal, R. 1989c. Agroforestry systems and soil surface management of a tropical alfisol: II. Water runoff, soil erosion, and nutrient loss. *Agroforestry Systems*, 8: 97-111.

Lal, R. 1990. Soil erosion and land degradation: the global risks. In: Lal, R. and Stewart, B.A. (eds.). *Soil Degradation*, pp. 129-172. Advances in Soil Science, Volume II. Springer-Verlag, New York, USA.

Lal, R. 1990b. *Soil Erosion in the Tropics: Principles and Management*. McGraw-Hill, Inc., USA.

Lal, R. 1991. Soil structure and sustainability. *Journal of Sustainable Agriculture*, 1(4): 67-92.

Lal, R. 1994. Soil erosion by wind and water: problems and prospects. In: Lal, R. (ed.). *Soil Erosion Research Methods*, pp. 1-9. Soil and Water Conservation Society, Ankeny Iowa and St. Lucie Press, Florida, USA.

Lal, R. and Stewart, B.A. 1990. Soil degradation: a global threat. In: Lal, R. and Stewart, B.A. (eds.). *Soil Degradation*, pp. xiii-xvii. Advances in Soil Science, Volume II. Springer-Verlag, New York, USA.

Laman, M., Sanders, H., Zaal, F., Sidikou, H.A., Toe, E., 1996. *Combating desertification – the Role of Incentives*. Centre for Development Cooperation Services, Vrije Universitiet Amsterdam, The Netherlands

Lang, R.D. 1992. Accuracy of two sampling methods used to estimate sediment concentrations in runoff from soil-loss plots. *Earth Surface Processes and Landforms*, 17: 841-844.

Larose, M., Oropeza-Mota, J.L., Norton, D., Turrent-Fernandez, A., Martinez-Menes, M., Pedraza-Oropeza, J.A. and Francisco-Nicolas, N. 2004. Application of the WEPP model to hillside lands in the Tuxtlas, Veracruz, Mexico. *Agrociencia*, 38(2): 155-163.

Larrea, S. 1997. *Experiencias y Lecciones de Agricultores Innovadores Sobre el Desarrollo Rural: Caso de Ginope, Honduras*. MSc dissertation, *Escuela Agrícola Panamericana*, Zamorano, Honduras.

Larson, W.E. and Pierce, F.J. 1994. The dynamics of soil quality as a measure of sustainable management. In: Doran, J.W., Coleman, D.C., Bezdicek, D.F. and Stewart, B.A. (eds.). *Defining Soil Quality for a Sustainable Environment* pp. 37-51. Soil Science Society of America Special Publication Number 35, Madison, Wisconsin, USA.

Lavelle, P., Spain, A.V., Blanchart, E., Martin, A. and Martin, S. 1992. Impact of soil fauna on the properties of soils in the humid tropics. In: Lal, R. and Sanchez, P.A. (eds.). *Myths and Science of Soils in the Tropics*, pp. 157-185. Soil Science Society of America Special Publication Number 29, Madison, Wisconsin, USA.

Lawrence, A. 1997. Contours, crops and cattle: participatory soil conservation in the Andean foothills, Bolivia. *Agroforestry Forum*, 8(4): 11-13.

Lawrence, M.B., Mayfield, B.M., Avila, L.A., Pasch, R.J. and Rappaport, E.N. 1998. Atlantic hurricane season of 1995. *Monthly Weather Review*, 126: 1124-1151.

Leonard, H.J. 1987. *Natural Resources and Economic Development in Central America: a Regional Environmental Profile*. International Institute for Environment and Development, Washington, D.C., USA.

Leslie, J. 2000. Running dry: what happens when the world no longer has enough freshwater? *Harpers Magazine*, July 2000: 37-52.

Lewis, L.A. 1992. Terracing and accelerated soil loss on Rwandian steeplands: A preliminary investigation of the implications of human activities affecting soil movement. *Land Degradation & Rehabilitation*, 3: 241-246.

Lewis, L.A. and Nyamulinda, V. 1996. The critical role of human activities in land degradation in Rwanda. *Land Degradation & Development*, 7: 47-55.

Ligdi, E.E. and Morgan, R.P.C. 1995. Contour grass strips: a laboratory simulation of their role in soil erosion control. *Soil Technology*, 8: 109-117.

Linden, D.R., Hendrix, P.F., Coleman, D.C. and van Vliet, P.C.J. 1994. Faunal indicators of soil quality. In: Doran, J.W., Coleman, D.C., Bezdicek, D.F. and Stewart, B.A. (eds.). *Definining Soil Quality for a Sustainable Environment*, pp. 91-106. Soil Science Society of America Special Publication Number 35, Madison, Wisconsin, USA.

Logan, T.J. 1990. Chemical degradation of soil. In: Lal, R. and Stewart, B.A. (eds.). *Soil Degradation*, pp. 187-221. Advances in Soil Science, Volume II. Springer-Verlag, New York, USA.

Loker, W.M. 1996. "Campesinos" and the crisis of modernization in Latin America. *Journal of Political Ecology*, 3: 69-88.

Long, N. 1992. From paradigm lost to paradigm regained: The case for an actor-oriented sociology of development. In: Long, N. and Long, A. (eds.). *Battlefields of Knowledge: The Interlocking of Theory and Practice in Social Research and Development*, pp. 16-45. Routledge, London, UK.

Lu, Y. and Stocking, M. 1998. *A Decision-Support Model for Soil Conservation: Case Study on the Loess Plateau, China*. Centre for Social and Economic Research on the Global Environment, Working paper GEC 98-04, University of East Anglia, UK.

Lutz, E., Pagiola, S. and Reiche, C. 1994. The costs and benefits of soil conservation: the farmers' viewpoint. *The World Bank Research Observor*, 9(2): 273-295.

Maass, J.M. 1992. The use of litter-mulch to reduce erosion on hilly land in Mexico. In: Hurni, H. and Tato. K. (eds.). *Erosion, conservation, and small-scale farming*, pp. 383-391. Geographica Bernensia, International Soil Conservation Organisation and World Association of Soil and Water Conservation.

Malley, Z.J.U., Kayombo, B., Willcoks, T.J. and Mtakwa, P.W. 2004. 'Ngoro: an indigenous, sustainable and profitable soil, water and nutrient conservation system in Tanzania for sloping land. *Soil & Tillage Research,* 77 (1): 47-58.

Marmillod, A. 1987. Farmers' attitudes towards trees. In: Beer, J., Fassbender, H.W. and Heuveldop, J. (eds.). *Advances in Agroforestry Research,* pp. 259-270. Centro Agronómico Tropical de Investigación y Enseñanza (CATIE) and Deutsche Gessellschaft für Technische Zusammenarbeit (GTZ), Costa Rica.

Marsden, D. 1994. Indigenous knowledge and power in development. In: Scoones, I. and Thompson, J. (eds.). *Beyond Farmer First: Rural People's Knowledge, Agricultural Research and Extension Practice,* pp. 52-57. Intermediate Technology Publications, London, UK.

Mastin, M.C. 2002. *Flood-Hazard Mapping in Honduras in Response to Hurricane Mitch.* U.S. Geological Survey Water-Resources Investigations Report 01-4277 [http://water.usgs.gov/pubs/wri/wri014277/pdf/WRIR01-4277.pdf] (Accessed April 2003).

Maxwell, S., Urey, I. and Ashley, C. 2001. *Emerging Issues in Rural Development.* An issues paper, Overseas Development Institute, London, UK

May, T. 1997. *Social Research: Issues, Methods and Process.* Oxford University Press, UK.

Mazzucato, V., Niemeijer, D., Stroosnijder, L. and Roling, N. 2001. Social Networks and the Dynamics of Soil and Water Conservation in the Sahel. Gatekeeper Series 101, International Institute for Environment and Development, London, UK.

McCown, S., Graumann, A. and Ross, T. 1998. Mitch: the deadliest Atlantic hurricane. http://www.ncdc.gov/ol/reports/mitch/mitch/html (Accessed November 1998)

McDonald M. and Brown K. 2000. Soil and water conservation projects and rural livelihoods: Options for design and research to enhance adoption and adaptation. *Land Degradation & Development,* 11: 343-361.

McDonald, M., Healey, J.R., Stevens, P.A., 2002. The effects of secondary forest clearance and subsequent land-use on erosion losses and soil properties in the Blue Mountains of Jamaica. *Agriculture, Ecosystems & Environment,* 92: 1-19.

McDonald, M.A., Stevens, P.A., Healey, J.R. and Prasad, P.V.D. 1997. Maintenance of soil fertility on steeplands in the Blue Mountains of Jamaica: the role of contour hedgerows. *Agroforestry Forum,* 8 (4): 21-25.

Mead, R., Curnow, R.N. and Hasted, A.M. 1996. *Statistical Methods in Agriculture and Experimental Biology.* Chapman and Hall, London, UK.

Meinzen-Dick, R., Adato, M., Haddad, L. and Hazell, P. 2003. *Impacts of Agricultural Research on Poverty: Findings of an Integrated Economic and Social Analysis.* Food Consumption and Nutrition Division Discussion Paper Number 164. International Food Policy Research Institute, Washington D.C., USA.

Meinzen-Dick, R., Adato, M., Haddad, L. and Hazell, P. 2004. *Science and Poverty: An Interdisciplinary Assessment of the Impact of Agricultural Research.* International Food Policy Research Institute, Washington D.C., USA

Mendoza, R.B. 1996. *Evaluación de Barreras Vivas de Gliricidia sepium (Jacq.) Sobre Pérdidas de Suelo, Agua y Rendimientos de Maiz, Frijol en Tres Sitios de la Cuenca El Pital.* Universidad Nacional Agraria, Facultad de Recursos Naturales y del Ambiente, Department de Uso y Manejo de Suelos, Managua, Nicaragua.

Mikhailova, E.A., Bryant, R.B., Schwager, S.J. and Smith, S.D. 1997. Predicting rainfall erosivity in Honduras. *Soil Science Society of America Journal,* 61: 273-279.

Miles, M. and Huberman A. 1994. *Qualitative Data Analysis: An Expanded Sourcebook.* Sage Publications, Thousand Oaks, California, USA.

Moehansyah, H., Maheshwari, B.L. and Armstrong, J. 2004. Field evaluation of selected soil erosion models for catchment management in Indonesia. *Biosystems Engineering,* 88 (4): 491-506.

Mokma, D.L. and Sietz, M.A. 1992. Effects of soil erosion on corn yields on Marlette soils in south-central Michigan. *Journal of Soil and Water Conservation,* 47(4): 325-327.

Moldenhauer, W.C. and Hudson, N.W. 1988. *Conservation Farming on Steep Lands.* Soil and Water Conservation Society and World Association of Soil and Water Conservation, Ankeny, Iowa, USA.

Moldenhauer, W.C., Kemper, W.D., Schertz, D.L., Weesies, G.A., Hatfield, J.L. and Laflen, J.M. 1994. Crop residue management for soil conservation and crop production: Present technology and acceptance in the United States. In: Napier, T.L., Camboni, S.M. and El-Swaify, S.A. (eds.). *Adopting Conservation on the Farm: An international Perspective on the Socioeconomics of Soil and Water Conservation,* pp. 37-46. Soil and Water Conservation Society, Ankeny, Iowa, USA.

Moran, E.F. 1987. Monitoring fertility degradation of agricultural lands in the lowland tropics. In Little, P.D., Horowitz, M.H. and Nyerges, A.E. (eds.). *Lands at Risk in the Third World: Local perspectives,* pp. 69-91. Westview Press, Boulder, Colorado, USA.

Morgan, R.P.C. and Rickson, R.J. 1995. *Slope Stabilisation and Erosion Control: A Bio-Engineering Approach.* E & F.N. Spon, London, UK.

Morris, S., Neidecker-Gonzales, O., Carletto, C., Munguía, M., Medina, J. and Wodon, Q. 2001. Hurricane Mitch and the livelihoods of the rural poor in Honduras. *World Development,* 30 (1): 49-60.

Mureithi, J. G., Tayler, R.S. and Thorpe, W. 1995. Productivity of alley farming with leucaena (*Leucaena leucocephala* Lam. De Wit) and Napier grass (*Pennisetum purpureum* K. Schum) in coastal lowland Kenya. *Agroforestry Systems,* 31: 59-78.

Murray, G.F. and Bannister, M.E. 2004. Peasants, agroforesters and anthropologists: a 20-year venture in income-generating trees and hedgerows in Haiti. *Agroforestry Systems,* 61-62(1): 383-397.

Mutchler, C.K., Murphree, C.E. and McGregor, K.C. 1994. Laboratory and field plots for erosion research. In: Lal, R. (ed.). *Soil Erosion Research Methods* pp. 11-37. Soil and Water Conservation Society, Ankeny Iowa and St. Lucie Press, Florida, USA.

Napier, T.L. 1989. Implementation of soil conservation practices: Past efforts and future prospects. In: Baum, E,E., Wolff, P. and Zöbisch, M.A. (eds.). *The Extent of Soil Erosion – Regional Comparisons.* Volume 1 in *Topics in Applied Resource Management in the Tropics,* pp. 9-34. German Institute for Tropical and Subtropical Agriculture, Witzenhausen, Germany.

Napier, T.L. 1991. Factors affecting acceptance and continued use of soil conservation practices in developing societies: a diffusion perspective. *Agriculture, Ecosystems and Environment,* 36: 127-140.

Napier, T.L., Camboni, S. M. and El-Swaify, S.A. 1994. A synthesis of adopting conservation on the farm: An international perspective on the socioeconmics of soil and water conservation. In: Napier, T.L., Camboni, S.M. and El-Swaify, S.A. (eds.). *Adopting Conservation on the Farm: An international Perspective on the Socioeconomics of Soil and Water Conservation,* pp. 511-517. Soil and Water Conservation Society, Ankeny, Iowa, USA.

Napier, T.L. and Sommers, D.G., 1993. Soil conservation in the tropics: A prerequisite for societal development. In: Baum, E.E., Wolff, P., Zöbisch, M.A. (eds.), *Acceptance of Soil and Water Conservation. Strategies and Technologies.* Volume 3 *in Topics in Applied Resource Management in the Tropics,* pp. 7-28. German Institute for Tropical and Subtropical Agriculture (DITSL), Witzenhausen, Germany.

Ndiaye, S.M. and Sofranko, A.J. 1994. Farmers' perceptions of resource problems and adoption of conservation practices in a densely populated area. *Agriculture, Ecosystems and Environment,* 48: 35-47.

Nearing, M.A. 1998. Why soil erosion models over-predict small soil losses and under-predict large soil losses. *Catena,* 32: 15-22.

Nearing, M.A. 2000. Evaluating soil erosion models using measured plot data: accounting for variability in the data. *Earth Surface Processes and Landforms,* 25 (9): 1035 - 1043.

Nehmdahl, H. 1999. Vetiver grass in a soil and water conservation trial - Tanzania. *Newsletter of the Vetiver Network,* 20: 47-49.

Nimlos, T.J. and Savage, R.F. 1991. Successful soil conservation in the Ecuadorian highlands. *Journal of Soil and Water Conservation,* 46 (5): 341 – 343.

Norman, D. and Douglas, M. 1994. *Farming Systems Development and Soil Conservation.* Food and Agriculture Organization of the United Nations, Rome, Italy.

Nyssen, J., Poesen, J., Moeyersons, J., Deckers, J., Haile, M., and Lang, A. 2004. Human impact on the environment in the Ethiopian and Eritrean highlands - a state of the art. *Earth Science Reviews,* 64 (3-4): 273-320.

O'Brien, K.L. 1998. Sacrificing the Forest: Environmental and Social Struggles in Chiapas. Westview Press, Boulder, Colorado, USA

O'Hara, S.L., Street-Perrott, F.A. and Burt, T.P. 1993. Accelerated soil erosion around a Mexican highland lake caused by prehispanic agriculture. *Nature*, 362: 48-51.

Oberthur, T., Barrios, E., Cook, S., Usma, H. and Escobar, G. 2004. Increasing the relevance of scientific information in hillside environments through understanding of local soil management in a small watershed in the Colombian Andes. *Soil Use and Management*, 20(1): 23-31.

Oldeman, L. R., Hakkeling, R.T.A. and Sombroek, W.G. 1991. *World Map of the Status of Human-Induced Soil Degradation: An explanatory Note.* International Soil Reference and Information Centre Wageningen, The Netherlands and United Nations Environment Programme, Nairobi, Kenya.

Oldeman, L.R. 1994. The global extent of soil degradation. In: Greenwood, D.J. and Szabolcs, I. (eds.) *Soil Resilience and Sustainable Land Use* pp. 99-118. CAB International, Wallingford, Oxon, UK.

Omoro, L.M.A. and Nair, P.K.R. 1993. Effects of mulching with multipurpose-tree prunings on soil and water run-off under semi-arid conditions in Kenya. *Agroforestry Systems*, 22: 225-239.

Oppenheim, A.N. 1986. *Questionnaire Design and Attitude Measurement.* Gower Publising Company, Aldershot, Hants, UK.

Osterman, D.A. and Hicks, T.L. 1988. Highly erodible land: farmer perceptions versus actual measurements. *Journal of Soil and Water Conservation*, 43(2): 177 – 181.

Ouedraogo, M. and Kaboré, V. 1996. The zaï: a traditional technique for the rehabilitation of degraded land in the Yatenga, Burkina Faso. In: Reij, C., Scoones, I. and Toulmin, C. (eds.). *Sustaining the Soil: Indigenous Soil and Water Conservation in Africa*, pp. 80-84. Earthscan Publications Ltd., London, UK.

Oxfam. 2002. *Rigged Rules and Double Standards: Trade, Globalisation, and the Fight Against Poverty.* Oxfam, Oxford, UK. Available at www.maketradefair.com

Pagiola, S. 1994. Soil conservation in a semi-arid region of Kenya: rates of return and adoption by farmers. In: Napier, T.L., Camboni, S.M. and El-Swaify, S.A. (eds.). *Adopting Conservation on the Farm: An international Perspective on the Socioeconomics of Soil and Water Conservation*, pp. 171-188. Soil and Water Conservation Society, Ankeny, Iowa, USA.

Pagiola, S. 1997. Land degradation under agricultural intensification in El Salvador: what do available data tell us? Paper presented at the Symposium on Sustainable Intensification in Fragile Lands XXIII International Conference of Agricultural Economists, Sacramanto, California, USA, August, 1997.

Painter, M. 1984. Changing relations of production and rural underdevelopment. *Journal of Anthropological Research*, 40: 271-292.

Pardey, P.G. and Beintema, N.M. 2001. *Slow Magic: Agricultural R&D a Century After Mendel*. Food Policy Statement, Number 36, International Food Policy Research Institute, Washington D.C., USA.

Paudel, G.S. and Thapa, G.B. 2004. Impact of social, institutional and ecological factors on land management practices in mountain watersheds of Nepal. *Applied Geography*, 24(1): 35-55.

Paulhus, J.L.H. 1965. Indian Ocean and Taiwan rainfalls set new records. *Monthly Weather Review*, 93: 331-335.

Pawar, P.B. 1998. Prospect and problems in use of vetiver for watershed management in sub mountain and scarcity zones (Maharashtra, India). *Newsletter of the Vetiver Network*, 19: 33-35.

Pawluk, R.R., Sandor, J. A. and Tabor, J.A. 1992. The role of indigenous soil knowledge in agricultural development. *Journal of Soil and Water Conservation*, 47 (4): 298-302.

Pellek, R. 1992. Contour hedgerows and other soil conservation interventions for hilly terrain. *Agroforestry Systems*, 17: 135-152.

Pereira, H.C. 1989. *Policy and Practice in the Management of Tropical Watersheds*. Westview Press, Boulder, Colorado, USA.

Perich, I., 1993. Umweltökonomie in der entwicklungspolitischen Diskussion. Ansätze und Perspektiven. – Berichte zur Entwicklung und Umwelt Nr.8, Gruppe für Entwicklung und Umwelt, University of Berne, Switzerland.

Perotto-Baldiviezo, H.L., Thurow, T.L., Smith, C.T., Fisher, R.F. and Wu, X.B. 2004. GIS-based spatial analysis and modeling for landslide hazard assessment in steeplands, southern Honduras. *Agriculture, Ecosystems & Environment*, 103 (1): 165-176.

Pierce, F.J. and Lal, R. 1994. Monitoring the impact of soil erosion on crop productivity. In: *Soil erosion research methods*. In: Lal, R. (ed.). *Soil Erosion Research Methods*, pp. 235-263. Soil and Water Conservation Society, Ankeny Iowa and St. Lucie Press, Florida, USA.

Pieri, C., Dumanski, J., Hamblin, A., and Young, A. 1995. *Land Quality Indicators*. World Bank Discussion Papers 315, The World Bank, Washington D.C., USA.

Pimentel, D., Harvey, C., Resosudarmo, P., Sinclair, K., Kurz, D., McNair, M., Crist, S., Shpritz, L., Fitton, L., Saffouri, R. and Blair, R. 1995. Environmental and economic costs of soil erosion and conservation benefits. *Science*, 267: 1117-1123.

Pollard, H.P. 1994. Prehispanic archaeology, ethnohistory, and soil erosion: A debate over modern agricultural sustainability. *Culture and Agriculture*, 49: 16-20.

Posner, J.L. and McPherson, M.F. 1982. Agriculture on the steep slopes of tropical America: current situation and prospects for the year 2000. *World Development*, 10(5): 341-353.

Pound, B. 1999. The appropriate use of qualitative information in participatpory research and development. What are the issues for farmers and researchers? *Agricultural Research and Extension Network Paper 92c,* Overseas Development Institute, London, UK.

Powers, S.P. 2001. Honduras suffered the brunt of Hurricane *Mitch.* [http://landslides.usgs.gov/Honduras/index.html] (Accessed April 2003).

Pretty, J. and Hine R. 2001. *Reducing Food Poverty with Sustainable Agriculture: a Summary of New Evidence. Final report from the SAFE-World Research Project.* University of Essex, Cochester, UK.

Pretty, J. and Koohafkan, P. 2002. *Land and Agriculture: A Compendium of Recent Sustainable Development Initiatives in the Field of Agriculture and Land Management.* Food and Agriculture Organization, Rome, Italy

Pretty, J.N. 1995. *Regenerating Agriculture: Policies and Practice for Sustainability and Self-Reliance.* Earthscan Publications Ltd, London, UK.

Pretty, J.N. 1995b. Crop yields rise with agricultural regeneration in Central America. IIED Perspectives 13, International Institute for Environment and Development, London, pp. 11-13.

Pretty, J.N. 1998. Furthering cooperation between people and institutions. In: Blume, H.-P., Eger, H., Fleischhauer, E., Hebel, A., Reij, C., and Stenier, K.G. (eds.). *Towards Sustainable Land Use: Furthering Cooperation Between People and Institutions.* Selected papers of the 9th conference of the International Soil Conservation Organisation, Bonn, Germany. *Advances in Geoecology,* 31: 837-849.

Pretty, J.N. 1998b. *The Living Land: Agriculture, Food and Community Regeneration in Rural Europe.* Earthscan Publications Ltd., London, UK.

Pretty, J.N. and Chambers, R. 1994. Towards a learning paradigm: new professionalism and institutions for agriculture. In: Scoones, I. and Thompson, J. (eds.). *Beyond Farmer First: Rural People's Knowledge, Agricultural Research and Extension Practice,* pp. 182-202. Intermediate Technology Publications, London, UK.

Pretty, J.N. and Shah, P. 1997. Making soil and water conservation sustainable: From coercion and control to partnerships and participation. *Land Degradation & Development,* 8: 39 – 58.

Pretty, J.N. and Shah, P. 1999. Soil and water conservation: A brief history of coercion and control. In: Hinchcliffe, F., Thompson, J., Pretty, J., Guijt, I. and Shah, P. (eds.). *Fertile Ground: The Impacts of Participatory Watershed Management* pp. 1-12. Intermediate Technology Publications, London, UK.

Pretty, J.N. and Shaxson, T.F. 1998. The potential of sustainable agriculture. *ENABLE* (Newsletter of the Association for Better Land Husbandry), 8: 4-21.

Pretty, J.N. Guijt, I., Thompson, J. and Scoones, I. 1995. *A Trainer's Guide to Participatory Learning and Action.* IIED Participatory Methodology Series. International Institute for Environment and Development, London, UK.

Pushparajah, E. 1989. *Soil Variability on Experimental Sites*. IBSRAM Technical Notes 3: 149-160. Bangkok, Thailand.

Quine, T.A., Walling, D.E., Chakela, Q.K., Mandiringana, O.T. and Zhang, X. 1999. Rates and patterns of tillage and water erosion on terraces and contour strips: evidence from caesium-137 measurements. *Catena*, 36: 115-142.

Quinton, J.N. 1997. Reducing predictive uncertainty in model simulations: a comparison of two methods using the European Soil Erosion Model (EUROSEM). *Catena*, 30: 101-117.

Rajasekaran, B. and Warren, D.M. 1995. Role of indigenous soil health care practices in improving soil fertility: Evidence from South India. *Journal of Soil and Water Conservation*, 50 (2): 146-149

Rao, C.N. and Rao, M.S. 1996. Effect of vetiver hedge on runoff, soil loss, soil moisture and yield of rainfed crops in Alfisol watersheds of southern India. *Newsletter of the Vetiver Network.*, 15: 21-22.

Rao, M.R., Ong, C.K., Pathak, P. and Sharma, M.M. 1991. Productivity of annual cropping and agroforestry systems on a shallow Alfisol in semi-arid India. *Agroforestry Systems*, 15: 51-63.

Rappaport, E.N. 1998. Atlantic Hurricanes. *Weatherwise*, 51 (2): 43-47.

Raun, W.R and Barreto, H.J. 1995. Regional maize grain yield response to applied phosphorous in Central America. *Agronomy Journal*, 87: 208-213.

Reij, C., Scoones, I. and Toulmin, C. (eds.). 1996. *Sustaining the Soil: Indigenous Soil and Water Conservation in Africa*. Earthscan Publications Ltd., London, UK.

Rhoades, R.E. 1991. *The Art of the Informal Agricultural Survey*. International Institute of Tropical Agriculture Research Guide 36, Ibadan, Nigeria.

Rist, S. 1992. Desarrollo y participación: experiencias con la revalorización del conocimiento en Bolivia. *Serie Técnica* No. 27. Agroecología Universidad Cochabamba, Bolivia.

Rivas, D. A. 1993. *Factors Affecting Soil Erosion in Maize (Zea mays L.) and Pineapple (Ananas comosus L.) Stands in Ticuantepe, Nicaragua: A Preliminary Evaluation of the Universal Soil Loss Equation using data from erosion plots.* MSc dissertation, Reports and Dissertations No. 15, Swedish University of Agricultural Sciences, Department of Soil Sciences, Sweden.

Robinson, C.A., Ghaffarzadeh, M. and Cruse, R.M. 1996. Vegetative filter strip effects on sediment concentration in cropland runoff. *Journal of Soil and Water Conservation*, 50 (3): 227-230.

Robinson, D.A. 1989. Motives and policies for soil conservation in colonial and early, post-independence Zambia. In: Baum, E,E., Wolff, P. and Zöbisch, M.A. (eds.). *The Extent of Soil Erosion – Regional Comparisons*. Volume 1 in *Topics in Applied Resource Management in the Tropics*, pp. 87-109. German Institute for Tropical and Subtropical Agriculture, Witzenhausen, Germany.

Rocha, J.L. and Christoplos, I. 2001. Disaster mitigation and preparedness on the Nicaragua post-Mitch agenda. *Disasters,* 25(3): 240-250.

Rodriguez, O.S. 1995. Hedgerows and mulch as soil conservation measures evaluated under field simulated rainfall. *Newsletter of the Vetiver Network,* 13: 26.

Roels, J.M. 1985. Estimation of soil loss at a regional scale based on plot measurements – some critical considerations. *Earth Surface Processes and Landforms,* 10(6): 587-595.

Röling, N. 2000. *Gateway to the Global Garden: Beta/Gamma Science for Dealing With Ecological Rationality.* Eighth Annual Hopper Lecture, October 24, 2000. University of Guelph, Canada.

Röling, N. and de Jong, F. 1998. Learning: shifting paradigms in education and extension studies. *Journal of Agricultural Education & Extension,* 5 (3): 143-161.

Romig, D.E., Garlynd, M.J., Harris, R.F. and McSweeney, K. 1995. How farmers assess soil health and quality. *Journal of Soil and Water conservation,* 50(3): 229 – 236.

Rose, C. W. 1993. Erosion and sedimentation. In: Bonell, M., Hufschmidt, M. and Gladwell, J. (eds.). *Hydrology and Water Management in the Humid Tropics: Hydrological Research Issues and Strategies for Water Management,* pp. 301-343. International Hydrology Series. Cambridge University Press, UK.

Rosecrance, R.C., Brewbaker, J.L. and Fownes, J.H. 1992. Alley cropping of maize with nine leguminous trees. *Agroforestry Systems,* 17: 159-168.

Rosegrant, M.W. and Cline, S.A. 2003. Global food security: challenges and policies. *Science,* 302 (5652): 1917-1919.

Ruaysoongnern, S. and Patanothai, A. 1991. Farmers' perceptions and the adoption of sustainable land management technologies – Thailand's experience. In: *Evaluation for Sustainable Land Management in the Developing World.* IBSRAM Proceedings 12(2), Bangkok, Thailand, pp. 491-507.

Ruben, R. 2001. Economic conditions for sustainable agriculture: a new role for the market and state. *LEISA,* 17 (3): 52-53.

Ruddell, E. and Beingolea, J. 1999. Increasing smallholder agricultural and livestock production in Andean mountain regions. Paper prepared for Bellagio Conference on Sustainable Agriculture. http:// www.rodaleinstitute.org/global/6 00 99.html (Accessed December 2004).

Runge, C.F., Senauer, B., Pardey, P.G. and Rosengrant, M.W. 2003. *Ending Hunger in Our Lifetime: Food Security and Globalization.* International Food Policy Research Institute and The Johns Hopkins University Press, Washington, D.C., USA.

Ruppenthal, M., Leihner, D.E., Hilger, T.H. and Castillo, F. J.A. 1996. Rainfall erosivity and erodibility of Inceptisols in the southwest Colombian Andes. *Experimental Agriculture,* 32: 91-101.

Rüttiman, M., Schaub, D., Prasuhn, V. and Rüegg, W. 1995. Measurement of runoff and soil erosion on regularly cultivated fields in Switzerland – some critical considerations. *Catena*, 25: 127-139.

Ryder, R. 1994. Farmer perception of soils in the mountains of the Dominican Republic. *Mountain Research and Development*, 14 (3): 261-266.

Sain, G.E. and Barreto, H.J. 1996. The adoption of soil conservation technology in El Salvador: Linking productivity and conservation. *Journal of Soil and Water Conservation*, 5 (4): 313 – 321.

Sanchez, P.A. 1987. Soil productivity and sustainability in agroforestry systems. In: Steppler, H.A. and Nair, P.K.R. (eds.). *Agroforestry a Decade of Development*, pp. 205-223. The International Institute for Research in Agroforestry, Nairobi, Kenya.

Sanders, D.W. 1988. Food and Agriculture Organization activities in soil conservation. In: Moldenhauer, W.C. and Hudson, N.W. (eds.). *Conservation Farming on Steeplands*, pp. 54-62. Soil and Water Conservation Society and World Association of Soil and Water Conservation, Ankeny, Iowa. USA.

Sanders, D.W. 1988b. Soil and water conservation on steep lands: A summary of workshop discussions. In: Moldenhauer, W.C. and Hudson, N.W. (eds.). *Conservation Farming on Steeplands*, pp. 275-282. Soil and Water Conservation Society and World Association of Soil and Water Conservation, Ankeny, Iowa. USA.

Sanders, D.W. and Cahill, D. 1999. Where incentives fit in soil conservation programs. In: Sanders, D.W., Huszar, P.C., Sombatpanit, S., Enters, T. (eds). *Incentives in Soil Conservation: From Theory to Practice*, pp. 11-24. Oxford & IBH Publishing, New Delhi, India.

Sanders, D.W., Huszar, P.C., Sombatpanit, S. and Enters, T. (eds), 1999. *Incentives in Soil Conservation. From Theory to Practice.* Oxford & IBH Publishing, New Delhi, India.

Scherr, S.J. 1999. *Soil Degradation: A Threat to Developing Country Food Security By 2020?* Food Agriculture and the Environment Discussion Paper 27, International Food Policy Research Institute, Washington, D.C, USA.

Scherr, S.J. and Yadav, S. 1996. Land Degradation in the Developing World: Implications for Food, Agriculture, and the Environment to 2020. Food, Agriculture, and the Environment Discussion Paper 14, International Food Policy Research Institute, Washington D.C., USA.

Schiller, J.M. Anecksamphant, C. and Sujanin, S. 1982. Development of areas of shifting cultivation in north Thailand: "Thai-Australia land development project". In: Kussow, W., El-Swaify, S.A and Mannering, J. (eds.). *Soil Erosion and Conservation in the Tropics*, pp. 77-96. American Society of Agronomy Special Publication Number 43, Soil Science Society of America, Madison, Wisconsin, USA.

Scoones, I. and Thompson, J. 1994. Introduction. In: Scoones, I. and Thompson, J. (eds.). *Beyond Farmer First: Rural People's Knowledge, Agricultural Research and Extension Practice*, pp. 1-12. Intermediate Technology Publications, London, UK.

Scoones, I. and Thompson, J. 1994b. Knowledge, power and agriculture – towards a theoretical understanding. In: Scoones, I. and Thompson, J. (eds.). *Beyond Farmer First: Rural People's Knowledge, Agricultural Research and Extension Practice*, pp. 16-32. Intermediate Technology Publications, London, UK.

Scoones, I., Reij, C. and Toulmin, C. 1996. Sustaining the soil: indigenous soil and water conservation in Africa. In: Reij, C., Scoones, I. and Toulmin, C. (eds.). *Sustaining the Soil: Indigenous Soil and Water Conservation in Africa*, pp. 1-27. Earthscan Publications Ltd, London, UK.

Secretaría de Recursos Naturales. 1994. *Manual Práctico de Manejo de Suelos en Ladera*. Secretaría de Recursos Naturales, Tegucigalpa, Honduras.

Segerros, K.H.M. and Kerr, B. 1996. Introduction. In: Chinene, V.R.N., Shaxson, T.F., Molumeli, P. and Segerros, K.H.M. 1996. *Guidelines to Better Land Husbandry in the SADC Region*, pp. 3-6. Environment and Land Management Sector, Southern African Development Community, Maseru and Agriculture Development Division, Commonwealth Secretariat, London, UK.

Shaxson, T.F. 1988. Conserving soil by stealth. In: Moldenhauer, W.C. and Hudson, N.W. (eds.). *Conservation Farming on Steeplands*, pp. 9-17. Soil and Water Conservation Society and World Association of Soil and Water Conservation, Ankeny, Iowa, USA.

Shaxson, T.F. 1992. Crossing some watersheds in conservation thinking. In: Tato, K and Hurni, H. (eds.). *Soil Conservation for Survival*, pp. 81-89. Soil and Water Conservation Society in cooperation with the International Soil Conservation Organisation and World Association of Soil and Water Conservation, Ankeny, USA.

Shaxson, T.F. 1993. Conservation-effectiveness of farmers' actions: a criterion of good land husbandry. In: Baum, E.E., Wolff, P., Zöbisch, M.A. (eds.), *Acceptance of Soil and Water Conservation. Strategies and Technologies*. Volume 3 *in Topics in Applied Resource Management in the Tropics*, pp. 103-128. German Institute for Tropical and Subtropical Agriculture (DITSL), Witzenhausen, Germany.

Shaxson, T.F. 1993b. Sustainability. *Journal of Soil and Water Conservation*, 48(4): 249-250.

Shaxson, T.F. 1994. Land husbandry's fifth dimension: Enriching our understanding of farmers' motivations. Paper presented at the 8th International Soil Conservation Organization Conference, New Delhi, India.

Shaxson, T.F. 1996. Land husbandry. In: Chinene, V.R.N., Shaxson, T.F., Molumeli, P. and Segerros, K.H.M. 1996. *Guidelines to Better Land Husbandry in the SADC Region*, pp. 21-40. Environment and Land Management Sector, Southern African Development Community, Maseru and Agriculture Development Division, Commonwealth Secretariat, London, UK.

Shaxson, T.F. 1997. Soil erosion and land husbandry. *Land husbandry: International Journal of Soil and Water Conservation* 2 (1): 1-14.

Shaxson, T.F. 1999. *New Concepts and Approaches to Land Management in the Tropics With Emphasis on Steeplands*. Soils Bulletin 75, Food and Agriculture Organization of the United Nations, Rome, Italy

Shaxson, T.F. 2004. Think like a root: the land husbandry context for conservation of water and soil. Submission to FAO e-conference on *Drought Resistant Soils: Optimization of Soil Moisture for Sustainable Plant Production*, November-December, 2004. http://www.fao.org/ag/agl/agll/soilmoisture/ (Accessed January 2005).

Shaxson, T.F. and Barber, R. 2003. *Optimizing Soil moisture for Plant Production*. Soils Bulletin 79, Food and Agriculture Organization of the United Nations, Rome, Italy.

Shaxson, T.F., Douglas, M.G. and Downes, R.G. 1997b. *Principles of Good Land Husbandry*. The Association for Better Land Husbandry, UK.

Shaxson, T.F., Hudson, N.W., Sanders, D.W., Roose, E. and Moldenhauer, W.C. 1989. *Land Husbandry: A Framework for Soil and Water Conservation*. Soil and Water Conservation Society and The World Association of Soil and Water Conservation, Ankeny, Iowa, USA.

Shaxson, T.F., Tiffen, M., Wood, A and Turton, C. 1997. Better land husbandry: re-thinking approaches to land improvement and the conservation of water and soil. *Natural Resource Perspectives* 19, Overseas Development Institute, London, UK.

Sheng, T.C. 1982. Erosion problems associated with cultivation in humid tropical hilly regions. In: Kussow, W., El-Swaify, S.A and Mannering, J. (eds.). *Soil Erosion and Conservation in the Tropics*, pp. 27-39. American Society of Agronomy Special Publication Number 43, Soil Science Society of America, Madison, Wisconsin, USA.

Sheng, T.C. 1990. Runoff plots and erosion phenomena on tropical steeplands. In: Ziemer, R. R., O'Loughlin, C. L. and Hamilton, L. S. (eds.) *Research Needs and Applications to Reduce Erosion and Sedimentation in Tropical Steeplands*, pp. 154-161. International Association of Hydrological Sciences Publication No. 192.

Shengluan, L. and Jiayou, Z. 1998. Experimental results of using vetiver grass hedgerows on China's red acidic soils. *Newsletter of the Vetiver Network*, 19: 29-32.

Sherwood, S. 1999. Personal communication (e-mail dated 7 January 1999).

Sherwood, S. and S. Larrea. 2001. Looking back to see ahead: Farmer lessons and recommendations after 15 years of innovation and leadership in Güinope, Honduras. *Agriculture and Human Values*, 18(2): 195-208.

Sherwood, S. and Uphoff, N. 2000. Soil health: research, practice and policy for a more regenerative agriculture. *Applied Soil Ecology*, 15: 85-97.

Sherwood, S., Nelson, R., Thiele, G. and Ortiz, O. 2000. Farmer Field Schools in potato: a new platform for participatory training and research in the Andes. *Newsletter of the Center for Research and Information on Low External Input and Sustainable Agriculture* (ILEIA), 16 (4): 24-25.

Siderius,W. 1987. Soil variability, experimental design, and data processing. In: Latham, M., Ahn, P. and Elliott, C.R. (eds.). *Management of Vertisols Under Semi-Arid Conditions*, pp. 233-245. International Board for Soil Research and Management, Proceedings No. 6, Bangkok, Thailand.

Siebert, S.F. and Lassoie, J.P. 1991. Soil erosion, water runoff and their control on steep slopes in Sumatra. *Tropical Agriculture* (Trinidad), 68 (4): 321-324.

Sikana, P. 1994. Indigenous soil characterization in Northern Zambia. In: Scoones, I. and Thompson, J. (eds.). *Beyond Farmer First: Rural People's Knowledge, Agricultural Research and Extension Practice*, pp. 80-82. Intermediate Technology Publications, London, UK.

Sillitoe, P. 1993. Losing ground? Soil loss and erosion in the highlands of Papua New Guinea. *Land Degradation & Rehabilitation*, 4: 143-166.

Sillitoe, P. 1998. The development of indigenous knowledge. *Current Anthropology*, 39(2): 223-252.

Sillitoe, P. and Shiel, R.S. 1999. Soil fertility under shifting and semi-continuous cultivation in the Southern Highlands of Papua New Guinea. *Soil Use and Management*, 15: 49-55.

Silverman, D. 1993. *Interpreting Qualitative Data: Methods for Analysing Talk, Text and Interaction*. Sage Publications, London, UK.

Sims, G.K. 1990. Biological degradation of soil. In: Lal, R. and Stewart, B.A. (eds.). *Soil Degradation*, pp. 289-330. Advances in Soil Science, Volume II. Springer-Verlag, New York, USA.

Sinclair, D.F. 1989. *Experimental Design for Land Clearing Experiments on Sloping Lands*. IBSRAM Technical Notes, No. 3: 327-335. Bangkok, Thailand.

Smith, J.E. 1997. *Assessment of Soil and Water Conservation Methods Applied to the Cultivated Steeplands of Southern Honduras*. MSc dissertation, Texas A&M University, USA.

Smyle, J.W. and Magrath, W.B. 1993. Vetiver grass – a hedge against eroison. In: Ragland, J. and Lal, R. (eds.) *Technologies for Sustainable Agriculture in the Tropics*, pp. 109-122. American Society of Agronomy Special Publication 56, Madison, Wisconsin, USA.

Sombatpanit, S., Sukviboon, S., Chiangmai, U.N. and Chinabutr, N. 1992. The use of plastic sheet in soil erosion and conservation studies. In: Hurni, H. and Tato. K. (eds.). *Erosion, Conservation, and Small-Scale Farming*, pp. 471-475. Geographica Bernensia, International Soil Conservation Organisation and World Association of Soil and Water Conservation, Ankeny, USA.

Sorrenson, W.J. 1997. *Paraguay: Financial and Economic Implications of No-Tillage and Crop Rotations Compared to Conventional Cropping Systems*. TCI

Occasional Paper Series No. 9, Investment Centre Division, Food and Agriculture Organization of the United Nations. Rome, Italy.

Sorrenson, W.J., Duarte, C. and Lopez-Portillo, J. 2001. *Aspectos Económicos del Sistemas de Siembra Directa en Pequeñas Fincas: Implicaciones en la Política y la Inversión.* Proyecto Conservación de Suelos MAG-GTZ, San Lorenzo, Paraguay.

Soto, B., Basanta, R., Perez, R. and Díaz-Fierros, F. 1995. An experimental study of the influence of traditional slash-and-burn practices on soil erosion. *Catena,* 24: 13-23.

Southgate, D. 1992. *The Rationality of Land Degradation in Latin America: Some Lessons from the Ecuadorian Andes.* IIED Gatekeeper Series GK 92-04, International Institute for Environment and Development, London, UK.

Southgate, D. 1994. The rationality of land degradation in Latin America: some lessons from the Ecuadorian Andes. In: Napier, T.L., Camboni, S.M. and El-Swaify, S.A. (eds.). *Adopting Conservation on the Farm: An international Perspective on the Socioeconomics of Soil and Water Conservation,* pp. 331-340. Soil and Water Conservation Society, Ankeny, Iowa, USA.

Stadel, C. 1989. The perception of stress by *campesinos*: A profile from the Ecuadorian Sierra. *Mountain Research and Development,* 9(1): 35-49.

Stocking M. and Abel, N. 1992. Labour costs: a critical element in soil conservation. In: Hiemstra, W., Reijntjes, E. and van der Werf (eds.). Let Farmers Judge: Experiences in Assessing the Sustainability of Agriculture, pp. 77-86. Intermediate Technology Publications, London, UK.

Stocking, M. 1987. Measuring land degradation. In: Blaikie, P. and Brookfield, H. (eds.). *Land Degradation and Society,* pp. 49-63. Methuen, London, UK.

Stocking, M. 1991. Seeds of change in soil conservation. *International Agricultural Development,* March/April: 7-9.

Stocking, M. 1993. The issues of a soil scientist in environment and natural resources research. In: Christiansson, C., Dahlberg, A., Loiske, V-M and Östberg, W (eds.). *A Discussion on Natural Resources Research in the Third World,* pp. 24-26. School of Geography, Stockholm University, Sweden.

Stocking, M. 1993b. The rapid appraisal of physical properties affecting land degradation. In: Christiansson, C., Dahlberg, A., Loiske, V-M and Östberg, W (eds.). *A Discussion on Natural Resources Research in the Third World,* pp. 20-23. School of Geography, Stockholm University, Sweden.

Stocking, M. 1994. Assessing vegetative cover and management effects. In: Lal, R. (ed.). *Soil Erosion Research Methods,* pp. 211-232. Soil and Water Conservation Society, Ankeny Iowa and St. Lucie Press, Florida, USA.

Stocking, M. 1995. Soil erosion and land degradation. In: O'Riordan, T. (ed.). *Environmental Science for Environmental Management,* pp. 223-242. Longman, Harlow, UK.

Stocking, M. 1995b. Soil erosion in developing countries: where geomorphology fears to tread! *Catena,* 25: 253-267.

Stocking, M. and Abel, N. 1989. Labour costs: a critical element in soil conservation. In: Tato, K and Hurni, H. (eds.). *Soil Conservation for Survival*, pp. 206-218. Soil and Water Conservation Society in cooperation with the International Soil Conservation Organisation and World Association of Soil and Water Conservation, Ankeny, USA.

Stocking, M. and Tengberg, A., 1999. Soil conservation as an incentive enough – experiences from southern Brazil and Argentina on identifying sustainable practices. In: Sanders, D.W., Huszar, P.C., Sombatpanit, S. and Enters, T. (eds), 1999. *Incentives in Soil Conservation. From Theory to Practice*, pp. 69-86. Oxford & IBH Publishing, New Delhi, India.

Stone, J.R., Gilliam, J.W., Cassel, D.K., Daniels, R.B., Nelson, L.A. and Kleiss, H.J. 1985. Effect of erosion and landscape position on the productivity of Piedmont soils. *Soil Science Society of America Journal*, 49: 987-991.

Stonich, S. 1991. Rural families and income from migration: Honduran households in the world economy. *Journal of Latin American Studies*, 23 (1): 131-161.

Stonich, S. 1993. *I Am Destroying the Land: the Political Ecology of Poverty and Environmental Destruction in Honduras*. Westview Press, Boulder, Colorado, USA.

Stonich, S. 1995. Development, rural impoverishment, and environmental destruction in Honduras. In: Painter, M. and Durham, W.H. (eds.). *The Social Causes of Environmental Destruction in Latin America*, pp. 63-99. The University of Michigan Press, USA.

Stonich, S. and DeWalt, B. 1989.The political economy of agricultural growth and rural transformation in Honduras and Mexico. In: Smith, S. and Reeves, E (eds.) *Human Systems Ecology: Studies in the Integration of Political Economy, Adaptation, and Socionatural Regions*, pp. 202-230. Westview Press, Boulder, Colorado, USA.

Sturm, L.S. and Smith, F.J. 1993. Bolivian farmers and alternative crops: some insights into innovation adoption. *Journal of Rural Studies*, 9 (2): 141-151.

Subudhi, C.R.; Pradhan, P.C. and Senapati, P.C. 1998. Effect of grass bunds on erosion loss and yield of rainfed rice – Orissa, India. *Newsletter of the Vetiver Network*, 19: 32-33.

Sukmana, S. 1995. Research and development on the application of vetiver for soil conservation in Indonesia. *Newsletter of the Vetiver Network*, 13: 26-28.

Suppe, F. 1988. The limited applicability of agricultural research. *Agriculture and Human Values*, 5 (4): 4-14.

Suresh R. 2000. *Soil and Water Conservation Engineering*. 3rd edition. Standard Publishers, Delhi, India.

Sutherland, A. 1999. Linkages between farmer-oriented and formal research development approaches. *Agricultural Research & Extension Network Paper* 92b, Overseas Development Institute, London, UK.

Tacio, H.D. 1993. Sloping Agricultural Land Technology (SALT): a sustainable agroforestry scheme for the uplands. *Agroforestry Systems,* 22: 145-152.

Tadesse, L.D. and Morgan, R.P.C. 1996. Contour grass strips: a laboratory simulation of their role in erosion control using live grasses. *Soil Technology,* 9: 83-89.

Tadesse, M. and Belay, K. 2004. Factors influencing adoption of soil conservation measures in southern Ethiopia: the case of Gununo area. *Journal of agriculture and rural development in the tropics and subtropics,* 105(1): 49-62.

TangYa, Zhang, Y.Z., Xie, J.S. and Hui, S. 2003. Incorporation of mulberry in contour hedgerows to increase overall benefits: a case study from Ningnan County, Sichuan Province, China. *Agricultural Systems,* 76 (2): 775-785.

Temu, A.E.M. and Bisanda, S. 1996. Pit cultivation in the Matengo Highlands of Tanzania. In: Reij, C., Scoones, I. and Toulmin, C. (eds.). *Sustaining the Soil: Indigenous Soil and Water Conservation in Africa,* pp. 145-150. Earthscan Publications Ltd., London, UK.

Tengberg, A., Stocking, M., and Dechen, S.C.F. 1998. Soil erosion and crop productivity research in South America. In: Blume, H.-P., Eger, H., Fleischhauer, E., Hebel, A., Reij, C., and Stenier, K.G. (eds.). *Towards Sustainable Land Use: Furthering Cooperation Between People and Institutions.* Selected papers of the 9th conference of the International Soil Conservation Organisation, Bonn, Germany. *Advances in Geoecology,* 31: 355-362.

Tenge A.J., De Graaff J. and Hella J.P. 2004. Social and economic factors affecting the adoption of soil and water conservation in West Usambara highlands, Tanzania. *Land Degradation & Development,* 15 (2): 99-114.

The Economist. 2004. Farming in Mexico: From corn wars to corn laws. *The Economist* September 25, 2004.

The World Bank Group. 2004. *Millennium Development Goals.* Available at http://www.developmentgoals.org/ Accessed December 2004.

The World Bank. 2000. *Implementation Completion Report on a Loan in the Amount of US$ million 33.0 to the Federative Republic of Brazil for Land Management II – Santa Catarina Project.* World Bank Report no. 20482. Washington D.C., USA.

The World Bank. 2001. *World Development Report 2000/2001: Attacking Poverty.* The World Bank, Washington D.C., USA.

The World Bank. 2003. *The CGIAR at 31: Celebrating its Achievements, Facing its Challenges.* World Bank Operations Evaluation Department Précis Number 232.

Thiele, G. and F. Terrazas (1998). 'The wayq'os (gullies) are eating everything!' Indigenous knowledge and soil conservation. PLA Notes 32: 19-23. IIED, London, UK.

Thirtle, C., Lin, L. and Piesse, J. 2003. The impact of research led agricultural productivity growth on poverty reduction in Africa, Asia and Latin America. Contributed paper for the 25th Conference of the International Association of Agricultural Economists, Durban, South Africa, August 2003.

Thompson, J. and Guijt, I. 1999. Sustainability indicators for analysing the impacts of participatory watershed management programmes. In: Hinchcliffe, F., Thompson, J., Pretty, J., Guijt, I. and Shah, P. (eds.). *Fertile Ground: The Impacts of Participatory Watershed Management*, pp. 13-26. Intermediate Technology Publications, London, UK.

Thompson, J. and Pretty, J.N. 1996. Sustainabilty indicators and soil conservation: The catchment approach in Kenya. *Journal of Soil and Water Conservation*, 51 (4): 265-273.

Thurow, T.L. and Smith, J.E. 1998. *Assessment of Soil and Water Conservation Methods Applied to the Cultivated Steeplands of Southern Honduras*. United States Agency for International Development, Soil Management Colloborative Research Support Program/Texas A&M University. Technical Bulletin No. 98-2.

Thurston, D.H. 1992. *Sustainable Practices for Plant Disease Management in Traditional Farming Systems*. Westview Press/Oxford & IBH Publishing, India.

Thurston, H.D., Smith, M., Abawi, G. and Kearl, S. (eds.). 1994. *Tapado - Slash/ Mulch: How Farmers Use It and What Researchers Know About It*. Cornell International Institute for Food, Agriculture and Development, Cornell University, USA.

Tiffen, M., Motimore, M. and Gichuki, F. 1994. *More People, Less Erosion: Environmental Recovery in Kenya*. John Wiley & Sons Ltd., Chichester, UK.

Tobisson, E. 1993. Changing role for rural sociologists. In: Hudson, N.W. and Cheatle, R. (eds.). *Working with Farmers for Better Land Husbandry*, pp. 59-63. Intermediate Technology Publications, London, UK.

Tolstoy, L.N. 1876. *Anna Karenin*. Translated and with an introduction by Edmonds, R. Penguin Books, 1969, UK.

Tomar, V.P.S., Narain, P. and Dadhwal, K.S. 1992. Effects of perennial mulches on moisture conservation and soil-building properties through agroforestry. *Agroforestry Systems*, 19: 241-252.

Toness, A.S., Thurow, T.L. and Sierra, H. 1998. *Sustainable Management of Tropical Hillsides: an Assessment of Terraces as a Soil and Water Conservation Technology*. United States Agency for International Development, Soil Management Colloborative Research Support Program/Texas A&M University. Technical Bulletin No. 98-1.

Tracy, F.C. 1988. The Natural Resources Management Project in Honduras. In: Moldenhauer, W.C. and Hudson, N.W. (eds.) *Conservation Farming on Steeplands*, pp. 265-268. Soil and Water Conservation Society and World Association of Soil and Water Conservation, Ankeny, Iowa, USA.

Tripp, R. 2001. *Seed Provision and Agricultural Development: the Institutions of Rural Change.* Overseas Development Institute, London, UK.

Tschinkel, H. 1987. Tree planting by small farmers in upland watersheds: Experience in Central America. *The International Tree Crops Journal,* 4: 249 – 268.

Turco, R.F.; Kennedy, A.C. and Jawson, M.D. 1994. Microbial indicators of soil quality. In: Doran, J.W., Coleman, D.C., Bezdicek, D.F. and Stewart, B.A. (eds.). 1994. *Definining Soil Quality for a Sustainable Environment,* pp. 73-90. Soil Science Society of America Special Publication Number 35, Madison, Wisconsin, USA.

Turkelboom, F., Poesen, J., Ohler, I., van Keer, K., Ongprasert, S. and Vlassak, K. 1997. Assessment of tillage erosion rates on steep slopes in northern Thailand. *Catena,* 29: 29-44.

UNAIDS. 2004. http://www.unaids.org/. Accessed December 2004.

Unwin, T. 1985. Farmers' perceptions of Agrarian change in North-West Portugal. *Jounral of Rural Studies,* 1 (4): 339-357.

Uphoff, N. 1996. *Learning From Gal Oya: Possibilities for Participatory Development and Post-Newtonian Social Science.* Intermediate Technology Publications, London, UK.

Valentin, C. 1989. *Surface Crusting, Runoff, and Erosion on Steeplands and Coarse Material.* IBSRAM Technical notes No. 3: 285-312. Bangkok, Thailand.

van Breeman, N. 1993. Soils as biotic constructs favouring net primary productivity. *Geoderma,* 57: 183-211.

Van Oost, K., Govers, G. and Van Muysen, W. 2003. A process-based conversion model for caesium-137 derived erosion rates on agricultural land: an integrated spatial approach. *Earth Surface Processes and Landforms,* 28 (2): 187-207.

Van Zyl, A.J., Smith, H.J. and Claassens, A.S. 1998. Sediment yield modelling in the catchments of the Lesotho Highlands Water Project. In: Haigh, M.J. Krecek, J., Rajwar, G.S and Kilmartin. M.P. (eds.). *Headwaters: Water Resources and Soil Conservation.* Proceedings of Headwater '98, pp. 329-333. Fourth International Conference on Headwater Control, Merano, Italy, April 1998. A.A. Balkema, Rotterdam, The Netherlands,

Versteeg, M.N. and Koudokpon, V. 1993. Participative farmer testing of four low external input technologies to address soil fertility decline in Mono Province (Benin). *Agricultural Systems,* 42: 265-276.

Vogel, H. 1988. Deterioration of a mountainous agro-ecosystem in the third world due to emigration of rural labour. *Mountain Research and Development,* 8 (4): 321-329.

Vosti, S.A. and Reardon, T. 1997. Introduction: The critical triangle of links among sustainability, growth and poverty alleviation. In: Vosti, S.A. and Reardon, T. (eds.) *Sustainability, Growth and Poverty Alleviation: A policy and Agroecological Perspective,* pp. 1-15. The John Hopkins University Press, USA.

Vosti, S.A. and Reardon, T. 1997b. Conclusion. In: Vosti, S.A. and Reardon, T. (eds.) *Sustainability, Growth and Poverty Alleviation: A policy and Agroecological Perspective*, pp. 339-346. The John Hopkins University Press, USA.

Wachter, D. 1994. Land Titling: Possible contributions to Farmland conservation in Central America. In: Lutz, E., Pagiola, S. and Reiche, C. (eds.). *Economic and Institutional Analyses of Soil Conservation Projects in Central America and the Caribbean*, pp. 150-157. World Bank Environment Paper no. 8., Washington, D.C., USA.

Wall, J. R. D. (ed.). 1981. *A Management Plan For the Acelhuate River Catchment, El Salvador: Soil Conservation, River Stabilisation and Water Pollution Control.* Land Resources Devlopment Centre, Overseas Development Administration, UK.

Warkentin, B.P. 1995. The changing concept of soil quality. *Journal of Soil and Water Conservation*, 50 (3): 226-228.

Warren, D.M. and Cashman, K. 1988. *Indigenous Knowledge for Sustainable Agriculture and Rural Development*. IIED Gatekeeper Series SA38, International Institute for Environment and Development, London, UK.

Warren, D.M., Slikkerveer, L.J. and Brokensha, D. (eds.). 1995. *The Cultural Dimension of Development: Indigenous Knowledge Systems*. Intermediate Technology Publications, London, UK.

Watkins, K. and von Braun, J. 2003. *Time to Stop Dumping On the World's Poor.* 2002-2003 IFPRI Annual Report Essay, International Food Policy Research Institute, Washington D.C., USA.

Wedum, J., Doumbia, Y., Sanogo, B., Dicko, G. and Cissé, O. 1996. Rehabilitating degraded land: Zaï in the Djenné Circle of Mali. In: Reij, C., Scoones, I. and Toulmin, C. (eds.). *Sustaining the Soil: Indigenous Soil and Water Conservation in Africa*, pp. 62-68. Earthscan Publications Ltd., London, UK.

Welchez, L.A. and Cherrett, I. 2002. The Quesungual system in Honduras: an alternative to slash and burn. *LEISA*, 18 (3): 10-11.

Wendt, R.C., Alberts, E.E. and Hjelmfelt, A.T., Jr. 1986. Variability of runoff and soil loss from fallow experimental plots. *Soil Science Society of America Journal*, 50 (3): 730-736.

Wenner, C.G. 1989. Soil and water conservation in the farming areas of Lesotho: A review and some proposals. In: Baum, E,E., Wolff, P. and Zöbisch, M.A. (eds.). *The Extent of Soil Erosion – Regional Comparisons.* Volume 1 in *Topics in Applied Resource Management in the Tropics*, pp. 57-86. German Institute for Tropical and Subtropical Agriculture, Witzenhausen, Germany.

Wiersum, K.F. 1994. Farmer adoption of contour hedgerow intercropping, a case study from east Indonesia. *Agroforestry Systems*, 27:163-182.

Wiggins, S. 1981. A case study in cost-benefit analysis: soil conservation in the Rio Acelhuate river basin, El Salvador. In: Morgan, R. (ed.). *Soil Conservation*, pp. 399-417. John Wiley and Sons, Chichester, UK.

Wilken, G.C. 1987. *Good farmers: Traditional Agricultural Resource Management in Mexico and Central America.* University of California Press, Berkeley, Los Angeles, USA.

Williams, B.J. and Ortiz-Solorio, C.A. 1981. Middle American folk soil taxonomy. *Annals of the Associatin of American Geographers*, 71 (3): 335-358.

Williams, J.D. and Buckhouse, J.C. 1991. Surface runoff plot design for use in watershed research. *Journal of Range Management*, 44 (4): 411-412.

Williamson, S. 2002. Challenges for farmer participation in integrated and organic production of agricultural tree crops. *Biocontrol News and Information*, 23 (1).

Wischmeier, W.H. 1976. Use and misuse of the Universal Soil Loss Equation. *Journal of Soil and Water Conservation*, 31 (1): 5-9.

Wischmeier, W.H. and Smith, D.D. 1965. *Predicting Rainfall Erosion Losses from Cropland East of the Rocky Mountains.* United States Department of Agriculture, Agriculture Handbook no. 282, Washington, D.C., USA.

World Bank. 1990. *Vetiver Grass: The Hedge Against Erosion.* The World Bank, Washington D.C., USA.

Wu, C. and Wang, A. 1998. Soil loss and soil conservation measures on steep sloping orchards. In: Blume, H.-P., Eger, H., Fleischhauer, E., Hebel, A., Reij, C., and Stenier, K.G. (eds.). *Towards Sustainable Land Use: Furthering Cooperation Between People and Institutions.* Selected papers of the 9th conference of the International Soil Conservation Organisation, Bonn, Germany. *Advances in Geoecology*, 31: 383-387.

Ya, T. 1998. Bioterracing and soil conservation: experience with contour hedgerow planting in parts of the Hindu Kush-Himalayan Region. *Issues in Mountain Development* 98/7. International Centre of Integrated Mountain Development, Kathmandu, Nepal.

Young, 1997. The effectiveness of contour hedgerows for soil and water conservation. *Agroforestry Forum*, 8(4): 2-4.

Young, 1997b. *Agroforestry for Soil Management.* CAB International, Wallingford, Oxon, in association with the International Institute for Research in Agroforestry, Nairobi, Kenya.

Young, A. 1989. The potential of agroforestry for soil conservation and sustainable land use. In: Baum, E,E., Wolff, P. and Zöbisch, M.A. (eds.). *The Extent of Soil Erosion – Regional Comparisons.* Volume 1 in *Topics in Applied Resource Management in the Tropics*, pp. 35-55. German Institute for Tropical and Subtropical Agriculture, Witzenhausen, Germany.

Zimmerer, K. 1992. Land-use modification and labour shortage impacts on the loss of native crop diversity in the Andean Highlands. In: Jodha, N. S.; Banskota, M. and Partap, T. (eds.) *Sustainable Mountain Agriculture*, pp. 413-422. Oxford and IBH Publishing, New Delhi, India.

Zimmerer, K. 1993. Soil erosion and labour shortages in the Andes with special reference to Bolivia, 1953-91: Implications for 'conservation-with-development'. *World Development*, 21 (10): 1659-1675.

Zimmerer, K. 1995. The origins of Andean Irrigation. *Nature,* 378: 481-483.

Zöbisch, M.A., Klingspor, P. and Oduor, A.R. 1996. The accuracy of manual runoff and sediment sampling from erosion plots. *Journal of Soil and Water Conservation,* 51 (3): 231-233.

Zougmore, R., Zida, Z. and Kambou, N.F. 2003. Role of nutrient amendments in the success of half-moon soil and water conservation practice in semiarid Burkino Faso. *Soil & Tillage Research,* 71 (2): 143-149.

Zúniga, E.A. 1990. *Las Modalidades de la Lluvia en Honduras.* Editorial Guayamuras, Tegucigalpa, Hopnduras.

Index

A-horizon 165
Abandonment 21, 23-25, 67-69, 156, 171, 182, 203
Africa 19, 25, 40, 56, 72, 74-76, 83, 87, 155, 173-174, 185, 205-206, 229
Agronomic practices 205, 215, 217
Andes 1, 40, 174, 225, 226
Animal feed 173-174, 200
Annual crops 8, 254, 256-257, 259, 260, 261
Anthropologists 39
Asia 5, 59, 83, 173, 201, 222, 229, 236
Association for Better Land Husbandry 208, 230
Australia 56

Bacteria 149, 152, 188
Bangladesh 45, 236, 237
Barrels 108-109, 111-114, 116, 122-123, 129, 139
Beans 48, 57, 70, 173, 185, 187-188, 193, 197, 200, 225, 232
Belgium 93
Benin 71, 173
Better land husbandry 1-12, 15, 17-20, 38, 40, 81-82, 139-141, 149, 159, 162, 164, 166, 168, 177-185, 196, 204-205, 208, 215, 127-220, 222-224, 228, 230, 233-234, 238, 242-244, 254, 261-266
Biological activity 10, 149, 150, 164, 165, 169
Biological methods 8
Biological processes 4-5, 150, 154, 156, 158, 163, 182
Bolivia 31, 40, 45, 58-60, 67, 70, 155, 200, 229, 231
Bulk density 74, 101, 137, 212
Burkina faso 205, 208
Burning 9, 11, 57, 65-67, 77, 102, 155-156, 168-170, 172, 174, 178, 195

Canícula 47, 101
Cape Verde Islands 25
Cash crops 3, 65
Cassava 1, 2, 57, 110, 176, 185
Catch-pits 4, 108-111, 114-116, 121, 129, 138, 212
Cation exchange capacity (CEC) 154
Central America 3-5, 19, 31, 40-41, 46-48, 74, 124, 126, 135, 174, 185, 192, 210-211, 232
Chemical fertiliser 9-10, 196,
China 2-3, 93, 108-110, 203, 220
Choluteca 48-50, 76, 100-101, 211, 214
Chronic Poor 220
Clay 3, 46, 75, 85, 101, 138, 149, 156-157, 207
Colombia 1-2, 5, 57, 110, 176, 203,
Consultative Group for International Agricultural Research 241
Costa Rica 45-46, 145, 147
Cover crops 8, 14, 20, 109, 135, 156, 164, 170, 173-174, 176, 177, 186-188, 191, 197, 200, 214, 245, 263
Credit 25, 29, 61, 71, 153, 191, 192, 204, 207, 222, 223, 227, 232, 234, 237
Crop production 2, 8, 12, 47, 58, 70, 74, 134, 155, 182, 220, 255, 256, 258, 259, 260
Crop residue 8-10, 17, 153-154, 167, 176, 186-187, 245, 255-256, 258, 259, 260

Debris flows 86, 211
Deforestation 3-4, 42, 65, 190
Developed world 6, 31, 43
Developing world 5-6, 34, 41, 43, 56, 68, 150, 220-221, 233, 235, 237
Direct incentives 22-25, 80, 170-171
Diversification 231, 238
Dominican Republic 23, 70, 76

Double-dug beds 208-210
Drainage channels 7, 107, 115, 226

Earth bunds 7
Earthworms 3, 75-76, 149, 151-154, 170, 188-190, 196
Ecosystem 23, 33, 149, 182
Ecuador 7, 37, 40, 58, 169, 228
El Niño 61, 115, 119-120, 134, 139, 160, 168
El Salvador 46, 57, 72, 145, 170
Elephant grass 1, 67
Empowerment 18-19, 38, 79, 199, 224-225, 227, 236-237
Enzymes 152
Erodibility 5, 13, 84-85, 89, 137-138, 142, 154, 179
Erosion pins 90, 99
Erosivity 5, 13, 84, 89, 101, 117-119, 142, 147, 167, 179
Ethiopia 5, 7, 45, 58, 134, 146, 238
Europe 5, 42, 69, 83, 87, 221
Evaporation 114, 164
Extension 9, 15, 17, 19, 35-38, 51, 55, 69-70, 72-73, 77-79, 81, 170, 172, 180, 183, 186, 191-192, 198-199, 207, 222, 224-227, 230, 234, 242, 265

Fallow 3, 5, 9-10, 46-47, 51, 65-66, 69, 75, 88, 150- 151, 155, 173-174, 179, 195, 255-256, 258-260
Farmer adaptation 199, 203
Farmer adoption 8-9, 18, 21, 23, 37, 45, 72, 169
Farmer experimentation 199, 204
Farmer Field Schools 225, 227
Farmer innovation 196, 204
Farmer-first 9, 19, 35-36, 79, 178, 182
Farmer-to-farmer extension 225-226
Farmers' realities 15, 19, 26, 32-35, 39, 42, 44, 49, 54-55, 76, 78-79, 183, 242
Fertilizer 74, 208, 221-222, 255-256, 258-260
Financial capital 28-29, 264-265
Firewood 194-195
Flooding 189, 210-211, 215-216

Food and Agriculture Organization of the United Nations (FAO) 6
France 5
Fruit trees 40, 149, 193-194, 203, 228
Fungi 149, 152-153, 188
Fungicides 153

Gerlach trough 5, 108, 138
Grazing 3, 9, 47, 100, 150, 255-259, 261
Green manure 9-10, 76, 135, 155, 164, 169, 173-174, 185-188, 191-192, 200, 202, 245, 255-256, 259-260
Green Revolution 220-222
Guatemala 2, 31, 69, 213, 232
Güinope 48-50, 62, 72, 109, 135, 195-197, 198-202, 204, 207
Gullies 56, 84, 90-91, 144, 213, 251, 253-254, 256-257, 259-261

Haiti 70
Herbicides 153, 187, 221
HIV/AIDS 69
Honduras 1, 3-8, 19, 23-24, 39, 43, 45-49, 52-53, 56, 59-62, 64-68, 70-74, 76, 82, 89, 95, 99-100, 107, 109, 111-115, 117-119, 124-126, 131, 135, 145, 155, 160-161, 168-172, 177, 184-185, 192-200, 202-203, 207, 210-216, 221, 234
Human capital 28, 38, 265, 267
Hurricane Mitch 8, 107, 114, 119-124, 129, 168, 177, 185, 210-217

Impact Monitoring and Assessment 9, 11, 244
Incentives 8-9, 20-25, 37, 58, 61, 67, 80, 170-171, 191, 228, 233
India 1-4, 6, 46
Indigenous knowledge 34, 39-40, 42, 62, 74, 184, 232
Indigenous soil conservation 46, 59, 69
Indonesia 2, 6, 72, 109-110, 203, 220
Infiltration 11, 14, 17, 25, 51, 57, 59, 75, 85, 94-95, 101, 114, 123, 137, 142-144, 147, 153, 157-158, 162, 164, 169, 175, 177, 181, 186, 188-189, 245
Innovation 16, 17, 21, 38-39, 180, 196, 199-200, 204, 227, 234, 241
Insecticides 58, 153, 188

Insects 149
Inter-rill erosion 84
ITDG 225-227, 237-238

Jamaica 67, 71, 76, 203, 217

Kamayoq 225-227
Kenya 4-6, 8, 39, 45, 56, 58-59, 61, 63, 67-68, 72, 109, 145, 168, 185, 208, 210, 230
Kinetic energy 84, 95, 117, 142

Labour costs 54, 66-69, 110, 170, 177, 207
Labour migration 206
Labour shortages 30, 62-63, 66-68, 82
Land degradation 1-7, 11, 13, 15-16, 18, 20, 25, 42, 46-48, 50-51, 54, 61-63, 65, 78, 82, 87, 91, 95, 99, 120, 140-142, 144-145, 147, 149-150, 185-186, 193, 195-196, 206, 218-219, 223-224, 229, 235, 237, 240, 243, 245, 251, 253
Land shortages 5, 62, 64-65, 76, 81, 155, 196
Land tenure 47, 61, 64, 71-72, 234
Landslides 8, 57, 84, 86, 90, 92, 210-217, 253
Laos 59
Latin America 31-32, 40, 64, 196, 225, 229, 236
Lesotho 59, 88, 93
Leucaena spp. 4
Live barriers 1, 6-8, 20-21, 25, 59, 67-68, 70, 77-78, 82, 95, 99-100, 102, 107, 109, 129-131, 133-139, 141, 146-147, 149-170, 172, 196-197, 200-204, 213, 242
Livelihoods 10, 21, 26-28, 30, 32-33, 35, 38, 51, 54, 57-58, 60-61, 82, 85, 88, 90, 120, 122, 125, 135, 142, 145, 149, 204, 219-220, 228-229, 236, 238, 241, 261-266
Local Agricultural Research Committees 225, 227

Macrofauna 152
Maize 1, 3-5, 30-32, 40, 46, 48, 57, 66, 70, 100, 107-108, 120-122, 125-127, 130-131, 133-134, 136, 139, 145-146, 155, 159-161, 167-168, 170, 172-174, 177, 185, 188, 190, 193, 196-197, 202, 207, 209-212, 221, 225, 232, 234, 241

Malaysia 1, 110
Mali 40, 45, 64, 74, 205, 207-208
Market opportunities 194, 227-228, 231, 236
Markets 3, 64, 180, 192, 197, 204, 219-220, 227-232, 234-238
Mass movements 84, 86, 90, 177, 185, 215, 217
Mechanical practices 14, 141
Mediterranean 5
Mesh bags 138
Mexico 5, 31, 40-42, 45, 47, 74, 168, 174
Models 87-90, 225
Mucuna spp. 8, 174, 177, 190, 214, 234
Mulch systems 169
Mulching 4, 174, 179

Napier grass 134, 200-202
Natural and social scientists 223
Natural capital 28-29, 261-262, 264-265
Natural resource management 22, 45, 88
Nematodes 152
Nicaragua 46, 109, 172, 174, 210, 213-214
Niger 7, 23, 59, 71, 109, 168, 205, 207
Nigeria 7, 23, 59, 109, 168
Nitrogen 2, 4, 9, 152-153, 155, 173-174, 187, 234
Nitrogen fixation 4, 173
Nutrient recycling 164

Off-site benefits 23
Onions 185, 188, 237
Organic fertiliser 155, 164, 196
Organic matter 2-3, 9-12, 40, 57, 65-66, 74-76, 85, 89, 141, 149-151, 153-156, 161, 164, 166, 170, 172, 174, 182, 188-189, 194-196, 200, 206, 208-209, 222, 234, 245
Outsiders 9, 25-26, 30-31, 33-38, 42, 44-45, 49, 54-55, 57, 59, 72-73, 77-80, 82, 85, 170, 176-177, 223-224
Outsiders' realities 35
Oxen 47, 58

Papa Andina project 232
Papua New Guinea 58, 89, 95, 166, 217
Paraguay 177, 187-188, 190-192

Participant observation   44-46, 50
Participation   8, 10, 15-17, 19, 35-39, 48, 60, 140, 149, 177-178, 181, 185, 195-199, 204, 207, 223-224, 230, 233, 266
Participatory Rural Appraisal   45
Perennial crops   85, 255, 257, 258, 260-261
Peru   7, 40, 43, 45, 58, 70, 74, 146, 225
Pesticides   62, 66, 71, 153, 182-183, 221-222, 229-230
Pests and diseases   51, 55, 60-62, 65, 81, 93, 154, 170, 178, 187, 193, 223, 226, 228
Philippines   6, 8, 23, 66-67, 69, 71, 134, 203
Phosphorous   2, 4, 6, 58, 152-153, 155, 185
Physical capital   28, 225, 265
Plant nutrients   2, 6, 13, 85, 145, 159-160, 164, 181
Plant residues   40, 85, 152, 164, 166, 170, 172, 174, 187
Plot size   4-6, 54, 102
Plough   47, 75-76, 86, 93, 158, 186-188, 190, 207-208
Policy Agenda   233
Population increase   65
Postrera   48, 107, 126, 130, 133-134, 161
Potassium   2, 153, 155
Potato   43, 58, 174, 185, 225, 232
Primera   48, 107, 126, 134, 160-161
Private sector   241
Protozoa   151
Public sector   222

Qualitative data   44
Quantitative data   44, 50
Questionnaire   44-46, 49-53, 60-62, 65, 75
Quezungual system   72, 169-170, 192-195

Radioactive tracers   91
Rain splash   84, 95
Rain-gauge   117-118, 122
Raindrops   8, 11, 13, 66, 84-86, 95, 142, 153, 158, 165-166, 168, 175, 186-187
Rainfall   2, 4-5, 7-8, 11, 13, 19, 30, 41-42, 46-47, 51, 56, 58, 61-62, 65, 84-86, 88-89, 93-94, 99, 101, 114-115, 117-120, 122-123, 126, 129-130, 133-134, 137, 139, 144-145, 147, 157, 159-160, 166-169, 182, 185-186, 189, 194, 196, 205-207, 210-211, 213, 215-216, 245
Rainfall depth   122-123, 130
Rainfall erosion indices   117
Rainfall intensity   84, 86, 88, 117, 119, 211, 216
Rainwater absorption   12
Rapid Rural Appraisal   45
Replication   4-7, 101-102, 107, 135-136, 138-139
Resource-poor farmers   9, 16, 26-27, 31, 36, 47-48, 61, 63, 66, 74, 82, 92, 95, 99, 120-121, 124, 126, 135, 139, 142, 160, 179-180, 183, 198, 208, 221-222, 224-225, 235, 238, 241, 263
Rills   56, 84, 91-92, 144, 251-252
Rock walls   7, 11-12, 147, 255-258, 260-261
Root growth   12, 14, 141, 147, 153, 156-159, 161, 176-177
Runoff   1-4, 6-8, 11-12, 14, 20-21, 51, 68, 76, 84-86, 90-95, 99-100, 108-114, 116, 122-123, 125, 127, 129-131, 137-138, 140-145, 158, 161-162, 164, 166, 168, 178-179, 181-182, 186, 188, 205-208, 215, 242, 245, 255, 257-261
Runoff Plots   90-92, 94, 99-100, 108-109, 116, 125, 138
Rural livelihoods   30, 38, 149, 220
Rural non-agricultural employment (RNAE)   235
Rural organisations   177
Rwanda   6, 21, 45, 59, 64, 92, 134

Salinization   5
Sand   5, 22-23, 40, 54, 71, 75, 79, 85, 89, 95, 101, 138, 144, 149, 156, 158, 237
Santa Catarina   174, 185-186, 188, 191
Santa Rosa   100-101, 107-108, 111-114, 116-117, 119-122, 124-127, 130-131, 133-139, 144, 160-161, 168, 210-211, 213, 215
Scouring   131, 134-135
Sealing   85, 154, 166
Self-help groups   209, 230
Semi-structured interviews   43, 45, 49-51, 61

Shifting cultivation 5, 47, 58, 150, 166
Sierra Leone 5
Silt 23, 89, 101, 138, 149, 156, 189, 237, 258-259, 261
Site heterogeneity 136-137
Slash and burn 6, 47, 65, 169, 195
Slope length 12, 123, 131, 144, 166
Slopes 4-6, 8, 14, 40, 46, 57-58, 82, 84-86, 89-95, 99-100, 110, 121-130, 132-133, 135-136, 138-139, 141, 160-161, 166-168, 170, 177, 193, 196, 207, 211-216, 240, 242, 258-260
Snails 151
Social capital 28-29, 38, 181, 204, 265
Soil accumulation 110, 131, 134
Soil aggregates 84, 153, 156, 158, 186, 188
Soil architecture 10, 12, 147, 154, 156-158, 160-161, 164, 176, 178, 182, 245
Soil augur 131
Soil biology 151, 170, 183
Soil characteristics 5, 94, 100-101, 136
Soil compaction 137, 158, 177
Soil conservation technologies 6-10, 18-21, 23-26, 28, 32, 38, 42, 45, 49, 51, 54-55, 59, 62-64, 66-72, 76-77, 79-81, 83, 87, 93-95, 120, 124-125, 127, 129-131, 135-136, 140, 142, 146-148, 159, 170, 172, 182, 196, 210, 213-218, 228
Soil depth 51, 85, 127, 131, 133, 148, 152
Soil Erosion 1-2, 5-7, 9-12, 18, 20, 26, 39, 42, 48, 51, 55-60, 67-68, 70 75, 77-78, 81-88, 91-94, 108-109, 117, 120, 124-125, 127, 129, 130-131, 133, 135-138, 140-142, 144-145, 147, 160-161, 164, 166-167, 178, 180, 182, 188, 195, 203, 205, 210, 214-215, 224, 240-242, 245, 251, 254
Soil fertility 4, 6, 17, 40, 56-58, 64, 76, 83, 134, 145, 151, 160, 186, 205-206, 209
Soil health 149
Soil loss 1-7, 9-11, 18-20, 26, 36, 39, 41-42, 46, 54-59, 61, 68, 76-77, 80-83, 85-95, 99, 102, 107-108, 110-112, 114-116, 120-122, 124, 127, 129-130, 133, 135-140, 145-147, 160-162, 167-168, 170, 172, 176-177, 185, 191, 212-213, 215, 217-218, 225, 240-241, 261, 265
Soil loss rates 46, 56, 91, 124
Soil loss tolerance 6, 83, 124

Soil Moisture 5, 12, 19, 56, 59, 74-75, 111, 122, 134, 147, 155-156, 158-160, 162, 164-165, 176, 180, 185, 194, 205, 208, 228, 245
Soil nutrients 9, 85
Soil organisms 3, 12, 17, 66, 141, 151-156, 162, 164, 166, 170, 172, 182, 195-196, 222
Soil particles 6, 12, 84, 86, 90, 131, 145, 153, 158, 159, 166, 252
Soil ph 2, 7, 10, 150, 152, 176, 181, 183, 187
Soil pores 157
Soil productivity 6, 19, 139, 145-147, 158-159, 182, 215, 221, 241, 245
Soil protection 12, 141, 166, 197, 222
Soil Quality 2, 8, 10-12, 15, 17-19, 48, 51, 65-66, 73-74, 80, 82, 127, 139-141, 149-151, 153-156, 159, 161-164, 166, 170, 172, 177-179, 181-184, 186, 194-197, 200, 204-206, 215, 218, 222-223, 233, 242, 245
Soil recuperation 166
Soil splash 86
Soil structure 14, 78, 85, 142, 144, 151-152, 154-158, 162, 166, 169, 176, 188, 194, 10
Soil technology 72
Soil tillage 92
Sorghum 3, 30, 48, 170, 193, 206-208
South America 1, 5, 7, 40, 69, 74, 145, 169, 185, 228
South East Asia 59
Soybean 187, 190
Spain 6, 31, 68, 76, 168
Sri Lanka 39, 93
Statistical test 53, 120
Steeplands 1, 3-6, 19-20, 40, 46-48, 62, 70, 85-86, 90, 94-95, 99-100, 120-121, 124, 141, 220, 228, 235, 240-241
Subsidies 8, 22-23, 39, 72, 222, 233
Subsoil formation 2
Sugar cane 149, 196, 203
Surface sealing 85, 154
Surface-flow 86
Sustainable agriculture 2, 18-19, 26, 79, 94, 150-151, 156, 169, 181, 199, 215, 219-220, 224, 229, 233, 243

Sustainable Livelihoods Framework 26-28, 262-265
Switzerland 109, 137
Syria 38

Taiwan 5, 89, 109
Tanzania 3, 6, 40, 45, 93
Temperate zones 1, 84, 154
Termites 152, 206
Terraces 1, 5, 7-8, 20-21, 32, 40-41, 47, 59, 64, 68-70, 78, 93, 168, 201, 228
Texture 3, 11, 74-75, 85, 92, 101, 137
Thailand 5-6, 23, 38, 55, 68, 71, 92-93, 109-110, 203
The Sahel 205
Tobacco 185
Tobago 94
Tropics 1-3, 20, 82, 84-85, 89-90, 121, 136, 141-142, 153-154, 170

United States 6, 56, 76, 82, 89, 93, 137-138, 145-146, 159, 210, 214, 221

Venezuela 2

Vetiver Grass 1-3, 5-6, 77-78, 82, 93, 95, 99-100, 102, 107, 120, 129-131, 134-135, 139, 149, 211, 242

Water harvesting 205-207
Water Loss 1, 5-6, 108-110, 112-113, 116, 118, 120-124, 129-130, 135-139, 145-146, 167-168, 176, 240, 242
Water stress 12, 157, 159
Water-holding capacity 2-3, 74, 85, 146, 150, 157, 188
Water-logging 55
Weeds 9, 55, 58-59, 65-66, 75, 93, 165, 172-173, 187, 191
World Bank 99, 130, 191, 220, 235, 240, 242
World Neighbors 68, 195-200, 203-204

Yemen 68

Zambia 5, 73-74
Zimbabwe 45, 74, 93
Zululand 56

Printed in the United States
by Baker & Taylor Publisher Services